Praise for *Critical Mass*!

Compelling ... as gripping as good fiction. Hydrick's book is important history well written.

Tony Hillerman
New York Times Best Selling Author
Retired Journalism Professor
Decorated Hero of D-Day and the Italian Front

Certainly leads the experienced physicist to believe.

Dr. Delmar Bergen, retired
Former Director, Nuclear Weapons Program
Los Alamos National Laboratory

*The assertion in **Critical Mass** that the uranium surrendered to U.S. authorities onboard the German submarine U-234 was enriched U^{235} [enriched uranium] is certainly a credible conclusion in view of the storage, containment and prevailing shipping conditions.*

Dr. Gary Sandquist
Former Instructor of Nuclear Engineering
United States Military Academy, West Point

***Critical Mass** brings to the surface defining new information, long hidden within archives, about the birth of the Atomic Bomb.... Should be in every library.*

D. Ray Smith
Oak Ridge Y-12 (uranium enrichment facility) Historian

This is a fascinating book ... with excellent primary source research.

Joe Sills
Former United Nations spokesperson

***Critical Mass** offers the scholar of modern history and the World War Two history buff important new information about the race for the atomic bomb. Its conclusions, based on primary sources, that the Manhattan Project used atomic bomb components received from Nazi Germany in the bombs dropped on Hiroshima and Nagasaki, appear plausible and logical.*

Hydrick's well-written account provides lucid understanding of hitherto unknown and important aspects of the birth of the Nuclear Age."

Dr. Anthony Stranges
Associate Professor of Modern Military Science and Technology
Texas A&M University

A coherent and well researched history of events that have been covered up for half a century – exciting and revealing!

Otis Maclay
Pacifica Radio Host

This book is a well-researched, well-reasoned, well-written persuasive argument for a revised interpretation of an important, perhaps even critical, chapter in our modern history. It deserves a careful reading and to be taken seriously by both scholars and laymen alike.

Dr. Douglas F. Tobler
Professor of Modern German History, Emeritus
Brigham Young University

The best primary source research I have seen in a long, long time.

Gordon Fowkes, Lt. Colonel, US Army (Ret),
University of Houston Military History Symposium

The facts that the uranium captured from Nazi Germany was: 1) stowed in gold-lined containers that, 2) were cylindrical in shape, 3) each possibly carrying half a critical mass, 4) that were described as becoming 'sensitive and dangerous' when opened, and 5) should be handled like TNT, certainly leads the experienced physicist to believe the material was enriched uranium. I cannot fathom anyone at the time taking such careful precautions, or claiming such danger, about comparatively harmless natural uranium.

Dr. Delmar Bergen, retired
Former Director, Nuclear Weapons Program
Los Alamos National Laboratory

Based on the information Mr. Hydrick presents, and my own knowledge of two World War Two SBR rubber plants, I find it hard to believe the traditional explanation that the Germans spent four fruitless years trying to bring a rubber plant on line, the technology for which they had previously developed, proven and used. I also cannot comprehend, nor do I believe, a buna plant of that time period consumed as much power as the eighth largest city in the world (Berlin – as stated by the directors of the plant).

George M. Ladzun, retired
Former Director, Process Development, Zeon Chemicals
Former Manager of two synthetic rubber plants
Former Process Engineer for buna plant start-up, BF Goodrich

The electrical consumption that I.G. Farben's directors described at their buna plant at Auschwitz is very much in line with the huge electrical requirements for electro-magnetically enriching uranium.

Dr. Delmar Bergen, retired
Former Director, Nuclear Weapons Program
Los Alamos National Laboratory

It was not a rubber plant. You can bet your bottom dollar on that.

Ed Landry
Former President and General Manager
Keystone Polymers, Inc.

For my mother, June Carsten, who instilled in me a love of learning and deep appreciation for truth.

And to those who believed when I had lost faith.

Above: *Critical Mass* researcher and author Carter Hydrick during his second presentation to scientists and technical personnel inside Oak Ridge National Laboratory July 13, 2005. (Photo courtesy Oak Ridge National Laboratory)
Below: Oak Ridge National Laboratory historian D. Ray Smith shows *Critical Mass* researcher and author Carter Hydrick a calutrons during a VIP tour prior to Mr. Hydrick's second presentation at Oak Ridge July 13, 2005. (Photo courtesy Oak Ridge National Laboratory)

Author's Note to the
Third Edition of Critical Mass

In mid-2015, two events occurred that drove me to pursue publishing this third edition of *Critical Mass* after having given up a decade earlier pursuing any meaningful success. The book had not achieved the exposure it deserved but had cost me personally in every currency conceivable, and I had determined to cut my considerable losses and leave it in the past. So I go forward with some fear of a repeat of the hazards, roadblocks, mistakes, and losses that have dogged this book since its inception. I do so with the sense this important history still must be correctly established no matter how checkered the past or process.

One of the events that changed my mind took place last year as I sat in a team leadership-building meeting at work. The team of 10 to 12 managers was getting to know each other using the time-worn activity, required by the team-building facilitator of each sharing something unique about ourselves the others didn't know. I shared I had written a book – and was immediately barraged by all the same responses of fascination and admiration and questions of what it's about that often come when people learn of *Critical Mass*. In the middle of all this conversation, one of the managers, who had not been involved previously but was busily thumbing his i-Pod instead, cried out as he stared at his screen, "Wow! That's an expensive book."

Perplexed, I responded, "It was only $19.95 for a paperback and $24.95 for hardcover."

He looked over the top of his reading glasses, "It's for sale on the internet for anywhere from $350 to $650," he said.

We gathered around and looked over his shoulder. Sure enough, to my utter amazement, hardcover used copies in very-good condition were being hawked for extraordinary prices. A few minutes later he actually found a copy in excellent condition for sale for $4,000.00 (I doubt it sold for that much, but would certainly like to know if it did). I was speechless and everyone else seemed just as stunned before excitement welled up from the others for me and how much money I must be making from the book. I explained that none of the money came to me, they were used books; the money went only to the owners of the books who were selling them.

For the first time I realized, however – a decade after giving up on *Critical Mass* because I thought all my hard work and investment had failed – that a groundswell had been developing over the past several years around its value as a valid, if alternative, history of the birth of the Atomic Age. For the first time in a long time, I was happy about something connected to this book.

The second event that compelled me to produce a new edition of *Critical Mass* resulted from a practice I had loosely followed during those years when the book was "in the wilderness" of once every six months checking Google to see what was being discussed about it. I was surprised to find a reference made in the United Kingdom's *Daily Mail* (13 July, 2011) regarding 126,000 barrels of spent uranium from the Nazi atomic bomb program having been found in a salt mine outside Hamburg. The reference then declared the discovery proved my research and findings correct. Which it does.

The discovery of the spent uranium was a stunning revelation to me. I had spent 10 years and tens of thousands of dollars of my own and other people's money researching *Critical Mass*, and another three years plus – and several thousand dollars more – promoting it wherever I could. I had had many successes, including invitations to present at Los Alamos National Laboratory, where the lion's share of the experimentation and design of the first bombs were performed, and Oak Ridge National Laboratory, where the uranium was enriched. Although my findings ran counter to the traditional histories of these august institutions, the reception was very good at both venues and I was invited back to present a second time at each. My second presentation at Los Alamos was standing-room only. And Oak Ridge historian D. Ray Smith pursued security clearances for me for the second presentation I made there, and I presented only to scientific and technical personnel within the classified section of the laboratory. I sold out all the books I had brought each time at both labs.

You would think with this kind of positive reception at the institutions my history attacked, success of *Critical Mass* was assured. Reception of my research and conclusions was always good when I had the opportunity to stand in front of any audience and present the documentation. At the first Oak Ridge presentation I had even presented before the director of technology for Oak Ridge during the war. He tried to argue nearly every assertion I made. But half-way through the presentation some of the scientists and technicians in the room boldly confronted him, insisting the documentation I was showing materially demonstrated a different history than what was understood broadly; and that my interpretations of these documents were correct; and that, in fact, these being the case, my conclusions of how these materials and programs would be used were exactly in line with what would have been done. After this point, I was no longer interrupted by the former director of technology.

One would think with this kind of support *Critical Mass* was a great success – but it sold fewer than 3,000 copies. Certainly not the volume it deserved to garner if it is the important history I think it is. Despite all of the evidence presented, there was one question I couldn't answer that kept many from believing.

These presentations at the labs were just four of a dozen presentations I made to science museums and university colloquia in many states. At the end of each of them, invariably, two challenges were fired at me: First, if the Nazis had the makings of an atomic bomb they would have made one and used it on the Allies. They did not, therefore they did not have a bomb and I was wrong.

This was easy to counter: By the time the Nazis had the components of a bomb, the Luftwaffe – the German air force – had lost superiority in the skies. The Nazis would not be willing to risk a $2 billion investment on the slim chance they could stealthily sneak the weapon to a strategic target worthy of the investment made, and successfully drop it on the target. Instead, I posited, the Germans would use it as a bargaining chip, which I believe the documentation in *Critical Mass* proves they did.

The other challenge I regularly faced was more difficult to parry. Those in the know would ask why, if the Germans had enriched enough uranium to make a bomb, there was no left-over spent uranium. To create a critical mass of enriched uranium would have left thousands of tons of uranium waste. Because there is no uranium waste, therefore, there could have been no Nazi uranium bomb.

I always had to admit that no massive stores of spent uranium had been found, and that perhaps they never would be, but I believed they existed somewhere. I had given up, however, on their ever being discovered. And the discovery, apparently, was needed in the minds of the masses to prove *Critical Mass*.

So it was a stunning relief when I learned the uranium deposit had been located. Uranium being the heaviest natural element on earth, each of the discovered 126,000 barrels would weigh several tons. Using the arbitrary number of 40 gallons per barrel, the total amount of spent uranium would be well over 300,000 tons: easily a quantity reflecting a mass production process had been in place, and immensely more than any experimental program would produce. The discovered uranium being from Nazi stores, my conclusions in *Critical Mass* are now proven.

Except historians are insisting the documentation found in the German record boxes stating the spent uranium is from the Nazis is in error, and the uranium actually came from later German power-generating programs. As their only proof of this, however, they say it is a well-known fact there can be no Nazi spent uranium because there is no evidence of a Nazi processing plant.

But there is. And has been since the late 1990's, in the form of my research as it has been shared in preliminary news reports and the first two editions of *Critical Mass*. It is time for the denial to end and all of the evidence to be brought together and become public. It is time for the true history of the first atomic bombs and the birth of the Nuclear Age to be correctly established. Driven to see the job to the end, I have decided to publish this third edition of *Critical Mass*. My thanks to everyone who participated in this long and challenging process – and especially to those who didn't let it die.

Carter Plymton Hydrick
May 2016

Foreword

I t is truly an honor to be asked by Carter Hydrick to write a forward to this third edition of his book, *Critical Mass*.

When I read Carter's first edition of *Critical Mass* some years ago, it became clear to me that there was more to the story of the race for the nuclear bomb then existed in the public domain. I thought I knew about everything associated with our nuclear weapons program. I certainly considered the story of the Manhattan Project to be well documented and that I knew the entire story.

I began my career at Los Alamos in the summer of 1957, directly involved in nuclear weapons work, which remained the case until my retirement. I started as a staff member working on nuclear weapon design, and eventually was promoted to Director of the LANL (Los Alamos National Laboratory) Nuclear Weapons Program. Other assignments included serving as a consultant to the Assistant Secretary of Defense for Atomic Energy, and as a consultant to the US delegation developing the protocols for the Short and Intermediate Range Missile Treaty between the U.S. and the U.S.S.R.

A few of my associates at the Los Alamos National Laboratory had been part of the Manhattan Project, and at cocktail parties and "water fountain talks" I picked up bits and pieces of the history, but always from the "Los Alamos" point of view. Occasionally we talked about the uranium enrichment program but by the time I came along the separation and enrichment programs were well developed and there was no scarcity of material. I was aware of the massive effort put forth by the Manhattan Project to produce highly enriched material but did not know just how scarce the material was in the spring of 1945, until I read the details in *Critical Mass*.

As you read this book I would like to stress two points that I believe materially change the important history of the birth of the Nuclear Age as we know it:

First, it is my view as a physicist, based on documentation provided here in *Critical Mass* that the effort the Germans put into preparing and shipping the 560 kilograms of uranium oxide surrendered on board the

German submarine U-234, was enriched in the isotope U235. In other words, it was enriched to create a nuclear weapon. This runs counter to the traditional history of these events but I am confident in my conclusion. I only am uncertain in my mind about what was the level of enrichment. But my belief is that the German scientists associated with this cache believed the amount of U235 carried by U-234 was sufficient to create a nuclear bomb.

Along these lines, it makes full sense that this material came from the so-called German buna plant at Auschwitz, described by Carter in the pages herein. The plant certainly was not a buna production facility. The electrical consumption alone – equal to that of Berlin, then the eighth largest city in the world – was preposterously high for making buna. The power expenditure was in line, however, with uranium enrichment requirements of an atomic weapon, based on my personal knowledge of its U.S. counterpart.

The traditional history of the buna plant explicitly states that, while it was in existence four years, it never created any buna. But the massive consumption of energy, whatever the plant was intended to create, suggests that it was to some significant degree operational and creating something. If not buna, what?

I was raised on a farm in western Kansas, and my daddy taught me that if a snake was the color of a rattlesnake and had rattles like a rattlesnake, treat it like a rattlesnake. The "buna" facility had the markings of a uranium separation production facility!

How did the Germans come to create so much enriched uranium before even the United States had achieved it? Unfortunately for the U.S., by far the most successful expert on uranium enrichment – in Germany or the United States – Manfred von Ardenne was captured and taken to Russia after the war. We know the Soviets almost immediately made Ardenne head of Institute A, their nuclear weapons program, and that he was the leader of the division for electromagnetic separation of isotopes. In 1953, Ardenne was awarded the Stalin Prize, first class, for contributions to the atomic bomb project, and became known as the Father of the Soviet Atomic Bomb.

The second material information you should take note of, I believe, is that the contributions to the Manhattan Project of the surrendered U-234 did not stop there. I believe the surrender of U-234 had impact on the development of the implosion device – the plutonium bomb dropped on Nagasaki – as well.

I was well acquainted with the development of the implosion device and the difficulties our scientists experienced in developing a detonation system that would give a proper spherical implosion. This concern kept the entire group on edge until the famous Trinity Test in New Mexico proved it to be successful.

Detonating the device, along with the scarcity of plutonium, led the scientists just prior to the Trinity test to build a large cast iron containment vessel in which they could place the device to contain it for recovery of the plutonium if the detonation system failed to create a nuclear explosion. Surprisingly, as the date of the test approached, last-minute improvements in the firing system reduced their concern enough about the reliability of the detonation system that the containment vessel was never used.

I was never told how the details of the improvements came about, but it was during this period that the passengers on board U-234 were debriefed and it was learned one in particular, the scientist Heinz Schlicke, had knowledge of fast operating energy transfer systems. The rapid and consistent release of electrical energy was a key part of the problem the LANL scientists were experiencing triggering the detonators with the simultaneity necessary to achieve a clean spherical implosion. There apparently is no written unclassified record available to provide us with what may have come from the debriefing of Heinz Schlicke but this we do know, over the summer months after his capture and the surrender of U-234 the confidence in the detonation system greatly improved, and the production of uranium for the gun weapon increased significantly.

I am convinced that Carter Hydrick presents in *Critical Mass* a credible theory backed by documentary evidence that the surrendered submarine U-234 contained important material to construct and possibly deliver both the uranium and plutonium atomic bombs. Carter has spent a significant part of his life sifting through the historical documents and technical information to give the reader some ideas as to just how close the race for a nuclear weapon was. In the process he has found information suggesting possible collusion between U.S. and German high officials that led to the surrender of U-234 to the U.S.

Delmar Bergen, PhD, (retired)
Former Director, Nuclear Weapons Program
Los Alamos National Laboratory

Acknowledgments

An undertaking of this size always requires a huge commitment that, paradoxically, due to myopia and naivety, usually is still too small to match the task at hand. The truth – that I was looking at the elephant's trunk but thought I was seeing the entire beast – was not discovered until the project was too far down the road to turn back or give up on.

This admission does not reflect a low level of enthusiasm for the work on my part. Working as a private researcher sans credentials, foundation funding, or assistants required a much higher level of commitment than originally anticipated. The effort ultimately cost over $20,000 and fifteen years of research, writing and marketing.

Those fifteen years and that $20,000 were not mine alone to give. My wonderful wife of nearly 45 years, Kris, and our children, Elias, Ashley, Yurii and Fleming, have by far sacrificed more than anyone in order to make this book possible. I must admit, sadly, and to my embarrassment, it was not always voluntary on their part. They all still love me, however, and have stuck by me and, for the most part I think forgiven me for the abuses of time and resources they had as much claim on as this history did – in fact more – but which they sacrificed. They are at the top of the list for a very personal and heartfelt "thank you," for giving up three different family vacations and countless additional hours with husband and father while I spent their time and money traveling to archives to do research or sitting alone in a closed room writing. It feels a very selfish thing for me to do. But understanding how important this history is, or at least how important *I believe* it is to our collective understanding of these important events, they supported me in the end; and even as this third edition is going to press have encouraged the process. There is nothing I can say or do that can express the intense appreciation I hold for each of them for their patience and support – often when it wasn't in their own interests. Thank you so much, guys.

For this third edition particularly I recognize a debt of gratitude once again to Dr. Delmar Bergen, retired director of the Nuclear Weapons Program at Los Alamos National Laboratory, for reaching deeper into his reserves of knowledge, courage and integrity to both add additional important technical insight, as well as having vetted the work and provided the forward to it, and ultimately having further endorsed and defended it.

Revisionist historian Dr. Joseph Farrell came to my attention while I once was surfing the internet to see what was happening with the two previous editions of *Critical Mass*. I found he had not only discovered and enthusiastically endorsed the book, but he led me to one of the compelling discoveries that swayed me to publish this third edition: the discovery of 126,000 barrels of spent uranium from the Nazi atomic bomb program, which proves the findings in *Critical Mass* are correct. He then in private encouraged me, despite some misgivings I had due to previous challenges with this project, to prepare a third edition. In the end, he led me to TrineDay, the publisher who agreed to take up the task. For all of these contributions I am grateful; including to Kris Milligen of TrineDay for his commitment to the work.

Henry Weissenborn, Jan Nizamani and Gene Lamb – all co-workers – deserve thanks for providing special services needed to help me prepare this edition.

I must say "thank you" too, to my sister, Christi Hydrick, who provided strong encouragement from the very beginning those 25-plus years ago – backed up by "putting her money where her mouth is" – when our emotional and financial reserves ran out. David Davis, a priceless friend since high school, did the same.

In November 1999, a Houston television station did a newscast about this research. In the segment, Dr. Anthony Stranges, PhD, associate professor of history at Texas A&M University, provided the opposing viewpoint in favor of the traditional history. I immediately made arrangements to show the research to Dr. Stranges, who kindly allowed me an opportunity. After three-and-a-half hours reviewing the documentation, Dr. Stranges agreed that my findings were, in fact, plausible and that critical review and additional research were appropriate. I very much appreciate Dr. Stranges, who not only had the courage to look unafraid at this evidence that flies in the face of the entrenched history, but who, over the course of the years since then, has also provided professional and moral support, along with opportunities for me to share my findings with the scholastic world.

During one of these events, while presenting to Phi Alpha Theta (the national collegiate historical society) at Texas A&M, for which Dr. Stranges was the faculty advisor, Dr. John Poston, Sr., head of the university's Nuclear Engineering Department, was in attendance. Dr. Poston provided valuable additional insight into atomic theory and technology that have further strengthened the conclusions forwarded within these pages.

Dr. Bernhard Wehring of the University of Texas J.J. Pickle Nuclear Research Center, and Drs. Harlow Russ, Edward Hammel and John Allred – all retired from the National Atomic Laboratory at Los Alamos, New Mexico – and the late Clarence Larsen – formerly director of the Oak Ridge calutrons – also provided valuable insight and information. Captain Carl Triebes (USN retired), former submariner and navigator extraordinaire, who helped me analyze U-234's war log, deserves recognition, as well. My son Eli went the additional mile by serving as my German translator for the war log and for many interrogation transcripts and original German documents.

Of prime importance in getting this work off the ground are Arthur Naujoks, the man who initiated this effort, and Rob Bell, who introduced me to Art. Rob was a friend and film producer who wanted me to write the screenplay for a motion picture about U-234. Art had proposed the film based on conclusions suggested by interpreting three or four pieces of evidence he had found that indicated there was an unknown but historically compelling account to tell. While these few pieces of information fit together well, they were very far from conclusive. My respect for properly recorded history led me to convince Art and Rob that the research should be completed first, proving or disproving the assumptions we were making about those few shreds of evidence. If the hypothesis proved true, we agreed, the documentation should be presented first in a book for critical review; if not, then we could go ahead and make a movie from the idea. Art asked how long it would take to do the research. I estimated no longer than a year.

After three or four years, Art and Rob gave up on me and we amicably agreed to go our separate ways. While I am sorry I disappointed them with my slowness, I am confident the right path was taken.

Cal Jochetz, Bret Reich, Rick Page and Ken Anderson deserve a note here for being the first to serve as editors for this work. That was many years and multiple drafts ago, but they set me on the right track with sound guidance both in terms of such things as structure, grammar and word usage, and, more importantly, in overall readability. Receiving their corrections was painful, so I suppose it is appropriate they are, respectively, a dentist and three lawyers. It has been many years since I have seen the last three, but I still consider them all good friends, and I appreciate immensely their willingness to hurt me in order to make this book better.

That having been said, any shortcomings are mine – they did everything they could to make me work through the pain I wanted to deny and get it right. They set the tone.

Finally, there are a number of others who helped drive this work forward. Robert Morin, of Birmingham, Alabama, was the first outside of my own family to contribute by providing means for me to work in the National Archives Southeast Region at East Point, Georgia, partially at his own expense. Archivists Barry Zerby and John Taylor lead a long line of specialists within the National Archives and Records Administration and elsewhere without whose help I could have completed this work. Archivist Charles Reeves gave great and enthusiastic assistance at the Southeast Regional Archives in East Point, Georgia.

The reader may note in the endnotes that citations are given from the Northeast Regional Archives in Waltham, Massachusetts, as well, but may notice no acknowledgements are credited to archivists from there. When I first started this project, all of U-234's documents were in the main building of the National Archives and Records Administration on Pennsylvania Avenue, Washington, DC. A year or so later, they were transferred to various satellite facilities, most notably NARA II on the University of Maryland campus. Many of the key documents that I had already reviewed in Washington, DC, however, were sent to the Northeast regional facility in Waltham. Rather than go to the expense of an additional trip just to re-identify locations of these documents in the northeast archive, I have used citations provided for them in Joseph Mark Scalia's book about U-234, *Germany's Last Mission to Japan: The Failed Voyage of U-234*. My thanks to Mr. Scalia, and by extension to the archivists who worked with him in the northeast archives.

Harry Cooper, founder and president of *Sharkhunters International*, a historical society dedicated to the preservation of submarine history, provided much assistance in connecting me to people close to events, as well as in giving me an open pass to the archives of *Sharkhunters*. Dr. Alan Bath, PhD, formerly of the Office of Naval Intelligence, and who wrote his doctoral thesis on Naval Intelligence, reviewed the manuscript for accuracy as relates to intelligence matters, most specifically ENIGMA/ULTRA and MAGIC/PURPLE cryptography, but also regarding American naval intelligence policy, processes and practices. Carina Lyne provided a great service in preparing the original documents for presentation within these pages.

There are many others who have helped with this work, I am sure, whom I have forgotten over these years. It takes so many people to make a project such as this happen – all people to whom I owe a deep debt of gratitude and to whom I express my profound appreciation. If I have forgotten you in these meager acknowledgements I am truly sorry. Please take compensation in knowing that you contributed to a work that will hopefully make a difference in how we understand this world in which we live.

Table of Contents

Introduction

This micro-history is the result of newly discovered, very significant events that occurred during the closing weeks of World War Two. As the story of *Critical Mass* unfolds, it questions the foundations of the traditional history of World War Two regarding the making and use of the first atomic bombs, as well as challenges our understanding of how the Nuclear Age was born. The facts reveal not only important new information about the race to produce the bomb, but the new information helps us understand how the sum of the history of man was combined in one brief moment to create a critical mass in humanity that exploded the old world forever and ushered in the Nuclear Era.

The previously secret – now declassified – unpublished military, governmental, intelligence and Department of Energy documentation cited throughout *Critical Mass* suggests the atomic bomb was not fully developed and built by American scientists and technicians, as traditional, long-standing history asserts. Instead, the evidence shows enriched uranium and other atomic bomb components developed by Nazi Germany were surrendered to United States forces during the final weeks of the war – probably according to prearranged surreptitious agreements – and were a vital part of the materials used to create the bombs that were dropped on Hiroshima and Nagasaki. The evidence indicates that without these materials the United States would have fallen short of achieving its nuclear weapons objectives. And it shows that without achieving these objectives, it is highly probable the Soviet Union would have become and remained for many decades the reigning super power on earth.

Interwoven into this story is provocative evidence that connects Hitler's behind-the-scenes right-hand man, Nazi Party Chief Martin Bormann, to Germany's very nearly successful effort to create an atomic bomb; and to Germany's last-ditch efforts to transfer that technology to Japan. Evidence also suggests that Bormann, at the latest possible moment, turned against his Asian ally and decided to hand the keys of world dominion – in the form of the atomic bomb – to any Allied country that would deal with him. Thus Bormann covertly negotiated a separate, and very secret, personal peace with the United States that allowed him to disappear from the front page of history and slide silently between the shadows of a murky past and a phantasmal future.

As I researched this history, the initial events, striking as they were, each led to astounding new revelations that had the net effect of continually, and, seemingly endlessly, expanding the scope of this book. As a private citizen who researched and wrote the book around the demands of a full-time job and who, with the aid of generous family and friends, financed the research and writing, generating unlimited resources to constantly expand the book's scope was impossible. Despite desires to throw light in every corner, proving the premises presented in *Critical Mass* has, of necessity, been circumscribed to proving the following three basic assertions:

1. That the Manhattan Project did not successfully produce all of the needed enriched uranium – isotope U^{235} – in time to fulfill its atomic bomb requirements, nor did it successfully create a triggering device for the plutonium bomb without the help of captured German components.

2. That uranium was enriched in Germany, despite the traditional history that states otherwise. Enrichment would have been in quantities that could have supplied the bomb-grade uranium needed by the United States to complete its atomic bomb project.

3. That enriched uranium (U^{235}) for the uranium bomb was obtained by the United States from Germany and was transferred into the possession of the Manhattan Project and ultimately used in the bombs dropped on Japan.

As a matter of sufficiently authenticating the above assertions, I have tried to obtain a minimum of two corroborating pieces of evidence to validate the theories presented. In almost every case, as will be seen, this has been accomplished. In many cases, three or more proofs are given. In a few instances only one piece of evidence is extant; but taken on the whole, the accumulated evidence is considerable if not incontrovertible.

The question may be asked that, with the hundreds if not thousands of books, articles and histories that have been written about the making of the first atomic bombs, how can any new and unpublished information be added to the chronicle. Remarkably, the answer, in part, is that very few of the writers of those histories ever saw any of the original records of the most seminal events that constituted the makings of the bombs. As far as I can tell, I was the first to review the actual uranium enrichment production records, the shipping and receiving records of materials sent from Oak Ridge to Los Alamos, the metallurgical fabrication records of the making of the bombs themselves, and the records and testimony regarding failure to develop a viable triggering device for the plutonium bomb. Of the 38 boxes of Oak Ridge records held in the Southeast Regional Archives in East Point, Georgia I had pulled for review, only four had been opened since their declassification in 1967 and 1978. I was the first to open and cull through the vast majority of boxes I was focused on, and within these containers I found many critical documents. And there are boxes that remain, their declassification seals yet unbroken.

Apparently, the authors described earlier have relied on personal accounts and the administrative, strategic and general records harbored in the National Archives in Washington for their research. The critical daily production records of Oak Ridge and elsewhere have been all but ignored, though they reveal important information not previously considered in other histories, and although they tell a different story than that presently believed. Even if those authors had read, assimilated and interpreted the available records, the discrepancies may have been considered anomalous and possibly would have been ignored when compared against the overpowering reputation of the traditional history. Most of that history can be traced in theme and content to Manhattan Project Commanding General Leslie Groves' book on the subject, *Now It Can Be Told*.

Now It Can Be Told presents the story of the making of the atomic bomb sthe United States government needed the world to hear at the time. There was, undoubtedly, justification for this guarded approach considering the exigencies of the era. The chronicle of history should, however, be corrected when opportunity allows – though it all too often is not – for the understanding and benefit of generations to come. And for the just recognition of all those who played a part, as well as the enlightenment of those who simply desire to know the truth.

Democracies especially depend on an informed citizenry to safeguard the proper use of power and appropriate oversight of important military and political policy. Certainly not all information and actions of a government at war or in conflict with another sovereignty can be reviewed on an open basis contemporaneously with the critical events. But as timely issues are resolved or neutralized by new events, it is incumbent upon that democratic society to carefully review and analyze the events and equitably judge the system and the people involved. In this manner we ensure the nation's best interests are preserved, and make whatever adjustments are necessary to provide a guide for future endeavors.

Other official and semi-official accounts of the Manhattan Project and the programs that competed against it have been written, the best among them being Richard Rhodes' exceptional Pulitzer Prize winning book, *The Making Of The Atomic Bomb*. *Critical Mass* attempts in no way to re-document the otherwise reliable historical elements of a very complex and detailed subject, other than to provide a basic understanding useful to the reader's analysis of the scenario forwarded within these pages. *Critical Mass* simply suggests the data recently found describe some very different events than are recounted in the presently accepted history.

As noted, many other authors' accounts are cited herein, but all of them, ultimately, either directly or indirectly, by default or design, have been molded by the man who presided over the project itself, General Groves. During the very process of the making of the atomic bombs, through compartmentalization and by mixing a high percentage of genuine data with innuendo – as well as judicious use of the occasional untruth – Groves was able to create a resilient and coherent self-perpetuating myth of the birth of the atomic age.

Although driven by primary, first-hand documentation, some of the information used to tell the story in *Critical Mass* comes from the writings of Groves and other authors. David Irving, Britain's controversial but document-dependant World War Two historian, has recorded much of the German effort to create a nuclear weapon in his book, *The German Atomic Bomb* – which Rhodes in turn depended upon for much of his account of the Nazi program. Irving's account alone, though he seems not to realize it, goes a long way toward impeaching the accepted history; which is that, because Germany failed to create plutonium, it therefore failed to build a weapon.

There are two ways to build an atomic bomb, one of plutonium, the other of uranium. Irving brings to light ample information that, when considered with other evidence newly discovered and revealed within *Critical Mass*, suggests the Germans produced the material for and all but assembled a uranium bomb.

In the traditional history of the bomb, Groves has positioned the German plutonium effort as the only nuclear initiative Germany seriously pursued. And he has magnified this misinformation, couched in a cushion of half-truths, to immense proportions – large enough to hide what appears to be a huge German uranium enrichment project behind it – and thus he has shielded the Nazi near-success from the view of the world. His motivations for doing so will be discussed in detail later.

One of many other authors quoted in *Critical Mass* is former World War Two intelligence officer Ladislas Farago, who documented Martin Bormann's escape from Nazi Germany at the end of the war and his ensuing life in semi-secret exile in South America in his book *Aftermath*. Farago was accused, with supporting evidence provided by the CIA, of having forged some documentation that he used to verify his claims about Bormann. There is evidence Farago did, in fact, use photographs that were not what he asserted; whether there was deception in his intent or not is uncertain. *Critical Mass* reviews the subject of the CIA and its predecessor the OSS and their possible involvement in negotiations with Bormann and eventual surrender of German-made nuclear bomb materials, as well as their protection of Bormann afterward. Suffice it to say here that, because the CIA appears to have participated in certain of these events, and because it then required controlling any leaks about these actions, the agency's involvement makes suspect any conclusions about the accusations against Farago.

Critical Mass goes on to quote other authors who have independently discovered documentation corroborating much of Farago's work. Among these authors is Paul Manning, former journalist for the *New York Times* and author of *Martin Bormann – Nazi In Exile*. Manning's credentials as a journalist are impeccable. Although he did not accept an offer immediately after the war to serve as the civilian deputy of the United States' occupation zone of Germany, the offer itself attests to the high regard in which he was held, as well as to the potential military intelligence and other resources he had available when researching his book.

Many other authors are quoted, as well, to highlight and validate the conclusions presented in *Critical Mass*. But the definitive body of evidence is the actual, seminal documents cited in this book that dispassionately record the numbers and weights and dates and times and places and people that constituted the real events that occurred.

The silent archives, in some cases long untouched, contain the remaining few pieces of the picture that had been painted over with duplicitous details and fraudulent facts. Exposing those lost data to the light of day is much like the art curator who takes a blacklight to a painting to ascertain its origin. Under scrutiny of light tuned only to see the original, the primary picture is exposed underneath as well as any revisions that may later have been made. So it is with the certifieds cited in *Critical Mass*. The light of day, "always a great disinfectant" as the saying goes, reveals through newly-disclosed documentation the true story of the Manhattan Project during the birth of its atomic offspring – with all its flaws, foibles and unholy alliances as well as its ultimate, although somehow twisted, success.

And even with those flaws and foibles it is, at once, a story of genius and perseverance as well as a lesson in man's own struggle to grow morally and spiritually at the same pace that he has grown intellectually and technologically. For, as social beings who must share this earth, we are all interdependent upon one another. When one such as Hitler rises to power, the only defense against the bully who insists on blood, when all reason has failed, is to be more the aggressor, or submit and perish. Such course devolves to a level of behavior differentiated from the instigator's only by the moral imperative of one's right to survive.

The sad fact is, we can rise as a race only to the level of our least enlightened. Until that time, the weight of our human frailties and flaws will at irregular intervals compress to critical mass and ignite a new explosion of chaos and distress, until we learn once and for all that our cumulative morality must meet or exceed our united intellects.

Prologue

Desire following accomplished cargo U-234 … and inventory list sent CNO (Commander Naval Operations) who will give shipping instructions and will control access to and disposition of all cargo due to vital importance to Pacific War.[1]
– U.S. Navy secret transmission #222115 from Commander and Chief and Commander Naval Operations to Portsmouth Naval Yard,

On 19 May 1945, eleven days after the surrender of Nazi Germany to Allied forces in Europe, a German U-boat was escorted into Portsmouth Naval Yard, New Hampshire. Underseacraft U-234 (as the Germans designated their submarines) was not the first U-boat to surrender at Portsmouth Naval Yard,[2] but it was by far the most enigmatic.

The massive size of the U-boat alone – the submarine was three times larger than the standard German submersible – was quite enough to draw attention to the vessel. But there were other hints that spoke of the craft's singular status: News reporters covering the surrender of U-234 had been ordered, contrary to all U-boat surrender procedures before and after its capture, to keep their distance from crew members or passengers of U-234 on threat of being shot by the attending Marine guards.[3] Strict orders had been given that the press and all other observers must stay at least eight feet away from the disembarking POWs, and ropes were used to cordon off a corridor so the prisoners coming off the boat could march unhindered through the crowd.

The first group of sequestered POWs that marched down the gangway revealed another clue to the mysterious vessel's mission. At the lead was Luftwaffe Lieutenant General Ulrich Kessler, former commander-in-chief of the German Air Force in Norway and the North Atlantic and newly-assigned air attaché to Japan.[4] The stern and monocled Prussian general was followed by an entourage; including Dr. Heinz Schlicke,[5] a high-frequency electronics expert; seven other military officers, all scientists, engineers or military staff members; and two mysterious civilians. Neither the reporters nor indeed any members of the United States armed services at Portsmouth would ever see or know about the passenger who it now appears had been on board but who was not among the prisoners taken off the boat. Nor would the reporters soon find out about the cargo still deep inside U-234's hull.

Reports did circulate of two Japanese officers on board who took their own lives in hara kari rather than submit to capture and the failure of their top-secret mission,[6] which was to guarantee the safe arrival of cargo and passengers to Japan. According to rumors later proved to be true, they were buried at sea. But it seems another, more foreboding guest also had been on board. He probably now was somewhere in Spain.

As the most important prisoners were whisked away by airplane to Washington D.C.,[7] and the reporters were complaining among themselves of their lack of opportunity to interview the cabalistic captives, the ultimate secret of U-234 still lay quietly in its sleeping hull. For jam-packed throughout the monstrous boat was the elite of Nazi Germany's newest weapons development; including a completely disassembled Messerschmidt 262 jet fighter[8] – the first jet aircraft to be used in combat – and the plans and material to build Germany's ultra-secret V-4 rocket, the long-distance version of the feared V-2; as well as a mysterious "stratosphere plane," apparently an aircraft that could reach and cruise in the rarefied atmosphere of near space. But most important, buried in the nose of the mammoth boat, sealed in cylinders "lined with gold,"[9] was 560 kilograms, 1,120 pounds, of uranium oxide[10] labeled "U235"[11] – the fissile material from which atom bombs are made.

Despite the cutting-edge technology, all of the other goods were dwarfed in significance by the uranium portion of U-234's cargo. The United States, dangerously short of its nuclear objectives, would spirit away the uranium, and other captured materials, for use in its own atomic bombs. Orders from the Commander Naval Operations in Washington D.C. to the Naval Yard at Portsmouth commanded:

> Desire following accomplished cargo U-234.... All material after rendering safe by mine disposal personnel placed safe stowage and inventory list sent CNO who will give shipping instructions and will control access to and disposition of all cargo due to vital importance to Pacific War.[12]

History seems to suggest U-234's killer cargo would soon be dropped on the U-boat's original destination – Japan.

How the uranium came to be in the U-boat, and the impact it would have on the outcome of the war against Japan and the future of the world, is a story of international intrigue to rival the most thrilling of espionage accounts. It is a story of unequaled scientific endeavor that toyed with the elementary building blocks of nature and of the universe and the staggering power with which it is held together; all being performed by the most brilliant minds of the age, under super-secret conditions, their projects swallowing staggering sums of money. It is the story of the struggle to control that power. It is a story of subterranean machinations unequaled in man's history, driven by selfish and evil men who were opposed by both honorable and suspect adversaries. And this chess-like struggle is counterpointed by raw power grabbing, achieved through such voluminous death and destruction as had never before been seen. To this day, it is the culminating story of power politics on earth.

The traditional history of the race for the atomic bomb – written by the victorious Allies – has taught us to believe the United States' atomic bomb program, the Manhattan Project, with help from Britain and mistakenly believing that Germany's nuclear program was striving to attain a bomb, had forged ahead and beaten the Germans to the development of a nuclear weapon. The authors of that history would have us believe that only later was it discovered the Germans never were successful in progressing beyond a very elementary level of the bomb's development.

Nothing could be further from the truth.

The arrival in the United States of U-234 and its stockpile of nuclear materials saved the United States' atomic bomb program. Because of shortages of time and materials within the American bomb project, unreported until now, the United States needed the German goods and technology not only to gain victory over Japan, the only surviving member of the German-Italian-Japanese triumvirate, and win the war, but to establish its objective, as well, of making its place once and for all among the top world powers.

Prior to World War Two, America, while considered an ambitious contender for a place in the upper echelon of Earth's geopolitical elite, was still relegated to the status of "also ran" among the countries of the world. The British Empire, though slowly being dismantled, could still brag the sun never set upon its holdings and still acted as the reigning world power. Germany, in the few short years before and during the early part of the war, was making an aggressive effort to unseat Britain's shaky imperialism. Japan hoped to do the same in the East. The Soviet Union, like the United

States, saw the war as an opportunity to make up distance in the international arena and fought for its place on the geopolitical globe. The potential introduction of an atomic weapon added a singular and determining dynamic to the equation. The bomb was the international equivalent of a sandlot free-for-all where one of the youngsters shows up with a machine gun: whoever had the weapon had control of the playground – if he could prove it was real and that he would use it.

Up to the time of the U-boat's surrender, America still had not been successful producing the weapon. True, the arrival of U-234's uranium on America's shore did not exactly replace the United States' atomic bomb program. To be sure, the Manhattan Project was enjoying levels of success on all fronts of the newly developing technology. Bomb-grade uranium was being enriched in Oak Ridge, Tennessee. Plutonium for nuclear weapons was being bred in huge reactors in central Washington State. Advances in the understanding of nuclear technology and how to utilize it were being achieved at the national laboratory in Los Alamos, New Mexico and elsewhere. The major problem was that the desired result – a usable bomb – was not going to be completed in time. While development of the bomb components were moving forward, a date had been set after which use of the bomb would be difficult to impossible; and even if it was used after that date, significant political potential would be lost.

To optimize atomic bomb technology politically required it be used to foreshorten the war in the Pacific by the time Russia declared war on Japan, which the Soviets planned to do in mid-August 1945. Any action significantly later would result in the parceling out of the Asia/Pacific region in the same way Europe was to be partitioned, leaving Russia with more than its fair share of the globe. Worse, once the Soviet Union was in the war in the Pacific, in order for the United States to use the bomb details of its existence and development would have to be shared with the Soviet Bear – an intimidating proposition indeed, giving nuclear war capabilities to Joseph Stalin. Russia's helping to foreshorten the fall of Japan through conventional warfare before the bomb could be used would also eliminate America's powerful (and arguably justifiable) rationalizations for use of the bomb against the Island Nation. The United States would not only lose the opportunity to save countless Allied and Japanese lives, but it would also lose powerful political advantage it hoped to gain psychologically and in real terms if it could demonstrate it had an atomic weapon and the resolve to use it.

Lacking enough enriched uranium to produce a uranium bomb (the type dropped on Hiroshima) at the time U-234 landed, and unable to complete the development of a reliable triggering device for a plutonium bomb (the type dropped on Nagasaki), the Manhattan Project greedily gobbled up the enriched uranium and detonation components available from the surrendered submarine. Three months later, the bomb materials finally reached Japan.

Why did Hitler fail to use the atomic bomb components before they fell into American hands? And how did Germany come to have the makings of an atomic bomb in the first place?

Substantial evidence exists that Germany led the race for an atomic bomb throughout the war. The Germans were unable, however, to use the weapon against their enemies in Europe because the weapon components were not completed until after the Luftwaffe (the German Air Force) had lost air supremacy in Europe. Germany then had no means to deliver the weapon on an enemy target. Following D-day, in June 1944, the Allies' close proximity to German troops on the European continent and in the Fatherland itself eliminated any potential attempt to use the weapon by Germany because of the improbability of successfully surreptitiously spiriting a nuclear weapon weighing several tons many miles behind enemy lines for detonation. There was no guarantee the weapon might not be captured and turned against them in one treacherous moment. As a result, arrangements were made at the highest level to export the technology and bomb components to Germany's ally, Japan. Germany presumably hoped a Japanese victory in the Pacific would mean a victory for the Triumvirate, which would still ultimately result in Aryan domination of Europe.

The masterminds behind development of the German nuclear weapon were not those men so often cited as the reigning scientific minds of the Third Reich – Heisenberg, Hahn, Weizsacker, Bothe and others. Heisenberg and his cronies are correctly characterized in the traditional history as having been so full of petty professional jealousies and pompous pride they could not productively work together to reach a common goal. The true masterminds of the German atomic bomb did not struggle under these conditions and the paltry budgets the traditional history suggests were provided the German scientists. The real masterminds of the German atomic bomb were not restricted by a lack of highly technical production capacity either, as the traditional history suggests was the German program. Nor was that production program under constant scrutiny by Allied nuclear reconnaissance efforts, as the traditional history asserts would have been the case had the Germans actually been more successful with their bomb project.

The true masterminds of the German atomic bomb program were experimental physicist Baron Manfred von Ardenne, his theoretical/experimental physicist colleague Fritz Houtermans, and an army of scientists and technicians. This intellectual armada was provided, as was the extravagant funding for the program, by one of the Third Reich's richest and most successful sources of research and development sustenance – the German postal service. As bizarre as this may seem, it is a documented fact the postal ministry supported with hundreds of millions of reichsmarks, possibly billions, an excellent technical branch that, among other remarkable achievements, successfully tapped and decrypted the complex technology of a dedicated transatlantic hot-line established early in the war between Franklin Roosevelt and Winston Churchill. All students of the race for the atomic bomb know Adolf Hitler himself once joked (as they, probably wrongly, characterized the comment) following a report by his Minister of Posts on the progress of the ministry's nuclear research and development program, that it would be his postal service, not his military, that would win the war for him with the secret weapon it was developing.[13]

Significant and substantial documentation exists that proves Ardenne and Houtermans worked on far-reaching nuclear weapons programs well advanced over Heisenberg's work, and funded and supported by the Ministry of Posts; and that Hitler on multiple occasions personally visited Ardenne's laboratories in Berlin Lichterfelde.[14] No record could be found by this author of Hitler ever having shown such support for Heisenberg's supposed atomic technology-leading nuclear reactor site, or that of any other nuclear scientist. Hitler insisted, nonetheless, that Germany would win the worldwide struggle using a secret weapon with destructive powers far superior to any previously known instrument of war.

Once Ardenne and Houtermans developed the technology, the massive and sophisticated mechanisms required for producing the fissile material had to be built and operated in a safe location, free from Allied detection and bombing. With United States aircraft screening German skies daily for emissions laced in xenon-133[15] and monitoring the Fatherland's electrical power usage for indications of inordinately high consumption of electricity and other characteristics that would signify existence of uranium enriching facilities, the Germans needed to find a location that would allow the work to go forward undisturbed. And big business needed to be brought in and tasked with building the secret, gargantuan industrial complex to enrich uranium. This business would be the German counterpart to DuPont, General Electric, Westinghouse, and Tennessee Eastman working on the Manhattan Project in the United States.

The safe location required for building this complex and the company retained to do the work were both found in I.G. Farben. During the war, this German conglomerate constituted the largest single chemical concern in the world, outsizing the top four United States chemical companies combined (all of which were working on the Manhattan Project). Farben had pioneered and owned the worldwide rights to the development and production of synthetic gasolines and rubbers – including those in the United States. Synthetic product technology required high-pressure, very demanding capabilities that were a good jumping-off point for the science needed to enrich uranium or build and operate a nuclear plutonium-breeding reactor.

In the early 1940s, the company had in fact started constructing a buna plant, a synthetic rubber product, at a location outside of Germany – perfect for the requirements of freedom from scrutiny by Allied reconnaissance. The buna plant construction would allegedly last twice as long as planned, eventually cost I.G. Farben 900 million reichsmarks[16] (approximately $2 billion), utilize over 25,000 laborers,[17] and devour more electricity than the entire city of Berlin, but it would never produce a pound of buna. All of these characteristics run counter to those one would expect to observe from an actual buna manufacturing plant. But they would have been perfectly expected of an electromagnetic isotope separation facility – the technology Ardenne and Houtermans had successfully pioneered – required to enrich uranium for nuclear bombs.

While the deadly purpose of the work done at the alleged buna facility has been virtually unknown to the world until now, the name of the place it was located carries an aura of condemnation nonetheless, earned for its other malignant purpose. The place is Auschwitz.

Located on the banks of the Vistula River in Poland, the Nazi concentration camp was the perfect site for Farben's facility. Vast coal reserves to provide the fuel necessary for allegedly heating buna and for use as buna feed stocks, but, in fact, for generating electricity, were mined nearby. The river provided additional energy and cooling for the plant's high-temperature processes. The major railroad spur located there provided easy transportation of equipment to, and product from, the plant. The vast pool of cheap labor available from the camp, which Farben infamously exploited throughout the war and for which the company paid the price dearly afterward when the Nuremberg Trials found it guilty of crimes against humanity and other human rights violations and ordered the dismantling of the conglomerate, were all components in selecting the location. But the best reason of all for selecting the site was the fact the proposed buna plant would provide the needed technological base and an excellent cover for what would actually be produced – enriched uranium.

Synthetic rubber technology was still so new few people would know what to look for in a buna plant – or a uranium enrichment plant. And even better, much of the high-end technologies, materials, skills and services required to deal with the high pressures, high temperatures and demanding performance of reactor vessels in a buna plant were also needed to construct and operate an electromagnetic separation facility.

Qualified Farben personnel performed the demanding aspects of construction, while almost all of the non-technical facilities at the "buna" plant at Auschwitz, requiring a veritable army to erect and maintain the buildings and simple structures that supported the sophisticated technology, were built and manned largely by the ill-fated inmates. The last of their physical energy – their very beings – was drained from them by the work and they were driven to the gas chambers while enriched uranium was being collected and stockpiled for its future, sinister purpose.

Standing silently at the center of this immense German atomic bomb program was one man. He had deep and powerful personal connections to each key player in the project. One of these relationships was old and sealed with blood – that he had with Rudolf Hoess, the Commandant of Auschwitz, with whom he helped murder a man, an enemy to the cause, when they were both young and unknown.

Another of these relationships was bound together by tremendous wealth. Billions, probably tens of billions of reichsmarks, were amassed when he colluded with Hermann Schmitz, the chairman of I.G. Farben. He and Schmitz camouflaged a huge fortune in securities and other portable properties and, with the war winding to a close, they removed the funds from Nazi Germany and carefully disseminated them into the markets of the free-world economy for their later use.

Still another of these relationships was bound together by power derived from the position he held as the most highly regarded lieutenant of arguably the most powerful man on Earth. His advocacy with the Fuehrer on behalf of Minister of Posts Wilhelm Ohnesorge fed the enormous research and development program Ohnesorge – formerly a mathematician and physicist – operated within the postal ministry.

The final, fateful, least secure but most important relationship was forged in the final days of his, and Nazi Germany's, power – with Grand Admiral Karl Doenitz. Doenitz was commander of the German U-boat navy, and was the man with whom he would trade the supreme leadership of all Germany for a place on one of the Admiral's fleeing U-boats, apparently having previously secretly agreed to pay to the United States the price of his own ransom – in the currency of atomic weaponry.

This man was Martin Bormann.

In the dismal days at war's end, buried in the bunker in Berlin that constituted the final seat of Hitler's government; with the world crashing in around them; with Hitler high on drugs and low on mental and emotional reserves; with the Nazi leadership panicked and breaking for cover; the mysterious Martin Bormann, true to his Machiavellian nature, weaved one of his most deadly, intricate and yet enduring webs. These machinations succeeded in deft strokes to dismember from the Nazi Party leadership both Hermann Goering, political heir apparent to Hitler's throne, and Heinrich Himmler, purportedly next in line of ascension. In their stead a seeming anomaly, and long misunderstood historical event, occurred, in which Admiral Doenitz, with no apparent political leanings and no political following, mysteriously took over the reins of German government from Adolf Hitler. Bormann had shrewdly negotiated with Doenitz the now hollow leadership of all Germany in exchange for a place on the U-boat he knew was leaving for Japan with its belly full of the power of the future, which to a large extent Bormann had made possible. From the bunker Bormann made his final arrangements to reach Doenitz at his U-boat headquarters. The web was being woven.

The fortune from I.G. Farben and additional funds that Bormann had camouflaged through a variety of financial instruments and then had surreptitiously exported into banks and businesses around the world silently awaited him once he was free of Germany. Agreement by the United States to allow his escape and to protect him from the justice sure to be demanded by Nazi victims following the war appears to have already been achieved. The arrangement would ensure his life-long freedom in the post-war world, paid for with the atomic bomb.

While traditional historians of the race for the atomic bomb have at times castigated and maligned or ignored these events over the past fifty years, this book documents and reestablishes them in their appropriate place in world events.

This rehabilitative work is based on facts newly uncovered that are contained in previously classified documents and uses other sources that prove the nuclear race was not only neck and neck – with the German program slightly ahead almost to the end of the war – but that, had it not been for other factors, the Germans or their remaining ally, Japan, may have won world domination using a German atomic bomb. Instead, using the cargo seized from U-234, the United States employed the new-won technology and dramatically altered the geo-political landscape of the future.

Finally, it is a curiosity that demands mentioning regarding the remarkable coincidence of the number designations of the captured U-boat, U-234, and its pestilent payload of uranium, U^{235}, being numerically sequential and both preceded by the same aberrant vowel. This coincidence is one of three of four haunting ironies that dangle obtusely throughout these appalling events like capricious cobwebs in the anteroom of a house of terrors; the presence of which should forewarn intrepid trespassers of the unsettling panoramas they are about to behold.

Endnotes – Prologue

1   U.S. National Archives, Northeast Region, Waltham, Massachusetts, *Navy Secret Dispatch #222115, U-234;Disposition of, 23 May 1945*, RG 181, box 531

2   U.S. National Archives, Northeast Region, Waltham, Massachusetts, *Log Of Public Relations - Restricted, 17 May 1945, Night Log* (Portsmouth Naval Yard), RG 181, box 531; U.S. National Archives II, *Memorandum For: Captain Herbster, Deputy Commander Northern Group, Subj: Publicity on Surrender of U-234, May 18, 1945*, RG 181, box 531; *Boston Globe*, May 19 and 20, 1945

3   U.S. National Archives, Northeast Region, Waltham, Massachusetts, *Navy secret dispatches #131509, Interviewing of Prisoners, 13 May 1945*, RG 181, Box 531, and *#151942, Disposition of U-234 Prisoners, 15 May 1945*, RG 181, box 531, (the exact same text in handwritten form, dated 16 May 1945, noting "this reenciphers COMINCH'S 151942," can be found at U.S. National Archives II, RG 38 – 370 15/09/04 box 13; *Boston Globe*, May 19 and 20, 1945

4   U.S. National Archives II, *Reports of Interrogation #5399, #1540: P/W Kessler, Ulrich, General der Flieger*, p. 4, Captured Personnel and Material Branch, Military Intelligence Division, War Dept., RG 38 – 370 15/09/01 box 2; U.S. National Archives II, Letter from General Ulrich Kessler titled *Personal Belongings* 28 May 1945, p. 2, RG 38 – 370 15/09/04 box 13

5   U.S. National Archives II, *Passengers and Crew of U-234 receipt, 15 May, 1945*, RG 38 – 370 15/09/04 box 13

6   *Boston Globe*, May 19 and 20, 1945; U.S. National Archives II, *Report On Interrogation Of The Crew Of U-234 Which Surrendered To The USS Sutton On 14 May, 1945 in Position 47°-07'N – 42°-25'W*, RG 38 – 370 15/09/01 box 2

7   U.S. National Archives II, *Navy Secret dispatch #151716, May 15, 1945*, RG 38 – 370

15/09/04 box 13; U.S. National Archives, Northeast Region, Waltham, Massachusetts, *Navy Secret dispatch #151942, Disposition of U-234 Prisoners, May 15, 1945*, RG 181, box 531, (the exact same text in handwritten form, dated 16 May 1945, noting "this reenciphers COMINCH'S 151942," can be found at U.S. National Archives II, RG 38 – 370 15/09/04 box 13

8   *Sharkhunters KTB 107*, p. 27; U.S. National Archives II, *Fragment of U-234 logbook*, RG 38 – 370 15/05/07 box 3-6; Robert K. Wilcox, *Japan's Secret War*, pp. 141, 142; U.S. National Archives II, transmission beginning *"unverified information received from Managing Director of Saudel Aircraft Works…"*, dated 8 May 1945, RG 38 – 370 1/4/7 box 113

9   U.S. National Archives, Northeast Region, Waltham, Massachusetts, *Navy Secret dispatch #26215, Mine Tubes, Unloading Of, 27 May 1945*, RG 181, box 531

10   U.S. National Archives II, *Manifest of Cargo For Tokio (sic) On Board U-234, Translated From German, 23 May 1945*, RG 38 – 370 15/05/07 box 3-6; U.S. National Archives II, original German loading manifest, RG 38 – 370 15/05/07 box 3-6

11   Wolfgang Hirschfeld and Geoffrey Brooks, *Hirschfeld: The Story of A U-boat NCO 1940-1946*, pp. 198,199

12   U.S. National Archives, Northeast Region, Waltham, Massachusetts, *Navy Secret dispatch #222115, U-234; Disposition of, 23 May 1945*, RG 181, box 531

13   David Irving, *The German Atomic Bomb*, p. 77; Robert Jungk, *Brighter Than A Thousand Suns*, p. 95

14   Dr. David Picker, *Hitler's Tabletalk*, (as quoted by Brooks and Hirschfeld in *Hirschfeld: The Story of a U-Boat NCO, 1940-1946* p. 230

15   Luis Alvarez, *Alvarez*, pp. 120, 121

16   Joseph Borkin, *The Crime and Punishment of I.G. Farben*, p. 116; Paul Manning, *Martin Bormann: Nazi In Exile*, p.153

17   Joseph Borkin, *The Crime and Punishment of I.G. Farben*, p. 3; Paul Manning, *Martin Bormann: Nazi In Exile*, p.153

The first atomic bomb used in war was dropped on Hiroshima Japan 6 August, 1945. It used uranium enriched in the isotope U^{235} as its nuclear fuel. Code named "Little Boy," this bomb killed between 90,000 and 146,000 people.

PART ONE

The Uranium Bomb

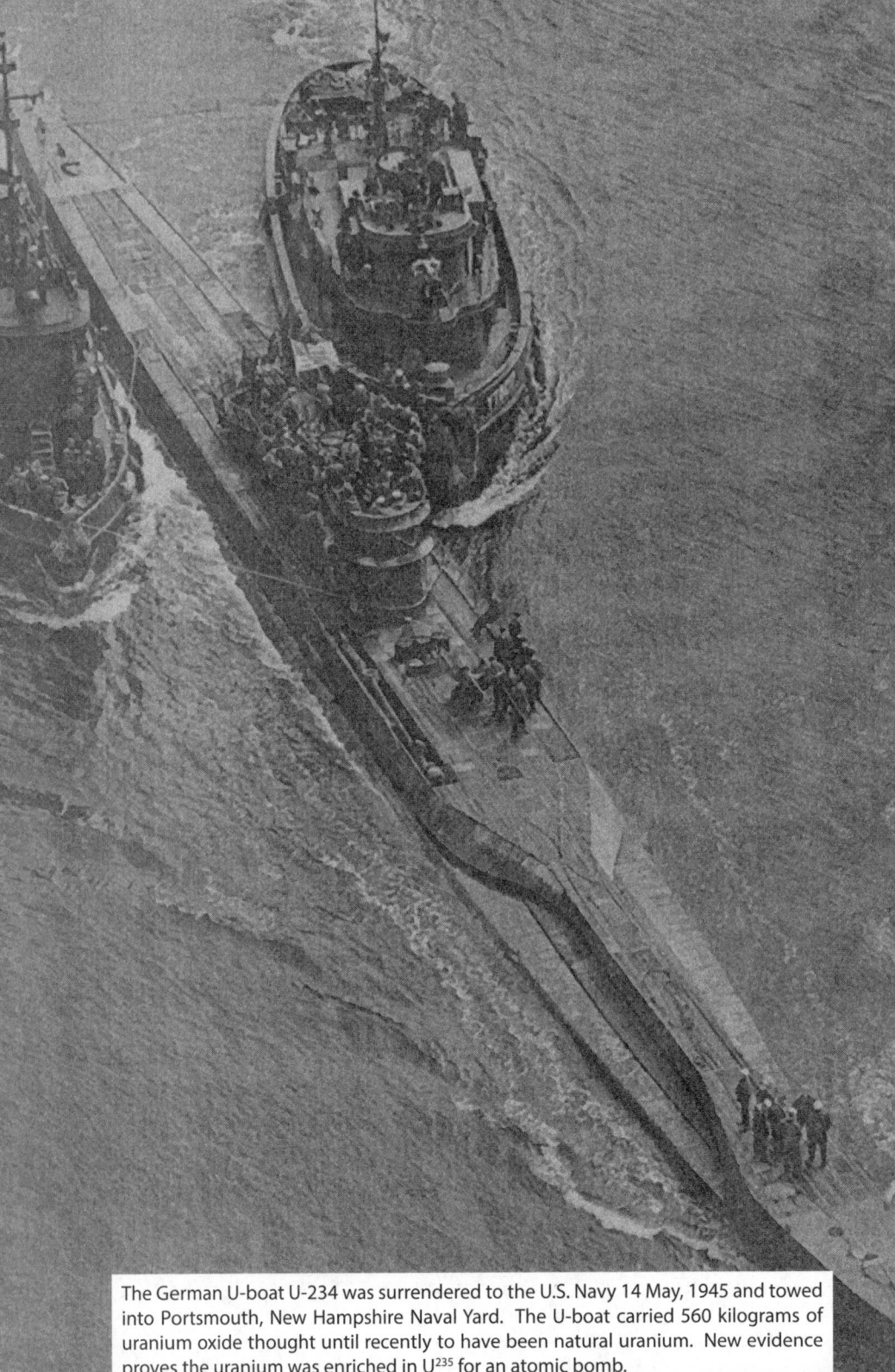

The German U-boat U-234 was surrendered to the U.S. Navy 14 May, 1945 and towed into Portsmouth, New Hampshire Naval Yard. The U-boat carried 560 kilograms of uranium oxide thought until recently to have been natural uranium. New evidence proves the uranium was enriched in U^{235} for an atomic bomb.

Chapter One

U-234/U²³⁵

"The most important and secret item of cargo, the uranium oxide, which I believe was radioactive, was loaded into one of the vertical steel tubes [of German U-boat U-234].... Two Japanese officers ... [were]... painting a description in black characters on the brown paper wrapping.... Once the inscription U235 had been painted on the wrapping of a package, it would then be carried over ... and stowed in one of the six vertical mine shafts."[1]
– Wolfgang Hirschfeld, Chief Radio Operator of U-234

"Lieut Comdr Karl B Reese USNR, Lieut (JG) Edward P McDermott USNR and Major John E Vance CE USA will report to commandant May 30ᵗʰ Wednesday in connection with cargo U-234."[2]
– US Navy secret transmission #292045 from Commander Naval Operations to
Portsmouth Naval Yard, 30 May 1945

"I just got a shipment in of captured material.... I have just talked to Vance *and they are taking it off the ship.... I have about 80 cases of U powder in cases. He* (Vance) *is handling all of that now."*[3]
– Telephone transcript between Manhattan Project security officers Major Smith
and Major Traynor, 14 June 1945.

The traditional history of the atomic bomb accepts as an unimportant footnote the arrival of U-234 on United States shores, and admits the U-boat carried uranium oxide along with its load of powerful passengers and war-making materials. The accepted history also acknowledges these passengers were whisked away to Washington for interrogation and the cargo was quickly commandeered for use elsewhere. The traditional history even concedes two Japanese officers were onboard U-234 and that they committed a form of unconventional Samurai suicide rather than be captured by their enemies.

The traditional history denies, however, the uranium on board U-234 was enriched and therefore easily usable in an atomic bomb. The accepted history asserted there was no evidence the uranium cargo of U-234 was transferred into the Manhattan Project (although recent admissions acknowledged this occurred). And the traditional history asserts the bomb components on board U-234 arrived too late to be included in the atomic bombs that were dropped on Japan.

The documentation indicates quite differently on all accounts.

Before U-234 landed at Portsmouth – before it even left Europe – United States and British intelligence knew U-234 was on a mission to Japan and that it carried important passengers and cargo.[4] A portion of the cargo, especially, was of a singular nature. According to U-234's chief radio operator, Wolfgang Hirschfeld, who witnessed the loading of the U-boat:

> The most important and secret item of cargo, the uranium oxide, which I believe was highly radioactive, was loaded into one of the vertical steel tubes one morning in February, 1945. Two Japanese officers were to travel aboard U-234 on the voyage to Tokyo: Air Force Colonel Genzo Shosi, an aeronautical engineer, and Navy Captain Hideo Tomonaga, a submarine architect who, it will be recalled, had arrived in France aboard U-180 about eighteen months previously with a fortune in gold for the Japanese Embassy in Berlin. I saw these two officers seated on a crate on the forecasting engaged in painting a description in black characters on the brown paper wrapping gummed around each of a number of containers of uniform size. At the time I didn't see how many containers there were, but the Loading Manifest showed ten. Each case was a cube, possibly steel and lead, nine inches along each side and enormously heavy.
>
> Once the inscription U235 had been painted on the wrapping of a package, it would then be carried over to the knot of crewmen under the supervision of Sub-Lt Pfaff and the boatswain, Peter Scholch, and stowed in one of the six vertical mineshafts.[5]

1	PACKAGE	2.5	CABLE	"
10	CASES	560.0	URANIUM OXIDE	JAP ARMY
17	PACKAGES	620.5	STOCK PARTS FOR PERCUSSION CAPS	"
1	CASE	25.0	HEMOGLOBIN	"
2	PACKAGES	59.0	DYES	".

Figure 1: 560 kilograms, almost 1,200 pounds of uranium oxide onboard U-234 destined for the Japanese army, was already enriched – possibly to bomb grade – based on evidence about the cargo, including the fact it was labeled U235.

Hirschfeld's straightforward account of the uranium being "highly radioactive" – he later witnessed the storage tubes being tested positively with Geiger counters[6] – and labeled "U235" provides profoundly important information about this cargo. U²³⁵ is the scientific designation of enriched uranium – the type of uranium required to fuel an atomic bomb.

While the uranium remained a secret from all but the highest levels within the United States until after the surrender of U-234, a captured German Enigma encoder/decoder had allowed the Western Allies to decode intercepted German radio transmissions. Some of these captured signals had already identified the U-boat as being on a special mission to Japan and even identified General Kessler and much of his cortège as likely to be onboard, but the curious uranium was never mentioned. The strictest secrecy was maintained, nonetheless, around the U-boat.

As early as 13 May, the day before U-234 was actually boarded by the *Sutton*'s prize crew, orders had already been dispatched that commanded special handling of the passengers and crew of U-234 when it was surrendered:

> Press representatives may be permitted to interview officers and men of German submarines that surrender. This message applies only to submarines that surrender. It does not apply to other prisoners of war. It does not apply to prisoners of the U-234. Prisoners of the U-234 must not be interviewed by press representatives.[7]

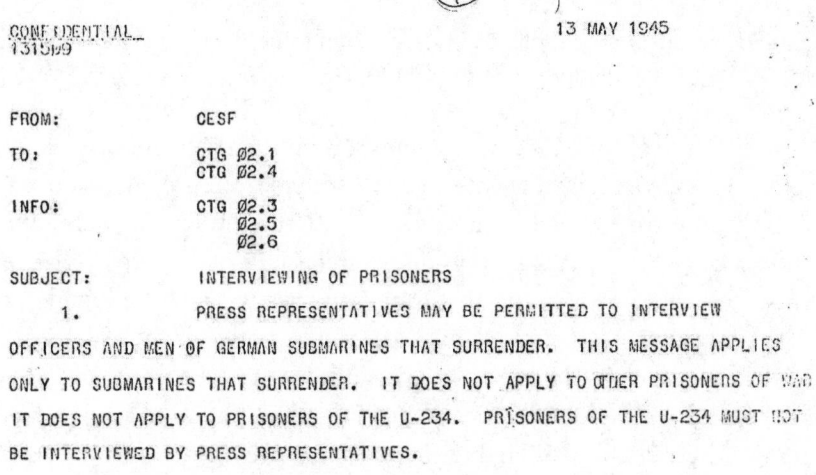

CONFIDENTIAL
131599

13 MAY 1945

FROM: CESF

TO: CTG Ø2.1
                CTG Ø2.4

INFO: CTG Ø2.3
                Ø2.5
                Ø2.6

SUBJECT: INTERVIEWING OF PRISONERS

    1.   PRESS REPRESENTATIVES MAY BE PERMITTED TO INTERVIEW
OFFICERS AND MEN OF GERMAN SUBMARINES THAT SURRENDER. THIS MESSAGE APPLIES
ONLY TO SUBMARINES THAT SURRENDER. IT DOES NOT APPLY TO OTHER PRISONERS OF WAR.
IT DOES NOT APPLY TO PRISONERS OF THE U-234. PRISONERS OF THE U-234 MUST NOT
BE INTERVIEWED BY PRESS REPRESENTATIVES.

Figure 2: Despite previous and later U-boat surrender practices, passengers and crew of U-234 were off limits to news media.

Two days later, while the Sutton was steaming toward Portsmouth with U-234 at her side, more orders were received. "Documents and personnel of U-234 are most important and any and all doubtful personnel should be sent here,"[8] the commander of naval operations in Washington, D.C. ordered. The same day, the commander in chief of the Navy instructed, "Maintain prisoners U-234 incommunicado and send them under Navy department representative to Washington for interrogation."[9]

S-E-C-R-E-T

151942

15 MAY 1945

ㄥ

FROM: COMINCH

ACTION: CESF

INFO: CINCLANT
                NYD PORTSMOUTH
                COM 1

SUBJECT: DISPOSITION OF U-234 PRISONERS

MAINTAIN PRISONERS U-234 INCOMMUNICADO
AND SEND THEM UNDER NAVY DEPARTMENT RE-
PRESENTATIVE TO WASHINGTON FOR INTERRO-
GATION.

Figure 3: The prisoners from U-234 were known to hold vital secrets, and thus required immediate, high-level interrogation.

The effort to keep U-234 under wraps was only partially successful. Reporters had been allowed to interview prisoners from previous U-boats, and, in fact, were allowed to interview captured crews from succeeding U-boats, as well. When the press discovered U-234 was going to be off limits, a cry and hue went up that took two days to settle. Following extended negotiations, a compromise was struck between the Navy brass and the press core.[10] The reporters were allowed to take photographs of the people disembarking the boat when it landed, but no talking to prisoners was permitted.[11] When they landed at the pier, the prisoners walked silently through the gawking crowd and climbed into buses, to be driven out of the spotlight and far from the glaring eyes of history.

On 23 May, the cargo manifest of U-234 was translated,[12] quickly triggering a series of events. On the second page of the manifest, halfway down the page, was the entry "10 cases, 560 kilograms, uranium oxide." Whoever first read the entry and understood the frightening capabilities and potential purpose of uranium must have been stunned. Certainly questions were asked. Was this the first shipment of uranium to Japan or had others already slipped by? Did the Japanese have the capacity to use it? Could they build a bomb?

Whatever the answers, within four days personnel from the Office of Naval Intelligence had brought U-234's second watch officer, Karl Pfaff – who had not been brought to Washington with the original batch of high-level prisoners but who had overseen loading of the U-boat in Germany – to Washington and interrogated him. They quickly radioed Portsmouth:

Pfaff prepared manifest list and knows kind documents and cargo in each tube. Pfaff states ... uranium oxide loaded in gold-lined cylinders and as long as cylinders not opened can be handled like crude TNT. These containers should not be opened as substance will become sensitive and dangerous.[13]

URANIUM OXIDE LOADED IN GOLD LINED CYLINDERS AND AS LONG AS
CYLINDERS NOT OPENED CAN BE HANDLED LIKE CRUDE TNT.
THESE CONTAINERS SHOULD NOT BE OPENED AS SUBSTANCE WILL BECOME
SENSITIVE AND DANGEROUS.

Figure 4: Stowing uranium in gold-lined containers all but proves the uranium was enriched. Gold is a very stable material against highly corrosive uranium and a suitable neutron shield, but is unnecessary and too expensive for transporting natural uranium.

The identification that the uranium was stowed in gold-lined cylinders and that it would become "sensitive and dangerous" when unpacked provides persuasive substantiation that this was U^{235} – enriched uranium – and not natural uranium. Uranium that has had its proportion of the isotope U^{235} increased compared to the more common isotope of uranium, U^{238}, is known as enriched uranium. When that enrichment becomes 70 percent or above it is bomb-grade uranium. The process of enriching uranium during the war was highly technical and very expensive – it still is.

Upon first reading the uranium on board U-234 was stored in gold-lined cylinders, this author tracked down Clarence Larsen, former director of the leading uranium enrichment process at Oak Ridge, Tennessee, where the Manhattan Project's uranium enrichment facilities were housed. In a telephone conversation I asked Mr. Larsen what, if anything, would be the purpose of shipping uranium in gold-lined containers.[14] Mr.

Larsen remembered the Oak Ridge program used gold trays when working with enriched uranium.

He explained that, because uranium enrichment was a very costly process, enriched uranium needed to be protected jealously; but because it is very corrosive, it is easily invaded by any but the most stable materials, and would then become contaminated. To prevent loss to contamination of the invaluable enriched uranium, gold was used. Gold is one of the most stable substances on earth. While expensive, Mr. Larsen explained, the cost of gold was a drop in the bucket compared to the value of enriched uranium. Would natural uranium, rather than enriched uranium, also be stored in gold containers? I asked. Not likely, Mr. Larsen responded. The value of natural uranium is, and was at the time, inconsequential compared to the cost of gold.

Assuming the Germans invested roughly the same amount of money as the Manhattan Project to enrich their uranium, which it appears they did,[15] the cost of the U^{235} on board the submarine was somewhere in the neighborhood of $100,000 an ounce – in 1945 dollars – by far the most expensive substance on earth. The fact the enriched uranium had the capacity to deliver world dominance to the first country that processed and used it made it priceless. A long voyage with the U^{235} stowed in anything but gold could have cost the German/Japanese atomic bomb program dearly in contaminated enriched uranium.

Further inquiries revealed gold was not only good for withstanding the corrosive characteristics of uranium but added a secondary benefit as a logical choice for protection from a nuclear chain reaction within the enriched uranium, as well.[16] While not the greatest of insulators from a chain reaction, gold absorbs neutrons at a relatively efficient rate. This is helpful especially for chain reaction-prone thermal neutrons that are often created when a neutron passes through water – a constant occurrence in the open mine tubes of a submarine, which is where the uranium tubes were stored. Neutrons passing freely through the cylinders could potentially have started a nuclear reaction. So besides shielding the uranium from contamination, the gold shield would also have helped insulate it from going critical.

The neutron-absorbing characteristic of gold, therefore, gave it an additional advantage as a deterrent from possible critical events within the U-boat.

Other substances, such as cadmium, are much better neutron absorbers but may not have been available to the Germans at the time.

In addition to the gold-lined shipping containers corroborating Hirschfeld's identification of the uranium as U^{235}, the description for handling the cargo and of the uranium's characteristics when its containers were opened also indicates the uranium was enriched and not natural uranium. Uranium of all kinds is not only corrosive, but it is toxic if swallowed. In its natural state, however, which is 99.3 percent U^{238}, the substance poses little threat to man as long as he does not eat it. The stock of natural uranium eventually was processed by the Manhattan Project originally had been stored in steel drums and was sitting in the open at a Staten Island storage facility.[17] Much of the German natural uranium discovered in salt mines at the end of the war also was stored in steel drums, many of them broken open. The material was loaded into heavy paper sacks and carried from the storage area to waiting trucks by apparently unprotected G.I.s.[18] Since then, more precautions have been taken in handling natural uranium, but at the time, caution was minimal and natural uranium was considered to be relatively safe.[19]

For the Navy, therefore, to note the uranium would become "sensitive and dangerous" when the containers were opened, and should be "handled like crude TNT," is another indicator the uranium was, in fact, enriched. Uranium enriched significantly in U^{235} produces a much higher level of alpha radiation than natural uranium. Alpha radiation alone, however, is not harmful; it is barely powerful enough to penetrate the dead layer of skin all humans carry. This, in and of itself, poses no threat.

But alpha radiation – which is merely neutrons being released by the U^{235} – in a large enough body of enriched uranium so dense a significant quantity of neutrons strike other atoms, causing them to split and release their nuclear energy, may potentially generate lethal gamma radiation.

Opening multiple containers each holding sub-critical amounts of enriched uranium in close proximity to one another, therefore, could raise the quantity of free neutrons available from the combined U^{235} to approach a critical state, and possibly initiate a slow chain reaction and the gamma radiation that would be released with it. Preventing such a potential health hazard would be handled with appropriate caution, such as warning against opening the containers, and that the contents would be "sensitive and dangerous" when opened, as the communiqué described.

Detractors to the suggestion the uranium on board U-234 was enriched argue that warning the uranium was "sensitive and dangerous" referred to it being pyrophoric, not radioactive. Pyrophoria describes substances that have the capacity of self combusting. The phenomenon occurs when uranium has been reduced to its metal form and then ground or otherwise converted to a fine grain, then exposed to oxygen.

But, as noted, this only occurs to uranium in its metallic state. Reduction to metal is the very last step that occurs in preparation for making a bomb. Uranium, raw or enriched, would not be stored or shipped in its metallic state for this reason.

The Manhattan Project intentionally converted its enriched uranium from tetraflouride form – its condition when processed through the beta calutrons – to tri-oxide for the express purpose of putting it in a stable ceramic state. The process is simple enough and certainly one the Germans would follow given the potential for disaster had they not. It is absurd to think they would ship 560 kilos of highly pyrophoric material of any kind anywhere, much less highly-valued uranium on a submarine loaded with near priceless cargo and intellectual capital in the form of important human beings, when this condition could be simply and inexpensively overcome. This is borne out by the fact literally every other reference to German uranium available – whether in the Salzburg mines, pitchblende from Joachimsthal or barrels bought from Union Miniére, states it was in oxide form.

Detractors to the enriched uranium idea forward other arguments against it, as well, such as if the uranium was highly enriched the gold lining in the cylinders would have melted. They assume this because of assumptions made in turn by Wolfgang Hirschfeld that the uranium was gamma radioactive.[20] Hirschfeld reported witnessing high radiation readings when the tubes were checked with Geiger counters. He reasoned the uranium was highly radioactive because it was exposed in a reactor, creating gamma radiation. Those who have tried to negate his testimony have nevertheless adopted his erroneous reasoning – and a resulting non sequitur. They denounce the high levels of radiation report by citing the fact Germany had no reactors, which is true, and therefore there could be no high radiation levels, which is untrue.

"That the gold would melt if enriched uranium was stored in it is a croc," responded Dr. Delmar Bergen, retired director of the Nuclear Weapons Program at Los Alamos. He explained that his program used highly neutron-absorbing cadmium, which is 40 times more efficient absorbing neutrons than gold, but that it did not produce enough heat to cause the cadmium to melt. "Gold would be even less prone to melt," assured Dr. Bergen.

But just as salient, there is no reason whatever the enriched uranium would have been exposed in a reactor. The two technologies are totally independent of one another. Although fueling a reactor with enriched uranium allows engineers to design a reactor that is much less complicated, a reactor will function without enriched uranium. On the other hand, reactors are not used in any way to create enriched uranium, and the fact Germany had no operating reactors is immaterial to the discussion. The traditionalists also forward the charge Germany had no production-level uranium enrichment plants, but, as shall be seen in a future chapter, this is also proven to be false.

In any case, while some gamma radiation results from natural alpha decay, this would not be life threatening and would not need shielding. On the other hand, risk of an inadvertent chain reaction is the key danger. Gold was a reasonable insulator to help shield neutrons, being 100 times more efficient at capturing thermal neutrons than lead.

While alpha radiation from enriched uranium is low-level radiation that would not melt the gold lining, it would violently excite a Geiger counter when generated at the high levels enriched uranium produces. This is what Hirschfeld witnessed.

And turning the tables, a question for the detractors is, if the uranium was not enriched, why use gold-lined cylinders in the first place? Detractors always avoid a reasonable explanation of their own for the gold, while trying to explain away the most logical conclusion made from it – that the uranium was enriched.

The reference to handling the containers of uranium in the same way TNT should be handled is apropos, as well. One of the basic rules for handling packaged TNT is to ensure large quantities are not stored in close proximity to one another, in order to avoid an accidental explosion of one batch igniting additional batches, and thus creating a catastrophic event. This aligns with one of the basic rules of handling enriched uranium, which requires that quantities approaching critical mass are not stored in close proximity to one another, in order to avoid creating a chain reaction – either causing irradiation or explosion. Other handling precautions for both substances have similar correlations.

By 16 June 1945, a cargo inventory of U-234 had been completed by the United States Navy.[21] The uranium was not on the list. It was not even marked as shipped out or having once been on board. It was never mentioned. It was gone – as if it never existed.

Where did the uranium go? Eleven days after U-234 was escorted into Portsmouth, and four days after Pfaff identified its location on the U-boat, a team was selected to oversee the offloading of U-234. Portsmouth received the following message:

Lieut. Comdr. Karl B Reese USNR, Lieut (JG) Edward P McDermott USNR and Major John E Vance CE USA will report to commandant May 30th Wednesday in connection with cargo U-234. It is contemplated that shipment will be made by ship to ordnance investigation laboratory NAVPOWFAC Indian Head Maryland if this is feasible.[22]

SECRET
292045 (P)   30 MAY 1945

FROM: CNO

TO: NYPORTS

INFO: NAV POW FAC INDIAN HEAD MD , INFORM OIL
                COMONE

SUBJECT: U-234, CARGO INFORMATION

REFERENCE: NYPORTS 281540

                LIEUT COMDR KARL B REESE USNR, LIEUT (JG)

EDWARD P MCDERMOTT USNR AND MAJOR JOHN E VANCE CE USA

WILL REPORT TO COMMANDANT MAY 30TH WEDNESDAY IN CONNECTION

WITH CARGO U-234.

Figure 5: Major John Vance of the Army Corps of Engineers, the parent organization of the Manhattan Project, was part of the otherwise all-Navy team that removed the captured cargo from U-234.

The order, dispatched by the commander of naval operations, is revealing if not outright startling for the selection of one member of its three-man team. Including Major Vance of the Army Corps of Engineers – and ultimately, if certain documents to be shared momentarily prove Major Vance to be a member of the Manhattan Project's intelligence team – in what was otherwise an all Navy operation, seems a telling selection. Someone, somewhere at a very high level, appears to have assured the Army was brought into the scavenging operation that had become U-234; not just any Army group, but the intelligence arm of the Manhattan Project.

If a telephone transcript taken from Manhattan Project archives refers to the same "Vance" as the Major assigned to offload U-234 – as it appears to – then Major John E. Vance was part of the intelligence team of America's super-secret atomic bomb project. The transcript is of a conversation between Manhattan Project intelligence officers Smith and Traynor and was recorded two weeks after "Major Vance" was assigned to the team responsible for unloading the material captured on U-234.

Smith: I just got a shipment in of captured material and there were 39 drums and 70 wooden barrels and all of that is liquid. What I need is a test to see what the concentration is and a set of recommendations as to disposal. I have just talked to Vance and they are taking it off the ship and putting it in the 73rd Street Warehouse. In addition to that I have about 80 cases of U powder in cases. He (Vance) is handling all of that now. Can you do the testing and how quickly can it be done? All we know is that it ranges from 10 to 85 percent and we want to know which and what.

Traynor: Can you give me what was in those cases?

Smith: U powder. Vance will take care of the testing of that.

Traynor: The other stuff is something else?

Smith: The other is water.[23]

TELEPHONE CONVERSATION BETWEEN MAJOR SMITH, WLO AND MAJOR TRAYNOR, 6/14/45.

S: I just got a shipment in of captured material and there were 39 drums and 70 wooden barrels and all of that is liquid. What I need is a test to see what the concentration is and a set of recommendations as to disposal. I have just talked to Vance and they are taking it off the ship and putting it in the 73rd Street Warehouse. In addition to that I have about 80 cases of U powder in cases. He (Vance) is handling all of that now. Can you do the testing and how quickly can it be done? All we know is that it ranges from 10 to 85% and we want to know which and what.

T: Can you give what was in those cases?

S: U powder. Vance will take care of the testing of that.

T: The other stuff is something else?

S: The other is water.

T: We can take a sample of each and have it analyzed. It will probably take two or three days. I will check around and let you know (probably tomorrow) just where the tests can be made and where the material can be stored. It may be a long distance but it will be safe.

S: That will be fine. Good-bye.

1. *Called Brodjon to see if Columbia had analyzing equipment set up. - B. will call me back.*

2. *Kusenbourn has lab demolished*

Figure 6: According to Manhattan Project telephone transcripts, an officer "Vance" was in charge of captured "U-powder" (undoubtedly uranium oxide) and other items matching descriptions in U-234's cargo manifest.

U-234's cargo manifest reveals that, besides its uranium, among its cargo were 10 "bales" of drums and 50 "bales" of barrels. The barrels are noted in the manifest to have contained benzyl cellulose, a very stable substance[24] that may have been used as a biological shield from radiation or as a coolant or moderator in a liquid reactor.[25] The manifest lists the drums as containing "confidential material." As surprising as it may seem, this secret substance appears to have been the "water" Major Smith noted in his discussion with Major Traynor, for nowhere on the German manifest is water listed. Why would water be described as "confidential material?" Why would Major Smith want the water tested? And what did he mean when he said that its concentration ranged "from 10 to 85 percent and we want to know which and what"?

The leaders of the German project to breed plutonium had decided to use heavy water, or deuterium oxide, as the moderator for a plutonium-breeding liquid reactor. The procedure of creating heavy water results in regular water molecules picking up an additional hydrogen atom. The percentage of water molecules with the extra hydrogen represents the level of concentration of the heavy water. Thus Major Smith's seemingly overzealous concern about water and his question about concentration is predictable, and even expected, if Smith suspected the material was intended for a nuclear reactor. In using heavy water as a major element of their plutonium breeding reactor project, it is easy to see why the Germans labeled the drums "confidential material." The evidence indicates that U-234 – if the captured cargo being tested by "Vance" was from U-234, which seems very probable given all of the details – carried components for making not only a uranium bomb, but a plutonium bomb, also.

Further corroborating the connection of the barrels and drums as those taken from U-234 is a handwritten note found in the United States' Southeast national archives held at East Point, Georgia.[26] Dated 16 June 1945, two days after Smith's and Traynor's telephone conversation, the note described how 109 barrels and drums – the exact total given in the Smith/Traynor transcript – were to be tested with Geiger counters to determine if they were radioactive. The note also included instructions that an "intelligence agent cross out any markings on drums and bbls. and number them serially from 1 to 109 and make note of what was crossed out." The note goes on to say this recommendation was given to and approved by Lt. Colonel Parsons, General Groves' right-hand man on the military side of the Manhattan Project. And lastly, the writer of the note had called Major Smith, apparently to report back to him, leading one to believe the note's author might have been Major Traynor.

Was the captured cargo discussed by Smith and Traynor from U-234? The presence of a Mr. "Vance" who was in charge of "U powder," almost certainly proves so. The documents under consideration and the conversation they detail are from Manhattan Project files and are about men who worked for the Manhattan Project. Using the letter "U" as an abbreviation for uranium was widespread throughout the Manhattan Project. That there could have been another "Vance" who was working with uranium powder – especially "captured" uranium powder – is improbable. And the fact the contents of the barrels listed on the U-boat manifest were identified as containing a substance likely to be used in a nuclear reactor, benzyl cellulose, and the barrels in the Smith/Traynor transcript and the untitled note – as well as the drums – were tested for radioactivity by Geiger counter, certainly links the "captured" materials to no other source than U-234.

Besides linking the "captured" uranium on board U-234 to the Manhattan Project through Major Vance, another striking and important detail is revealed in the Smith/Traynor telephone transcript. The uranium is spoken of as being in "about 80 cases." Assuming these are the same containers the uranium was shipped in, and there is good reason to believe they are, those cases would have been the same gold-lined cylinders mentioned in Navy secret cable #262151, referred to earlier. To have distributed up to 560 kilograms of natural uranium into 80 cases – whether as stowed on U-234 by the Germans or after offloading by the Americans – does not make sound economic sense. In terms of monetary cost or in terms of space, which was at a desperately high premium inside U-234, such a distribution is perplexing. The logical option would have been to transport natural uranium in as large a quantity as reasonable for lifting and stowage, saving space and cost. To have divided 560 kilograms into 80 cases means each case weighed about seven kilograms (15.5 lbs.), not an efficient volume for shipping and handling at all.

But if the uranium was enriched there would be a requirement to keep the powder separated into sub-critical quantities to avoid creating a critical mass – and the devastating nuclear chain reaction that would follow. In optimum conditions, critical mass was 15 kilograms. Stowing half of critical mass in each container, about seven kilograms – which just happens to be the amount of 560 kilos divided into 80 cases – would be a safe, and logical, quantity.

But in the U-boat, due to space constraints, the containers of enriched uranium would be packed in close proximity to one another, still supporting a critical scenario. The solution: make the containers cylindrical so when tightly packed the walls would not touch on all surfaces, breaking up the mass and reducing the chance of a reaction.

The new-found evidence taken en mass demonstrates that, despite the traditional history, the uranium captured from U-234 was enriched uranium that was commandeered into the Manhattan Project more than a month before the final uranium slugs were assembled for the uranium bomb. The Oak Ridge records of its chief uranium enrichment effort – the magnetic isotope separators known as calutrons – show that a week after Smith's and Traynor's 14 June conversation, the enriched uranium output at Oak Ridge nearly doubled – after six months of straight-line output.[27] Edward Hammel, a metallurgist who worked at the Chicago Met Lab and Los Alamos, where the enriched uranium was fabricated into the bomb slugs, corroborated this report of late-arriving enriched uranium. Mr. Hammel told the author very little enriched uranium was received at the laboratory until just two or three weeks – certainly less than a month – before the bomb was dropped.[28]

The Manhattan Project had been in desperate need of enriched uranium to fuel its lingering uranium bomb program. The cumulative evidence seems very persuasive that U-234 provided the enriched uranium needed, as well as components for a plutonium breeder reactor.

Endnotes: Chapter One – U-234/U^{235}

1   Wolfgang Hirschfeld and Geoffrey Brooks, *Hirschfeld: The Story of A U-boat NCO 1940-1946*, pp. 198,199

2   US National Archives, Northeast Region, Waltham, Massachusetts, Navy secret dispatch #292045, U-234, Cargo Information, 30 May 1945, RG181, box 531

3   US National Archives Southeast Region, East Point, Georgia, telephone transcript titled Telephone Conversation Between Major Smith, WLO and Major Traynor, 14 June, 1945

4   US National Archives II, extract of intercepted transmission sent from Chief Inspector in Germany to Bureau of Military Operations and Military Affairs, #165, 15 April, 1945, RG 38 – 370 01/04/07 box 113

5   Wolfgang Hirschfeld and Geoffrey Brooks, *Hirschfeld: The Story of A U-boat NCO 1940-1946*, pp. 198,199

6   Wolfgang Hirschfeld and Geoffrey Brooks, *Hirschfeld: The Story of A U-boat NCO 1940-1946*, Appendix

7   US National Archives, Northeast Region, Waltham, Massachusetts, Navy confidential dispatch #131509, Interviewing of Prisoners, 13 May 1945, RG 181, box 531

8   US National Archives II, Navy secret dispatch #151716, (no title/subject given), 15 May 1945, RG 38 – 370 15/09/04 box 13

9   U.S. National Archives, Northeast Region, Waltham, Massachusetts, Navy Secret dispatch #151942, Disposition of U-234 Prisoners, May 15, 1945, RG 181, box 531, (the exact same text in handwritten form, dated 16 May 1945, noting "this reenciphers COMINCH'S 151942," can be found at U.S. National Archives II, RG 38 – 370 15/09/04 box 13

10   US National Archives, Northeast Region, Waltham, Massachusetts, Log of Public Relations – Restricted, by Commander N.R. Collier, 17 May 1945; transcript, Telephone Conversation Between Capt. V.D. Herbster, USN (Ret.), and Commodore Kurtz, U.S.N. E.S.F.,

18 May 1945; second telephone conversation transcript Captain Herbster and Commodore Kurtz, 18 May 1945, all can be found in RG 181, box 531

11   US National Archives, Northeast Region, Waltham, Massachusetts, Log of Public Relations – Restricted, by Commander N.R. Collier, 17 May 1945; transcript, Telephone Conversation Between Capt. V.D. Herbster, USN (Ret.), and Commodore Kurtz, U.S.N. E.S.F., 18 May 1945; second telephone conversation transcript Captain Herbster and Commodore Kurtz, 18 May 1945, all can be found in RG 181, box 531

12   US National Archives II, Manifest of Cargo For Tokio On Board U-234, translated from German, 23 May, 1945, RG 38 – 370 15/05/07 box 3; original German loading manifest, RG 38 – 370 15/05/07 box 3

13   US National Archives, Northeast Region, Waltham, Massachusetts, secret dispatch #262151, Mine Tubes, Unloading of, 27 May, 1945, RG 181, box 531

14   Personal telephone conversation between the author and Clarence Larsen, Director of Y-12 calutrons operations at Oak Ridge, no date recorded

15   Joseph Borkin, *The Crime and Punishment of I.G. Farben*, p 116; Paul Manning, Martin Bormann: Nazi In Exile, p.153; compare to Chapter Four, page 82

16   Dr. John Poston, Sr., Nuclear Engineering Department Head, Texas A&M University, correspondence with author, December 7, 8, 11, 2000

17   Richard Rhodes, *The Making of the Atomic Bomb*, p. 427

18   Richard Rhodes, *The Making of the Atomic Bomb*, pp. 608, 609

19   Richard Rhodes, *The Making of the Atomic Bomb*, p. 461

20   Wolfgang Hirschfeld and Geoffrey Brooks, *Hirschfeld: The Story of A U-boat NCO 1940-1946*, pp. 228, 229

21   US National Archives II, Manifest of cargo for Tokio (sic) on Board U-234 – forwarding of; original German loading manifest, RG 38 – 370 15/05/07 box 3

22   US National Archives, Northeast Region, Waltham, Massachusetts, Navy secret dispatch #292045, U-234, Cargo Information, 30 May 1945, RG181, box 531

23   US National Archives Southeast Region, East Point, Georgia, telephone transcript titled Telephone Conversation Between Major Smith, WLO and Major Traynor, 14 June, 1945

24   Personal telephone conversation between the author and Dr. Susan Frost, PhD, Associate Professor of Biochemistry and Molecular Biology, College of Medicine, University of Florida, 30 August 1999, also Dr. Wentworth, University of Houston

25   Interscience Publishers, *Concise Encyclopedia of Nuclear Energy*, p. 688

26   US National Archives Southeast Region, East Point, GA, untitled handwritten note dated 6/16/45

27   US National Archives Southeast Region, East Point, GA, Beta Oxide Transfer Report, RG 194 – 69 A 406 section 326 box 17; see also chart on page __

28   Personal telephone conversation between the author and Edward Hammel, Manhattan Project metallurgist, 14 May, 1996

General Leslie R. Groves oversaw the Manhattan Project, the United States' program that developed the first nuclear weapons. In his previous assignment he built the Pentagon.

Chapter Two

The Two Billion Dollar Bet

"A study of the shipment of (bomb-grade uranium) for the past three months shows the following... : At the present rate we will have 10 kilos about February 7 and 15 kilos about May 1."[1]
– From a memo written by chief Los Alamos metallurgist Eric Jette, 28 December, 1944.
The uranium bomb required 50 kilos by July 24.

By mid-May of 1945, as U-234 was being escorted into Portsmouth, almost two billion dollars had been spent on the Manhattan Project, making it the greatest wager ever to that point in time. The man who threw the dice, and was about to lose it all, was Brigadier General Leslie Richard Groves.

In the course of just three years, using taxpayers' money unbeknownst to them, Groves had built a secret industry that outstripped any other enterprise on earth. He had purchased vast tracts of land in Washington state, Tennessee, New Mexico and elsewhere, engulfing hundreds of thousands, if not millions, of acres. On these reservations he built huge factories that contained the most advanced technology on the face of Earth. He made multi-million dollar deals with many of the globe's top companies – companies like DuPont, Westinghouse, and Raytheon.

To support these contracts and newly constructed facilities, he built whole towns, complete with roads, schools, postal services, banks, unions and everything else necessary to maintain a community. And he manned these municipalities with hundreds of thousands of workers and their families, including many of the greatest intellects alive. No fewer than thirteen of the physicists and chemists involved in the Manhattan Project either had already won, or later would go on to win, the Nobel Prize.

All of this had been assembled and focused on one task – to make an atomic bomb. Now the effort seemed to be exploding in his face.

The construction of an atomic bomb requires two things: enough fissile material to achieve critical mass and explode, and a trigger to start the explosion. Despite the immense investment, progress was remarkably slow on both requirements. Contrary to presently accepted history, by mid-May of 1945, neither requirement had been obtained. According to recently uncovered information from contemporaneous Manhattan Project documents – enriched uranium production charts and memos on metallurgical progress – and other never-before-revealed sources, including first-hand information revealed to the author during interviews with Manhattan Project personnel, the objectives still had not been achieved. And Groves had a third requirement that was about to make the other two points moot. Time was a factor, and it was running out.

Germany, the chief rival in the atomic bomb race according to intelligence reports,[2] – notwithstanding its now-surrendered status – planned to provide its Asian ally, Japan, with an atomic bomb[3] to use in the Pacific. U-234 had not been the only U-boat scheduled to voyage to Japan.[4] At least one other vessel, possibly more, apparently also carried in its belly enriched uranium intended for Tokyo.

Apparently, the race for the atomic bomb was much closer than most would have supposed – possibly even closer than Groves thought. After all, the General had spy Paul Rosbaud, code named "Griffin," keeping the Allies informed of German progress and possibly even of shipments to the Island Nation. There seems to have been no such counterpart in Japan to serve Groves as a conduit. If uranium had been sent to Japan, as appears probable, Groves most likely knew through Rosbaud, but what was happening to it once it arrived there, he could only guess.

Groves was not pressured by this threat alone, he also had to worry about the fact that, should the Allies' war effort survive the German/Japanese conspiracy, in July, Truman, Churchill and Stalin were scheduled to meet in Potsdam to partition the remnants of Europe that the Third Reich had left behind. The result of the Potsdam Conference would go a long way toward deciding the balance of power in the post-World War Two Era.[5] Additionally, in the Yalta Conference held earlier in February, Stalin had already declared his intent to go to war with Japan as soon as he became unencumbered in Europe, a promise he kept in mid-August.[6]

This was the deciding factor: the impetus for much of what would follow regarding U-234 and the conclusion of the war.

If Russia joined the war in Japan before the United States completed its operations to overtake the nation, all the Soviets had to do was jump across the Sea of Japan and overwhelm the tiny nation. The United States, still thousands of miles away and unable yet to mount an offensive on

Tokyo, would be powerless to stop them. "To the victor go the spoils," and so the entire Asian landmass with all its people, resources, industry, and commerce, as well as Eastern Europe, would fall into the hands of the Soviet Union. The United States would be relegated, once again, to "also ran," but in a much more precarious position.

A demonstration of the power of "the bomb" to end the war with Japan – while circumventing Stalin's plans in Asia and displaying to the rest of the world the United States possessed this awful weapon – would establish America as the military leader of all nations. And it would certainly impact any negotiations and the resulting geo-political complexion of the modern age.

But here stood Groves, as yet unsuccessful, with the sands of time slipping through his hands. Despite massive, sometimes reckless, always all-out spending; despite playing all the odds, even those with the slimmest chance of winning; despite assembling the greatest braintrust ever brought together in the United States; and even despite Groves' own expansive experience and unquestioned self-confidence, the gamble appeared to be a bust.

Almost $2 billion to produce just over 100 pounds of fissile material for the uranium bomb and about 30 pounds for the plutonium bomb, and a way to detonate them, had not bought enough brainpower to meet the deadline. The cost, had the effort been successful, equaled almost $100,000 per ounce of enriched uranium – in 1945 dollars.

Although the great effort had been successful enriching uranium and reducing it to its explosive metallic form, it appears over half of the hard-earned material never saw a uranium bomb; it was secretly being used to fuel the huge plutonium-breeding reactors at Hanford, Washington. The reactors, fueled by the enriched uranium, would produce several orders of magnitude of more explosive plutonium than the enriched uranium they consumed, promising quicker, easier, less expensive bombs. And many more plutonium bombs could be produced than the single uranium bomb that could have been produced with the amount of enriched uranium consumed in the reactors.

The end result for the uranium enrichment effort was less than half of the enriched uranium metal required for a nuclear device was available by mid-May, according to calculations made by top Manhattan Project metallurgist, Eric Jette[7] and with which later information agrees.

> A study of the shipment of (bomb-grade uranium) for the past three months shows the following …: At the present rate we will have 10 kilos about February 7 and 15 kilos about May 1.[8]

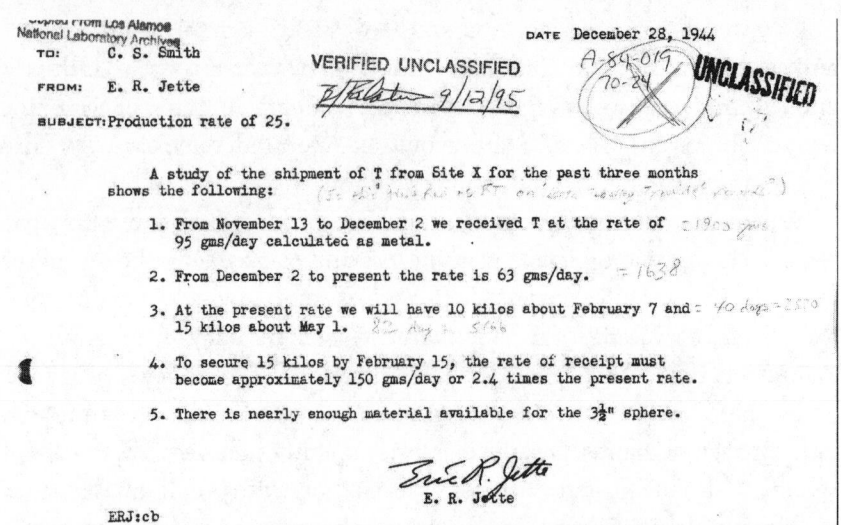

Figure 1: The Manhattan Project's chief metallurgist predicted enriched uranium output of only one critical mass within three months of the uranium bomb being dropped on Hiroshima. The uranium bomb used three critical masses.

Even doubling that rate of output, the program would fall far short of the amount required for a bomb to have been dropped in early August to pre-empt Stalin. And yet the bomb dropped on Hiroshima is known to have been at least a 50-kilogram uranium bomb.

Jette's calculations correspond almost precisely with and are validated by information supplied in Richard Rhodes' book *The Making Of The Atomic Bomb*, in which Rhodes sets the amount of enriched uranium metal available for a uranium bomb by April 1945 as "a near critical assembly."[9] According to Rhodes' calculations, which are based on information recorded at the time by James Bryant Conant, one of the scientific advisors on the Manhattan Project and president of Harvard University, 42 kilograms, or 92.4 pounds, of enriched uranium is equal to 2.8 critical masses.[10] One critical mass therefore, the amount barely available in mid-April with only three months of production time left, is exactly 15 kilograms, or 33 pounds, the amount Jette predicted would be available by 1 May and which Rhodes validates.

In theory, one critical mass was all that was needed to make a bomb; but in reality, due to inefficiencies caused by impurities still mixed throughout the enriched uranium, the bomb actually required over three critical masses in order to achieve the level of explosion desired. Robert Serber, who wrote *The Los Alamos Primer*, gives the total figure for the uranium bomb at "about 50 kilograms,"[11] over three times critical mass.

The point is, in mid-April, after almost a year of processing enriched material, because of the demand to use enriched uranium to produce the much more practical and powerful plutonium bomb, the uranium program had barely one-third the processed uranium required to make a weapon.

The loss of the uranium bomb option would have been inconsequential with a valid plutonium bomb, but as already noted, two months earlier at Yalta, Stalin had agreed to enter the war against Japan "two or three months after Germany has surrendered."[12]

Serious issues triggering the plutonium bomb had not been resolved in April, and it became apparent if they were not resolved, Soviet entrance into the war in Japan before a United States atomic bomb had been developed and demonstrated, had the potential to be catastrophic.

The American strategy for defeating Japan was to avoid facing Japan at her strongpoints where she had amassed huge armies to overrun and occupy important countries throughout the Pacific Rim. She held China, Korea, Singapore, Hong Kong, and Southeast Asia. The United States instead had followed the strategy of island hopping to the Japanese homeland. This gamble avoided facing Japanese strength on Japanese terms, and instead allowed the Americans to fight smaller – but still costly – and still important strategic battles on her own terms at places like Midway, Guadalcanal, Iwo Jima, and Guam as it made its way back across the Pacific to eventually attack Honshu, the Japanese main island. Honshu contained Tokyo, the Japanese government, and the greatest population of Japanese people. When Honshu was taken, the war against the Japanese would be won.

But if the Soviets entered the war with Japan, they potentially could move their four million-man army across Asia on the Trans-Siberian Railroad and force a single, overwhelming attack on Honshu, jumping just a few hundred miles from Soviet soil across the Sea of Japan onto the main island. United States troops would have looked on helpless from Tinian and Guam and the other, now irrelevant, rocks they held throughout the Pacific, still over a thousand miles away, as the entire Asia/Pacific Rim and all its riches and resources fell into Soviet hands.

Had this happened, the post-war world would have been a decidedly different place than we know it. The Cold War, which had left Russia consigned to controlling a few Eastern European countries – with the specter of nuclear retaliation from the United States hanging over its head if it tried to expand from there – would instead have been much more chilling for Americans and the democratic world in general.

Instead of holding only Eastern Europe, the Soviet Union could have controlled all of the major landmass, commercial- and trade-centers, and populations in the traditional "Old World," with the exception of Western Europe.

Without a demonstrated nuclear deterrent to stop them, with the United States an ocean away, it is probable the African Continent would have fallen without a struggle. The question is what would have happened to the remaining Allies in Western Europe – Britain, France and some of the smaller countries? It seems probable, with these exhausted countries already spent from their efforts in the war, with the American public tired of the war as well and unwilling to come to their aid again from across the Atlantic, they would quickly have fallen under Russian influence, too.

How would the Soviet Union have looked at America from there, alone across the Atlantic and unable to keep it at bay – at least initially – without a nuclear weapon? Would the Soviets have come after the United States? Or would they have been satisfied to leave it alone?

Or, once the United States had successfully completed its nuclear weapon, which it assuredly would have, would it attack the Soviet Union and start a new, eminently greater conflagration while it still had the atomic upper hand? The answers, probably fortunately, we will never know.

Now, with enriched uranium stocks depleted by plutonium demand and the plutonium bomb, in turn, so far undetonatable, the entire enormous nuclear enterprise appeared destined for defeat.

Yet even now, both Groves and his superiors knew the gamble had been a strategic imperative. To sit on the sidelines of international influence, when America was just coming into its own; to allow fascist, communist or imperialistic governments to control the destiny of the world – especially of the free nations – was inconceivable. The wager was essential no matter how small the chance of winning.

For the opportunity even to sit at the table and bet, knowing the stake was world dominion, Roosevelt had anted-up $2 billion, and, with foreknowledge some say, had allowed Pearl Harbor to be bombed. Thus the United States entered the war for a chance to play the nuclear game. Now the deck almost had been played out and the clear winner was going to be its greatest future opponent.

Groves from the very beginning when he took over the Manhattan Project from Colonel J.C. Marshall in September of 1942, [13] despite all his later efforts, had given the improbable scheme a small chance of success.[14] Marshall had been the Manhattan, New York district engineer for the Army Corps of Engineers. He was assigned to the project shortly after Roosevelt, in late 1939[15], received the famous letter written by Albert Einstein that explained the destructive realities of nuclear energy and that the Germans were working feverishly on its unleashing. Einstein had written the letter at the behest of two renowned Hungarian physicist émigrés, Eugene Wigner and Leo Szilard. The letter was delivered personally to the president by economist and Roosevelt confidant Alexander Sachs, who read it to the president aloud in the Oval Office.

Roosevelt by his own native genius seems quickly to have understood the full implications of the development. Before Sachs left the White House that day, the President had established a committee for pursuing nuclear energy.

But despite Roosevelt's quick reflexes, the work moved slowly. Responding to a report by aide Vannevar Bush two years later, in the early Spring of 1942, Roosevelt – who seemed to understand the urgency of the atomic initiative better than most of his nuclear advisors – wrote emphatically, "The whole thing should be pushed not only in regard to development, but also with due regard to *time*. This is very much of the essence."[16] The President seems to have been the only one who understood the full gravity of the circumstances.

When James B. Conant reported in mid-1942 that Germany might be ahead in the arms race by as much as a year[17] – and despite traditional history there is evidence this was so – impetus was finally given to the program, but it still took until September of that year to recruit Groves.

The colonel who in the early part of the decade had overseen the construction of the great symbol of United States military might, The Pentagon, had been made a brigadier general responsible for the development of the weapon ultimately destined to guarantee that power. Groves' response to learning the project for which he was being recruited could single-handedly win the war speaks volumes about the size of his ego and the extent to which his experience building the Pentagon and handling a $10 billion budget as the number two man in the Corps of Engineers had alienated him from feelings of mere human dimensions. He said simply: "Oh."[18]

The one thing Roosevelt did not need to worry about with Groves was wasted time. The general went to work immediately, criss-crossing the country to familiarize himself with the theory and processes and all of the research and development programs presently in progress. What he found was discouraging.

First, uranium, at least at the time, was relatively rare. Experts in the United States knew of only a few light deposits of the very heavy element but were doing little to mine it. Up to that point, there had not been a lot of use for uranium except in ceramic glazes. To get what it needed, the Manhattan Project would have to go outside of the sovereign borders of the United States, or so it seemed.

In a quirk of circumstance, over 1,000 tons of raw uranium ore had been sent to New York and was sitting in open steel drums in a warehouse on Staten Island.[19] The uranium had come from what Groves later identified, wrongly, as the richest uranium reserves in the world – those of the Belgian Congo – by way of Belgium and the Brussels-based company that owned the mines, Union Miniére. Union Miniére had provided rare-earth minerals for radiation studies performed by the famous French Curie family.

Groves' misstatement that the Belgian Congo held the richest uranium reserves is the leadoff in a long litany of hidden or half-truths, shaded assertions and outright lies later employed to paint a public picture decidedly different than the events that actually transpired. The details of this deception will be outlined later. Simply put, the mischaracterization is a single brushstroke – among a multitude – that makes up part of a larger picture created after-the-fact to hide the evidence that the Third Reich already had in its possession far more raw uranium than it would ever need for its purposes. The Reich held within its hands total control of the largest and most high-grade uranium ore deposit in the world, that at Joachimsthal, Czechoslovakia.

The president of Union Miniére, M. Edgar Sengier, having been approached previously by agents of the German government to buy the valuable mineral stocks, carefully avoided closing a deal with the German emissaries. Sengier knew of uranium's ultimate possibilities. Through his dealings with the Curies he had been invited by Frederic Joliot-Curie in 1939 to help build an atomic bomb in the Sahara desert, according to General Groves's book, *Now It Can Be Told*.[20]

Such a fascinating revelation from Groves demands a question: Build an atomic bomb for whom? Certainly Joliot-Curie was not planning it for personal world dominion. He must have known such a project could only be accomplished at enormous cost and effort, if it was possible at all. Given later accusations regarding Joliot-Curie that show every indication of having been true, and despite his reported membership in the French resistance, it is possible he planned on consorting with the Germans. At any rate, Sengier appears to have declined that offer, as he presently did the agents' bid for the bulk uranium stores.

Instead, right under the Germans' noses, he shipped the uranium to the United States for safekeeping. Once having made such a prudent and noble move at the potential cost of the loss of great profit for himself and his company, not to mention the threat to his physical safety that defying the Nazis could mean, he tried to make a deal with the United States to cover his lost investment. But the old Manhattan Project regime, for whatever reason, had not responded.

Groves now snapped it up. Over twelve hundred tons of uranium might be enough to harvest the 110 pounds of U^{235} needed to make a bomb. But raw uranium ore is only the basest form of uranium. From the ore, full of a variety of polluting elements and minerals, pure uranium must be refined; a considerable process in and of itself. Then the real challenge begins: Uranium atoms, like most elements, exist in various versions called isotopes. These different versions of the atom contain the same numbers of protons and electrons, which define the element and create its characteristics, but have a different number of neutrons, which, while not changing the element's characteristics, alter the atom's structure and weight.

The vast majority of uranium is the more stable isotope identified as U^{238} (U for uranium, 238 for this particular isotope's atomic weight), which constitutes 99.3 percent of all of the uranium on earth. The remaining less-than-one percent is mostly U^{235} – the fissile form of uranium. When a U^{235} nucleus absorbs a passing neutron energy is absorbed, offsetting to a degree the nuclear binding power and causing the nucleus to become unstable. Often the nucleus will fracture and divide, leaving two sub-uranic elements behind. At the same time, it releases additional neutrons along with a portion of the energy that had kept the uranium nucleus bound together. This nuclear energy is by far the strongest force known to man. Although because of each atom's minuscule measurements the energy released on a per atom basis seems like an infinitesimal force, actually, the power discharged is proportionally enormous.

To appreciate the truly diminutive size of an atom, journalist Chapman Pincher has given the following scale against which the minuteness of atoms can be measured. Envision a straight pin magnified so large that its head lay in London, England and its point terminates in the country of Bangladesh, on the far side of India – a distance covering approximately one-third the circumference of the earth. The atoms of such a needle would be the size of golf balls.[21] Yet according to real-world examples cited in Richard Rhodes' book, *The Making of the Atomic Bomb*, the strength of the nuclear force in a single atom contains enough energy to make a grain of sand jump, a mass hundreds of thousands if not millions of times greater than that of the atom. Rhodes adds that there is enough power in one cubic meter of uranium to lift one million million kilograms (or 2.2 million million pounds) 27 miles into the air. Put another way, one pound of uranium can produce nine million kilowatt hours, for which New York City would pay about $1.2 million.

Almost as soon as the first atom was split, physicists the world over realized if these great forces could be systematically released and controlled in large quantities of atoms, an enormous source of energy would be made available. On the heels of this realization came the revelation that if this energy could all be released in an instant, a super powerful explosion would occur, the likes of which had not been experienced on earth.

Calculations and experiments soon proved that in properly prepared uranium, for each neutron that split a nucleus, of the many neutrons that would be released an average of two-and-a-half would hit and split other nuclei, which would split yet two more each, and so on – creating a chain reaction that theoretically could sustain itself until the nuclear fuel ran out. In reality, the heat generated from the reactions expands the assembly until the chain reaction is broken – but not before huge amounts of energy are released. The knowledge the atom was split, along with the fact Nazi Germany was the first to uncover these cosmic secrets, is what caused Einstein, Szilard and Teller to write their famous letter of warning to Roosevelt.

The great challenge of this task for all warring factions was in accumulating enough uranium that was predominantly pure U^{235}, and whose atoms were closely enough positioned together, so released neutrons could reach the surrounding U^{235} atoms and create a chain reaction. A method had to be found to virtually pluck U^{235} atoms one at a time from the average of 140 U^{238} atoms surrounding each one of them and gather them together in a single body of U^{235}. Given the acutely minute, super-submicroscopic media to be meddled with and the overwhelming ratio of U^{238} to U^{235}, the prospects were surely daunting.

When Groves was given the assignment to oversee this Draconian task in the fall of 1942, however, he had been told by his superior that the project was well in hand. He was stunned to find upon review that so little had in fact been accomplished.

For starters, almost no one in the United States had been able to technically devise how to separate U^{235} from raw uranium. Thus far everything was theory – with one small exception. Nobel Laureate Dr. Ernest Lawrence at the University of California in Berkeley was just in the process of developing an electro-magnetic mass separator that, using mammoth-sized magnets and hundreds of thousands of volts to power them, could separate U^{235} from U^{238} to at least a nominal degree of enrichment. Groves presumably was encouraged when he heard about the breakthrough.

Traveling to Berkeley, the General entered Lawrence's laboratory and was brought to where he could see the enriched uranium product – he was led to a microscope. Undoubtedly dumbfounded and disappointed, Groves bent over the lens to see a spec of uranium that measured 75 micrograms of only 30 percent enriched uranium.[22] For comparison, a dime weighs 2.5 million micrograms. He knew by this time the amount needed for a bomb was still a matter of theory. But estimates ranged anywhere from five pounds to 600 pounds (Manhattan Project scientists would ultimately conclude the bomb would need to be about 110 pounds) of from 80 to 90 percent enriched material. Compared against the meager offering he was staring at through the microscope lens, the chances of producing the roughly calculated critical quantities in bulk production amounts, and in a usable time frame, were so astronomical as to be meaningless.

Despite Groves' disappointment, the perennially optimistic Lawrence assured the General what he had seen represented great strides, and that from this feeble foundation he could build a device capable of separating uranium in mass production quantities – tens of grams at a time. Groves was nonplused. They were still talking in fractions of ounces. But Lawrence's process was the best chance he had – for everyone else so far, any kind of serious isotope separation had been impossible.[23]

While in Berkeley, the new-formed cradle of American nuclear research, the General also took the time to visit several other researchers, experimenters and theoreticians, and this proved to be fortuitous. He met J. Robert Oppenheimer, the man Groves would choose to direct the laboratory that would develop the United States atomic bomb. Robert Serber, a close friend and co-worker of Oppenheimer's, in his preface to the post-war version of *The Los Alamos Primer*, which he originally published during the war at Oppenheimer's request to orient newly arriving Manhattan Project personnel into the program, described Groves' ego-emanating entrance the first time they met.[24] Apparently Groves had no more than entered the room when he removed his jacket and handed it to a colonel he had "in tow," and curtly ordered the high-ranking officer to find a laundry and get his tunic cleaned.

Oppenheimer, on the other hand, was quite a different personality. He was young, ascetic, wealthy, and seemingly frail, although later events would prove him to have a great capacity for bearing physical, psychological, emotional and intellectual abuse. Oppy, as he was affectionately known by friends, was scientifically and clinically critical while at the same time embracing Far Eastern metaphysical mysticism. The paradox made him an astonishing choice for project director. The greater half of the astonishment was that Oppy was a theoretician, not an experimentalist. The new laboratory was, of necessity, going to be nothing if not overwhelmingly experimental.

Oppenheimer's lack of experimental experience caused many who coveted the position, or who otherwise had what appeared to be legitimate concerns, to cry foul. Groves would have none of it. He had quietly grasped Oppenheimer's unique genius, his brilliantly quick analytical and intuitive facility and a talent for exciting people about the work, and was not about to let him go.

What concerned Groves more was the future lab director's leftist connections. Not that Groves felt they were much of a hindrance to Oppy's doing the job. But security checks had to be performed and they soon revealed that not only had Oppenheimer once been a registered member of the American Communist Party, but his wife, brother and ex-fiancé, as well, were presently members or had been members at one time.

The endless pursuit by military security to rectify this apparent security breach kept Groves almost continually in a position of having to protect his chief deputy. His willingness to do so is surely a strong endorsement of Groves' belief and confidence not only in Oppenheimer but also in his own extraordinary ability as a judge of people. The results Oppenheimer brought forth stand as an undeniable testament to the General's ability to measure a man. What is most remarkable is that although he had considered others, Groves was 99 percent decided Oppy was his man after only one or two meetings.

A month later, in November 1942, Groves and Oppenheimer, with a handful of others, were at a boys' ranch standing atop a 7,200-foot-high plateau in New Mexico. Oppenheimer, who owned property in New Mexico and loved the vast, scenic expanses, had suggested the location over several rivals, some close by, others as far away as Utah and Washington State. As they stood under the cottonwood trees – for whose Spanish appellation the boys' school had been named, Los Alamos – Groves consented to purchase the property as the sight for America's new atomic bomb laboratory.[25]

A full four months after that, in the end of March 1943,[26] the small group would finally return, accompanied by a nucleus of scientists that would ultimately grow to be one of the greatest collections of intellects ever concentrated on one task. Enrico Fermi, Emilio Segré, Hans Bethe, Otto Frisch and many others, all Los Alamos personnel during the war, were just a few of several scientists at the project who had already won or would go on to win the Nobel Prize and other top awards of science. Along with them they brought equipment commandeered from laboratories across the United States[27] and a support force of almost 5000 people, many with their families.

Despite the thin chance the American effort could achieve making a bomb General Groves made an early and full commitment to the project. Before he had pinned the new general's star on his collar (an inducement to get him to accept the Manhattan Project assignment over his preference to serve in a theater of war), before he even ran to Berkeley to find what level of scientific talent was available, Groves signed the directive that began the purchase of 59,000 acres of mostly undeveloped land in Eastern Tennessee. The complex built there would soon come to be known as Oak Ridge, and it would house most of the technologies to enrich production quantities of bomb-grade uranium[28] – many of which would fail or only achieve nominal success during the war.

On the site eventually would be established a gaseous diffusion isotope separation plant that would utilize hundreds of thousands of stacks of pipes in an all-but-failed effort to enrich uranium before the war was over. This plant would enclose almost 42 acres under a single roof and cost one-half billion dollars, the greatest single expenditure of the wartime program. A liquid thermal diffusion plant under the operation of the Navy would be constructed as well. By far the most successful form of isotope separation would be the electro-magnetic isotope separators pioneered by Ernest Lawrence. Groves would one day brag that every gram of U^{235} produced for the Manhattan Project had been processed through Oak Ridge's magnetic isotope separators – called calutrons, after the California State University (Cal. U.) at Berkeley, where it was developed. But even with the calutrons, none of these processes were close to being viable at production-level quantities at the end of 1943. And the famous claim that all of the uranium enriched passed through the celebrated calutrons during that process has now become doubtful, based on recently discovered information.

Five days less than a year after the bombing of Pearl Harbor, on 2 December 1942, Italian émigré physicist Enrico Fermi and his research team, working in an old squash court under the University of Chicago's Stagg Field grandstand, opened another door leading to an atomic bomb. They produced the first man-made self-sustaining nuclear chain reaction.[29] The experimental reactor pile, built of over 400 tons of graphite and uranium, proved that a slow chain reaction could be achieved and controlled. The reactor also provided the means to learn more about uranium bombarded by neutrons and how it morphs into the new element plutonium.

Plutonium, as created in a reactor, fissions more readily than U^{235}. The bomb makers at first counted this a blessing. And plutonium had chemical characteristics that were different from other substances.[30] By finding these differentiating properties, the plutonium could be separated from its parent, uranium, by chemical means, a far less expensive and comparatively easy process than the impossibly demanding physical separation procedures required to harvest one atom at a time, as was necessary to enrich uranium. There was now a second, much better option for developing an atomic bomb.

Hopes were high. Everyone from Groves and Oppenheimer to Fermi and Lawrence enthused over the plutonium prospect.[31] In fact, the whole object of creating a reactor pile changed from creating heat to make steam for industrial power to breeding plutonium for a bomb. Groves immediately went to work establishing a plutonium-breeding pilot reactor at Oak Ridge for experimentation, as well as beginning the procurement of property in the state of Washington for the purpose of constructing a series of plutonium breeding production reactors.

The scientists, however, soon found problems with the plutonium option. Previous plutonium breeding experiments had been performed in a cyclotron that could bombard target uranium with only very small amounts of neutrons. The result was the expected transmutation of U^{238} to plutonium 239 (Pu^{239}). The comparative flood of neutrons released in a chain reacting pile needed to produce bomb-size volumes of plutonium, however, placed the parent U^{238} awash in free neutrons. Because of the much greater density of free neutrons, there was a small percentage, perhaps as much as 10 percent of the final product, that morphed the next step to plutonium 240 (Pu^{240}). The much faster fission rate of Pu^{240} – 100 million times faster than that of U^{235} or Pu^{239} – made the triggering mechanism of the uranium bomb impractical for the plutonium weapon.

The latter two isotopes fission slowly enough that to assemble a critical mass one needed simply shoot one sub-critical piece of material into another piece with a high-velocity cannon. The pieces would come together to achieve critical mass at about 3,000 feet per second. The result, a nuclear explosion.

This "gun bomb" method of assembling the fissile material was so slow, however, that with Pu^{240} premature nuclear explosion was almost certain – to obviously disastrous results.

Groves and his cadre of scientists now had a challenge creating a plutonium bomb as perplexing and problematic as the original isotope separation assignment. They must find a way to trigger a critical assembly, in other words, to move multiple blocks of matter at velocities no human, for any reason, had ever envisioned attempting. The plutonium option was now just as much a long shot as the original uranium bomb.

Endnotes: Chapter Two – The Two Billion Dollar Bet

1   US National Archives Southeast Region, East Point, Georgia, E.R. Jette to C.S. Smith memorandum: Production rate of 25, December 28, 1944, A-84-019-70-24

2   Arnold Kramisch, *The Griffin*

3   Robert Wilcox, *Japan's Secret War*, pp. 15,16

4   Sharkhunters KTB 103, p. 7 and KTB 110, p. 10

5   Arnold Kramisch, *The Griffin*

Robert Serber, *The Los Alamos Primer*, p. xvii

6   Richard Rhodes, The Making Of The Atomic Bomb, p. 691

7   US National Archives Southeast Region, East Point, Georgia, E.R. Jette to C.S. Smith memorandum: Production rate of 25, December 28, 1944, A-84-019-70-24

8   US National Archives Southeast Region, East Point, Georgia, E.R. Jette to C.S. Smith memorandum: Production rate of 25, December 28, 1944, A-84-019-70-24

9   Richard Rhodes, *The Making Of The Atomic Bomb*, p. 612

10   Richard Rhodes, *The Making Of The Atomic Bomb*, p. 601

11   Robert Serber, *The Los Alamos Primer*, p. xv

12   Protocol Proceedings of the Crimea Conference, Agreement Regarding Japan, 11 February, 1945

13   Herbert Childs, *An American Genius*, p. 335

14   Leona Libby, *The Uranium People*, p. 213

15   Richard Rhodes, *The Making Of The Atomic Bomb*, pp. 313, 314

16   Richard Rhodes, *The Making Of The Atomic Bomb*, p. 406

17   Richard Rhodes, *The Making Of The Atomic Bomb*, p. 406

18   Leslie Groves, *Now It Can Be Told*, p. 15

19   Richard Rhodes, *The Making Of The Atomic Bomb*, p. 427; Leona Libby, The Uranium People, p. 83

20   Leslie Groves, *Now It Can Be Told*, p. 33,34

21   Chapman Pincher, *Into The Atomic Age*, p.7

22   Stephen Groueff, *Manhattan Project*, p. 36

23   Leslie Groves, *Now It Can Be Told*, p. 96

24   Robert Serber, *The Los Alamos Primer*, introduction

25   Richard Rhodes, *The Making Of The Atomic Bomb*, pp. 450, 451

26   Robert Serber, *The Los Alamos Primer*, p. ix

27   Leona Libby, *The Uranium People*, p. 194

28   Richard Rhodes, *The Making Of The Atomic Bomb*, p. 427; David Irving, The German Atomic Bomb, p.150

29   Richard Rhodes, *The Making Of The Atomic Bomb*, pp. 436 - 442

30   Leona Libby, *The Uranium People*, p. 77; Richard Rhodes, The Making Of The Atomic Bomb, pp. 388, 389

31   Herbert Childs, *An American Genius*, p 324, 325; Leona Libby, *The Uranium People*, p. 79; Richard Rhodes, *The Making Of The Atomic Bomb*, pp. 368, 416, 431

Chapter Three

Uranium

Until 14 May 1945, the day U-234 surrendered to the United States, Germany had always held the lead in the race for the atomic bomb – even before anybody knew there was a race being run. Way back in 1789, 150 years before the pernicious purpose of uranium was conceived, Martin Klaproth discovered this last, and heaviest, of the elements found in nature. Appropriately, given later physics history – or perhaps inevitably – Klaproth was German. In the century and a half between Klaproth's discovery and the splitting of the first atom – a uranium atom – little happened with the element. In the small amounts it could be found, uranium was considered relatively rare, although it has since been discovered in varying quantities in almost all geographic regions on earth. Prior to the effort to build a bomb, however, uranium was used almost exclusively as a pigment in ceramic glazes; no one could devise any other practical use for it. But when the first atom was split at the end of 1938, the whole world changed.

Advances in physics, particularly the effort to understand the make-up of the atom, had physicists and radiochemists across the globe experimenting with uranium, the natural world's largest atom. As a result, the first atom was split, quite by accident, by Otto Hahn and Fritz Strassmann, two Germans at the Kaiser Wilhelm Institute of Physics in Berlin.

Hahn and Strassmann – both radiochemists not physicists – did not immediately realize what they had achieved. They had been bombarding uranium with slow neutrons expecting its transmutation to other isotopes of uranium or other heavy elements. But the result of their experiment showed, along with isotopes of uranium, of which U^{238} is the most common, evidence of traces of barium were present as well, which has an atomic mass slightly larger than half of uranium's mass.

At first neither scientist could reckon how the atomic weight had been cut in half. The cleaving of an atom, with its powerful internal force holding it together, was considered impossible and splitting the atom had never crossed their minds. The pair assumed they had not carried out their experiments correctly; but careful checks using control samples they knew were pure proved they had not contaminated the experiment with material already containing barium. Only then did they consider the impossible may have happened. Hahn wrote his former co-worker, Lise Meitner, an Austrian-born Jew who, now in her 60s, had over 40 years experience in radiochemistry and a native genius for diagnosing chemical and nuclear puzzles.

On Christmas Eve, while contemplating during a holiday at the seaside in Sweden the remarkable events described in Hahn's letter, her nephew and fellow researcher Otto Frisch visited Meitner. Frisch would later be the one who coined the term "fission"[1] – borrowed from the microbiology lexicon and which describes the dividing of living cells – as the moniker for the splitting of atoms. He would also shortly immigrate to the United States and perform the famous, and very dangerous, critical mass experimental studies on uranium at Los Alamos known as "tickling the tail of the dragon."

Meitner and Frisch discussed how it could be possible barium should come from uranium. In the course of considering several possibilities, they contemplated the puzzle in the light of Niels Bohr's new model of the nucleus – not a collection of tightly bound neutrons and protons, but "freely" bound neutrons and protons. They reasoned that, although the nuclear force holding these components together is undoubtedly the strongest on earth – even though active for extremely small distances only – each proton in the nucleus contains a small electrical force of its own that counters, to a degree, that nuclear force. As the nucleus of each element in ascending order contains one or more additional protons than the previous element, by the time uranium – the natural element with the most protons of all, at 92 – is reached, the countering force of the cumulative protons is barely less than the total nuclear force. The scientists realized this would explain why there are no more natural elements beyond uranium – because the accumulated electrical force of the extra protons in an atom larger than uranium would counter the atomic force to a point where the nucleus is no longer able to hold itself together. Any elements beyond uranium must have disintegrated to other elements earlier in earth's history.

But the uranium nucleus holds together barely, the opposing forces causing the sub-nuclear particles to float "loosely" around one another in a fluid form. The unstable geometric construction of a U^{235} atom, particularly, when struck by the energy of a neutron, theoretically starts the nucleus "wobbling," becoming narrower in the middle. The nuclear force in each of the two outer lobes immediately takes control and parses off the lobes into independent, non-uranic spheres of their own – one of them barium.

Thus Meitner and Frisch explained, and therefore validated, Hahn's and Strassmann's discovery – and set in motion with their explanation the fearful, surreal absurdity that would become man's future. For Meitner also calculated the nuclear reaction after the split, caused by the repulsion of the protons in each nucleus pushing away from each other at one-thirtieth the speed of light, would generate about 200 million electron volts of energy per atom.[2] In comparison, the strongest of chemical reactions such as a dynamite explosion, produces a very paltry five electron volts.

Hahn had written not only Lise Meitner on that fateful December night, he had also contacted Paul Rosbaud, the editor of Germany's foremost scientific publication, *Naturwissenschaften*.[3] Rosbaud would soon come to be known in Allied intelligence circles as *The Griffin*, the codename assigned him upon joining the ranks of Germans spying for the Allies. Those he spied upon almost certainly included two unheralded scientists, Baron Manfred von Ardenne and Fritz Houtermans. Ardenne and Houtermans later would develop a uranium enrichment process that, by all accounts, was superior to the Manhattan Project's calutrons and was surely the basis for Germany's successful uranium enrichment efforts. Rosbaud was particularly close to Houtermans. Many of Rosbaud's activities are recorded in Arnold Kramisch's excellent book, *The Griffin*. But despite Rosbaud's close friendship with Houtermans', in his book he never mentions Ardenne – or Houtermans' multi-year employment at Ardenne's laboratory – which certainly seems telling considering what the two men accomplished.

Presumably, General Groves received Rosbaud's reports through the United States/British intelligence master, Sir William Stephenson, and therefore known on an ongoing basis what was the condition of his nemesis' program. Statements the General made during the war indicating he often thought the enemy was a year or two ahead of the United States' program can, therefore, probably be considered accurate. If they were, assertions made by General Groves after the war indicating he had been wrong in this conclusion were surely designed to divert attention from the German isotope separation program.

The purpose of the feint would be that if the existence of the German uranium enrichment program could be hidden, the cover story could be established that Germany's atomic bomb effort consisted only of failed efforts to create a reactor pile to breed plutonium. This would appear to demonstrate the Germans had failed and the Americans alone succeeded in the nuclear race.

Rosbaud's complete excising from his book of Ardenne's and Houtermans' substantial uranium enrichment efforts probably served a similar purpose, testifying to the Griffin's post-war service to his Anglo-American masters.

At any rate, on Hahn's request, Rosbaud had agreed to hold space in the next issue of his journal for an upcoming paper Hahn promised to prepare by print time. The article not only ran in early January 1939, quickly spreading the news throughout the global scientific community, but Frisch returned to work with Niels Bohr in Copenhagen after his Christmas holiday with Meitner and told "The Great Dane," as he was affectionately called, of their theory.[4]

Bohr responded before Frisch had hardly finished explaining, gasping, "Oh what idiots we have all been! Oh but this is wonderful! This is just as it must be." The Great Dane left Denmark within a week of this revelation on a previously planned trip to the United States to work for a short period at the Institute for Advanced Study. Once there, he was instrumental in disseminating the news to the rest of the world. Then the new discovery's ultimate outcome was calculated – that a nuclear chain reaction might be created. Szilard and Teller, quickly recognizing the unthinkable possibilities, contacted Einstein, who wrote his famous letter to Roosevelt in response to such a prospect.

The chain reaction conclusion also made Hahn consider an action he had never before contemplated. Upon realizing the likely outcome of his discovery would be the loss of tens- or hundreds-of-thousands of lives – possibly millions – Otto Hahn seriously considered taking his own life.[5]

The taking of one life would have been a small matter and a futile action, however. The nuclear door had been opened and could never be closed again. Despite later and persistent claims that Germany put little effort – and that erring – into the development of an atomic bomb, quite the opposite actually occurred. As a nation with a disciplined, precise and loyal nationalistic character and a tradition of cultivating the ultimate in technology, under the rule of a dictator with a fetish for innovative armaments and a commitment to using them, Germany was already on the verge of waging war using the most technically-advanced fighting machine ever. The airplanes, tanks and submarines of Blitzkrieg were unsurpassed and it would be years before the Allies equaled the armaments of the Third Reich. During the course of the war, Hitler added rocketry, silent electric torpedoes and jets to his arsenal, none of which were matched by any other belligerent nation during the course of the conflict. In truth, on the whole, German weaponry was probably never equaled during the war. Many experts maintain Germany lost World War Two directly because of strategic blunders committed by Adolf Hitler and little else.

With a superior technical culture, a lead on the field, and many of the best scientists available – all at the behest of a madman well-established to have a penchant for ingenious and decisive weaponry – it certainly would be expected that Germany would run hard in the nuclear arms race and would break out of the gate first. The idea accepted wholesale in the traditional history, that German efforts to produce *the* deciding weapon of the war, an atomic bomb, were vapid, poorly executed, uninspired projects, runs wholly counter to the character of the regime and the Germanic race and culture. To this day, in a world of global technological parity, Germany is still looked up to as a scientific and technical leader of the world.

According to author/historian David Irving, in his book, *The German Atomic Bomb*, the post-war criticism of Germany's supposedly insipid effort to create an atomic bomb is both inaccurate and unwarranted.[6] Irving cites some 50 German scientists[7] toiled night and day throughout the war, in both plutonium breeding and uranium separation efforts, many of which achieved high levels of success. And Irving adds that those who spread the misinformation about these scientific efforts should have known better; they knew the story and had all of the documentation. This contrasts with the official history stating only a handful of half-hearted German scientists working on an impotent reactor pile intended, but failed, to breed plutonium. Such is the story promoted by General Groves and the Manhattan Project's intelligence arm, Alsos (Greek for "grove," Alsos was the codename given the Manhattan Project's enemy information gathering function).

By the summer of 1939, scant months after Hahn's and Strassmann's discovery had been published, the German Army had established a uranium project in Gottow, near Berlin, with Dr. Kurt Diebner at its head.[8] By the time war broke out a few months later, Germany was the only country studying the use of atomic power for military means, and it pushed forward with vigor. By contrast, the United States' efforts stalled and were not to be purposefully pursued until General Groves was appointed head of the program almost two and one-half years later, near the end of 1942.

A first secret conference on atomic power was held in Berlin on September 16, 1939.[9] Most of the Reich's top nuclear scientists soon afterward were inducted into the army – an action Groves would later seriously consider for the American program but was convinced otherwise by Oppenheimer – and assigned to laboratories throughout the Fatherland to study nuclear fission for military uses. The first laboratory, in Dahlem, near Berlin, was established and called "The Virus House,"[10] a name concocted as a ruse to cultivate an atmosphere of fear around the facility and thus drive off unwanted observers.

Despite later assertions, the Third Reich very soon had on hand copious amounts of raw, as well as very highly refined, uranium, and controlled a great deal more – almost a limitless supply for its needs. The first ton of "extremely pure" uranium oxide was delivered in the first weeks of 1940.[11] This had already been refined from the raw uranium ore and was, for all intents and purposes, ready to be used for experimentation – or for enriching to bomb grade as soon as the technology could be developed.

As early as the Summer of 1941, according to historian Margaret Gowing, Germany had already refined 600 tons of uranium to its oxide form, the form required for ionizing the material into a gas, in which form the uranium isotopes could then be magnetically or thermally separated or the oxide could be reduced to metal for a reactor pile.

In fact, Professor Dr. Riehl, who was responsible for all uranium throughout Germany during the course of the war, says the figure was actually much higher, which it proved to be. In addition, the Nazi program was extracting one ton per month of uranium oxide from separate ore stocks left over from a commercial venture following a previous extraction of radium to be used in German toothpaste!

To create either a uranium or a plutonium bomb, at some point uranium must be reduced to metal. In the case of plutonium, U^{238} is metalicized; for a uranium bomb, U^{235} is metalicized. Because of uranium's difficult characteristics, however, this metallurgical process is a tricky one. The United States struggled with the problem early and still was not successful reducing uranium to its metallic form in large production quantities until late in 1942. The German technicians, however, true to their whiz-kid reputations, by the end of 1940 had already processed 280.6 kilograms of uranium into metal, over a quarter of a ton.

From June of 1940 to the end of the war, Germany seized 3,500 tons of uranium compounds from Belgium – almost three times the amount Groves had purchased from Union Miniére – and stored it in salt mines in Stassfurt, Germany.[12] Groves brags that on 17 April 1945, as the war was winding down, Alsos recovered some 1,100 tons of uranium ore from Stassfurt and an additional 31 tons in Toulouse, France, as well as eight tons of refined oxide from the Stassfurt mines.[13] And he claims the amount recovered was all Germany had ever held; asserting, therefore, that Germany had never had enough raw material to process the uranium either for a plutonium reactor pile or through isotope separation techniques.

If Stassfurt once held 3,500 tons and only 1,130 were recovered, however, some 2,370 tons of uranium ore was unaccounted for. Thus twice the amount the Manhattan Project possessed and is assumed to have used throughout its entire wartime effort – and a quantity certainly far in excess of the amount Germany would have used for experimental needs – was unaccounted for.

This has been borne out in spades by the history-shattering discovery in the summer of 2011 of more than 126,000 barrels of Nazi *spent* uranium – the used uranium left over from an enrichment process – announced by Germany's ASSE regulatory commission.[14] Not even mentioned in the U.S. press, the news caused major headlines throughout Germany as its Greenpeace counterpart raised the concern of nuclear contamination. The uranium resides over 2,000 feet deep in an abandoned and "crumbling" salt mine outside of Hanover.

"Our association sank radioactive wastes from the last war, uranium waste, from the preparation of the German atomic bomb," reads a 1967 statement from the ASSE II nuclear fuel dump archives.

One gallon of uranium weighs approximately 150 pounds. If each barrel was 40 gallons, a logical size to assume, each barrel of uranium would weigh about 6,000 pounds. Multiplied by 126,000 barrels, the store reasonably contains 756,000,000 – yes, million – pounds of spent uranium equaling 378,000 tons.

This volume blasts to insignificance the 2,370 tons missing from Groves' account and validates the unending supply of uranium the Nazi's held at Joachimsthal in Czechoslovakia.

It also blasts the long-asserted proclamation the Nazi program never enriched production quantities of uranium. And with the other revelations revealed herein, even while historians try to claim it cannot be true because "everyone knows" the Nazis had no means of enriching uranium, it proves the chief assertion of *Critical Mass*, that the Nazis provided enriched uranium for the United States' Atomic Bomb.

Dr. Werner Heisenberg headed the plutonium bomb effort for Germany. As with the United States program, the Germans early realized the benefits of a plutonium bomb over a uranium explosive.[15] They knew plutonium could be bred from uranium and separated chemically much easier, faster and less costly than the isotopes of uranium could be separated from one another. In addition, because the plutonium fission process was three times more powerful than uranium's, theoretically, to make an equal-size bomb only one-third the amount of plutonium was required.

Heisenberg's efforts ran into a roadblock, however, when, in 1940, his co-worker Dr. Walther Bothe seriously miscalculated the neutron absorption rate of graphite,[16] which the researchers thought to use as a moderator to prevent any experimental chain reaction from becoming ungovernable and causing a meltdown. The error proved to have a profound impact on the success of the German plutonium project. In want of an alternate moderator, the scientists turned to deuterium oxide[17] – heavy water – an isotope of common water but with an additional neutron. The new requirement for heavy water, a rare substance not found in nature but requiring long amounts of time to process, would ultimately resign the German plutonium effort to – not failure, a chain reaction was eventually achieved – but to second place behind the American plutonium project.[18]

The carbon miscalculation combined with the shortage of heavy water constituted the failure of the Germans to build a plutonium bomb, which proved later to be the perfect screen behind which General Groves hid Germany's other atomic bomb effort, uranium isotope separation. As seems to have happened at almost every serious juncture, the two nations' programs followed parallel thinking and parallel processes. But General Groves and others – for reasons probably necessary at the time – have buried the history of the German uranium enrichment effort. Desiring after the war to destroy the evidence of German uranium isotope separation, for reasons to be reviewed later, the General de-emphasized the Nazi uranium enrichment effort until its historic profile was small enough to be hidden safely behind the failed plutonium picture.

General Groves does not appear to be the only person after the war to distort the facts of this episode to suit his purposes. Professor Heisenberg and others, purportedly desiring to divest themselves of what they said was the undeserved stigma of working on an atomic bomb for the Nazis, but in reality desiring to hide their failure to build a nuclear reactor despite great and earnest efforts, decided to inculcate the fantasy, as well – and successfully did so, possibly in collusion with Groves.

Heisenberg later contended he and others of his staff had innocuously but bravely resisted their fascist government. He insisted he did not believe at the time the making of an atomic bomb to be a possibility at all, but had acted as though it were in order to keep the Nazis happy and distracted.[19] The professor assured those who would listen he had been resisting and subverting the objectives of the Nazi regime by monopolizing the invaluable services of some of the Reich's greatest men of science, who might otherwise have been forced to put their efforts to use for Hitler in projects more productive to the Fuehrer's purposes.

In reality Heisenberg, like most scientists of his bent and professional stature, not only could not resist the pursuit of his science for the sheer inducement of discovering what lay around the next cosmic corner, but he did indeed believe a nuclear blast initiated by man was possible. He had admitted to Manfred von Ardenne[20] and to Niels Bohr, before the latter had escaped Denmark upon its occupation by the Nazis, that he thought an atomic bomb was possible[21] – even though Bohr, himself, at this time, did not believe such an explosion would ever be achieved. Heisenberg tried to explain away this statement after the war as having been misunderstood by the Danish Nobel Laureate; but the Great Dane was certainly convinced he had understood correctly what had been said. And the fact Heisenberg had asserted the same thing to Ardenne supports Bohr's claim.

Furthermore, Dr. Heisenberg was in the forefront from February to June of 1942 in an effort to get party leadership to more fully appreciate the value atomic explosives could serve in the war.[22] In June, he estimated a bomb could be built in as little as two years.[23]

Developers of the American plutonium project realized relatively late-in-the-game they had a problem with triggering the plutonium bomb. Up to that time, they had given the plutonium program their prime effort and resources. On the other hand, serious doubts about the success of the German plutonium program came early because of the heavy water crisis, forcing the Nazis from almost the very beginning to concentrate their efforts, resources and expectations on isotope separation to enrich uranium. By virtue of this fact alone, one would expect the German isotope separation program demanded more focus and would have been more successful than the plutonium effort, and would not have been left completely unpursued, as is asserted.

At about this time in mid-1942, American James B. Conant, one of the civilian administrators of the Manhattan Project and a personal confidant of Roosevelt, reported to the president the Germans "might be ahead of us by as much as a year."[24] Considering British spy Paul Rosbaud's position in the midst of the German effort, one can assume Conant got this estimate from good sources.

In fact, the estimate may have understated Germany's lead. By this time, Germany already had at least five, and possibly as many as seven, serious isotope separation development programs underway. From among these devices, three very innovative technologies were being pioneered, beginning with Dr. Erich Bagge's "isotope sluice" and a similar machine constructed by a Dr. Korsching. Before the middle of 1944, Bagge's isotope sluice would enrich uranium on a single pass to four times that achieved in the United States using gaseous diffusion,[25] which is alleged to be the United States' saving technology. While earlier Ernest Lawrence's experimental cyclotron at the University of California in Berkley produced only 100 micrograms of "partially separated" uranium,[26] Bagge's experimental isotope sluice had yielded 2.5 grams of "much enriched" uranium.[27] More than likely, the German program was well ahead of the Manhattan Project's efforts.

Had the Germans actually enriched uranium on a large-scale basis, and there now is ample evidence they did, they may have used a multi-stage technique, as the Americans had. The American effort used a procedure that passed already-enriched uranium through enrichment processes a second or third time to further increase the level of U235 concentration to levels up in the high eighty and low ninety percentiles required for a bomb. In America, gaseous diffusion and thermal diffusion are supposed to have saved the bomb enrichment program in the waning days of the separation effort by providing those needed, partially enriched feedstocks to Lawrence's beta calutrons in the final hour. Oak Ridge records discovered by the author and reviewed later in this book, however, contradict this assertion; although the calutron process did run product through the system multiple times.

One may assume the German effort followed a similar obvious path, as so often happened between the two programs, and that the product of the isotope sluice – or any of the other separation technologies – might therefore have been used as feedstocks for one of the other four separation techniques the Germans were refining.

The isotope sluice was not the strongest of the Nazis' separation efforts. A stronger performer was the centrifuge, and then its progeny, the ultracentrifuge. A special alloy called "Bondur" had already been developed in 1941 specifically designed to handle the harsh, corrosive uranium compounds used in the ultracentrifuge.[28] The United States' isotope separation effort, on the other hand, struggled to find a similar material that would serve well against the corrosive uranium gases.

By May 1944, the American production efforts operating at their highest efficiency resulted in enriching uranium from its raw state of .7 percent to about 10 to 12 percent on the first pass. By comparison, the first German experimental ultracentrifuge trials resulted in enriching the material to seven percent.[29] The experimental result was less than American production efforts and what had been predicted by its German inventors, but it was a good showing for its first experimental outing compared to what the Manhattan Project produced from its already-tweaked production model calutrons.

Ultracentrifuge initial output was so impressive, in fact, that following its very first experimental run funding and authority were established to build ten production model ultracentrifuges in Kandern, a town in the southwest of Germany far from the fighting. When Allied bombing became continuous in the north, many separation processes had been moved south; Bagge's isotope sluice went to Hechingen and the 10 ultracentrifuges went to Kandern, located near the juncture of the borders of Germany, Switzerland and France. The Nazis were now committed in a big way to ultracentrifuge production – and therefore to enriching uranium.

True to his objectives, however, after the war Groves once again warped the truth, downplaying the production plants by mentioning only that "U235 separation experiments" were being conducted in Celle and Freiburg[30] – never anything of the success of those experiments or of the ten ultracentrifuge production plants or of Ardenne's efforts at Lichterfelde.

Despite such subjugation of the truth, David Irving in his book *The German Atomic Bomb*, identifies what, at least for a time, were thought by the Allies to be fourteen isotope separating facilities being built in the region.[31] Groves himself admitted concern these plants were being erected to enrich uranium. According to Groves, he saw patterns similar to Oak Ridge in these plants; but quick intelligence analysis suggested the facilities were crude and inefficient factories for synthetically converting shale to oil. Such a revelation hints at their actually being a cover for nuclear weapons activity. After all, synthetic processing was the cover given the "buna plant" at Auschwitz that actually appears to have been electro-magnetically enriching uranium. And there existed a "gentlemen's agreement" between I.G. Farben and Allied forces[32] not to bomb synthetic processing plants. Despite the "shale oil" plants' seeming inconsequence as ultimately described by Groves, compared to the important schedule of non-nuclear strategic targets needing attention, Allied bombers were diverted from some of their important missions to destroy the chain of plants. Surely the bombing was counter to the "gentlemen's agreement" unless there was something that justified their destruction beyond the fact they were allegedly synthetic processing plants.

The converting of shale to oil is a synthetic gasification process pioneered by I.G. Farben and its technology is in many ways similar to that of producing synthetic rubber, also called buna. Given events related later in this chapter and elsewhere, it would not be surprising to find these plants had, indeed, been enriching uranium.

Even the impressive successes of the ultracentrifuge do not match up to the "most far reaching" achievements attained in isotope separation by Baron Manfred von Ardenne. Ardenne and his associate, Fritz Houtermans, as early as 1941, had already calculated the critical mass[33] of U^{235} and had begun construction of "a magnificent laboratory" underground – safe from the bombing of Allied airplanes – in Berlin Lichterfelde.[34] The laboratory contained a two million-volt electrostatic generator and a cyclotron – at the time there was only one other cyclotron throughout the Reich, that of the Curies, which had been commandeered in France. By April 1942, Ardenne also had in his laboratory a completed magnetic isotope separator[35] not unlike the calutrons of Ernest Lawrence, which General Groves would not deploy at Oak Ridge for another year-and-a-half. Ardenne had designed the separator in 1940, barely on the heels of the discovery of a possible fission explosion.

And so, supplied with his million-volt generator to provide the copious amounts of power needed to operate the magnetic isotope separator, Ardenne seems to have been ahead of everybody else in the field of uranium enrichment. In addition, the ion plasma source Ardenne had designed for his isotope separator to sublime the uranium compound was far superior to that provided for the calutrons – a profound distinction considering the calutron's sublimation process was one of its key weaknesses. Calutron efficiency for sublimation ran up to 75 percent at best.[36] Ardenne's invention apparently was much more efficient – and has come to be the premiere source world-wide for emitting particle rays; and is known to this day as "The Ardenne Source."

One other important distinction separated Ardenne's and Houter-mans' work from the other German efforts. All of the other programs were under the direction and were part of the German Army, supplied by and accountable to the military. By contrast, all of Ardenne's facilities – the bomb-proof lab, the million-volt generator, the cyclotron, and the magnetic isotope separators themselves – were provided by, and ongoing funding made available through, the patronage of one man, Reich Minister of Posts and member of the Reich President's Research Council on Nuclear Affairs, Wilhem Ohnesorge. Like the Manhattan Project scientists, Ardenne and Houtermans worked within the more intellectually free environment of a civilian, not military, organization.

Production for the German isotope enrichment projects, once the experimental and design work were completed by Ardenne and the others, appears to have been undertaken by the I.G. Farben company under orders of the Nazi Party. The company was directed to construct at Auschwitz a buna factory,[37] allegedly for making synthetic rubber. Following the war, the Farben board of directors bitterly complained that no buna was ever produced despite the plant being under construction for four-and-a-half years; the employment of 25,000 workers from the concentration camp, of whom it makes note the workers were especially well-treated and well fed; and the utilization of 12,000 skilled German scientists and technicians from Farben. Farben also invested 900 million reichsmarks (equal to approximately $2 billion of today's dollars) in the facility. The plant used more electrical power than the entire city of Berlin yet it never made any buna, the substance it was "intended" to produce.

When these facts were described to an expert on polymer production (buna is a member of the polymer, or synthetic rubber, family), Mr. Ed Landry,[38] Mr. Landry responded directly, "It was not a rubber plant, you can bet your bottom dollar on that."

Landry went on to explain that while some types of buna are made by heating, which requires using relatively large amounts of energy, this energy is invariably supplied by burning coal. Coal was plentiful and well-mined in the area and was a key reason for locating the plant at Auschwitz when it was still intended to be a buna facility.[39] The heating-of-buna process, to Landry's knowledge, was never attempted using electricity, nor could he envision why it would have been. Landry totally dismissed the possibility a buna plant, had it tried an electric option, would ever use more electricity than a large city, as was described by the Farben directors. And the investment of $2 billion is, "A hell of a lot of money for a buna plant" even these days, according to Mr. Landry.

The probability of the Farben plant having been completed to make buna appears to be slim to none. The facility had all the characteristics of a uranium enrichment plant, however, which undoubtedly it would never have been identified as. Instead, it would be given an appropriate cover story to camouflage it – such as it being a buna plant. In fact, buna was an excellent cover because of the high level and similar types of technology involved in both buna production and uranium enrichment. Indeed, as has been noted previously, General Groves and his intelligence analysts had already mistakenly identified as possible enrichment plants what he later alleged were production facilities for a synthetic process similar to buna. And do not forget the protection provided by the gentlemen's agreement, which forbade the Allied bombing of Farben synthetic rubber plants – another great reason to use the buna cover.

One last detail of interest regarding this phantom factory: I.G. Farben had close ties with and often financed or otherwise served directly the clandestine purposes of Adolf Hitler – usually working through the Fuehrer's top aide, Martin Bormann, or through Bormann's bureaucracies.

The various components of the German isotope separation efforts easily could have been implemented with a high degree of secrecy, even from other high-level Nazis, given Bormann's close-knit relationships with Ohnesorge; Schmitz, who was the chief of I.G. Farben; Hoess, the commandant of Auschwitz; and Heinrich Mueller, who, among his many other duties as head of the Gestapo, oversaw the supplying of forced laborers to Auschwitz.[40]

Endnotes: Chapter Three – Uranium

1   Richard Rhodes, *The Making Of The Atomic Bomb*, p. 263

2   Richard Rhodes, *The Making Of The Atomic Bomb*, p. 259

3   Richard Rhodes, *The Making Of The Atomic Bomb*, p. 253

4   Richard Rhodes, *The Making Of The Atomic Bomb*, p. 261

5   David Irving, *The German Atomic Bomb*, p. 45

6   David Irving, *The German Atomic Bomb*, p. 302

7   David Irving, *The German Atomic Bomb*, p. 179

8   David Irving, *The German Atomic Bomb*, pp. 41, 42

9   David Irving, *The German Atomic Bomb*, p. 44

10   David Irving, *The German Atomic Bomb*, p. 56

11   David Irving, *The German Atomic Bomb*, p. 51

12   David Irving, *The German Atomic Bomb*, p. 71

13   Richard Rhodes, *The Making Of The Atomic Bomb*, pp. 607 - 610

14   Allen Hall, *Mail Online*, 13 July 2011

15   David Irving, *The German Atomic Bomb*, p. 222

16   David Irving, *The German Atomic Bomb*, p. 85

17   David Irving, *The German Atomic Bomb*, p. 58; Leona Libby, *The Uranium People*, pp. 73, 74

18   David Irving, *The German Atomic Bomb*, p. 211

19   McGeorge Bundy, *Danger and Survival: Choices About The Bomb In The First Fifty Years*, pp. 15 - 18

20   David Irving, *The German Atomic Bomb*, p. 77

21   David Irving, *The German Atomic Bomb*, p. 102

22   Richard Rhodes, *The Making Of The Atomic Bomb*, pp. 402, 403

23   Albert Speer, *Inside The Third Reich*, p. 21

24   Richard Rhodes, *The Making Of The Atomic Bomb*, p. 406

25   David Irving, *The German Atomic Bomb*, p. pp. 89, 90 and 284

26   Richard Rhodes, *The Making Of The Atomic Bomb*, p. 487

27   David Irving, *The German Atomic Bomb*, p. 234

28   David Irving, *The German Atomic Bomb*, p. 91

29   David Irving, *The German Atomic Bomb*, pp. 91, 173, 229

30   Leslie R. Groves, *Now It Can Be Told*, p. 337

31   David Irving, *The German Atomic Bomb*, p. 253

32   Joseph Borkin, *The Crime and Punishment of I.G. Farben*, p. 130

33   David Irving, *The German Atomic Bomb*, p. 92

34   David Irving, *The German Atomic Bomb*, p. 290; Richard Rhodes, *The Making Of The Atomic Bomb*, p. 371,

35   David Irving, *The German Atomic Bomb*, p. 76 - 78, 116, 235

36   Telephone conversations, (dates), between the author and Gerald Rice, beta calutrons operator at Oak Ridge

37   Paul Manning, *Martin Bormann: Nazi In Exile*, p. 153

38   Ed Landry, personal interview with the author, May 22, 1996; President and General Manager of Keystone Polymers, Inc. of Houston, Texas. Mr. Landry holds a degree in chemistry with emphasis on polymer science, earned on a two-year fellowship at the University of Akron, the home of the Goodyear Rubber Company and the leading school on polymers in the United States.

39   Joseph Borkin, *The Crime and Punishment of I.G. Farben*, p. 115

40   Yisrael Gutman and Michael Berenbaum, *Anatomy of the Auschwitz Death Camp*, p. 39

59   David Irving, *The German Atomic Bomb*, p. 77

Baron Manfred von Ardenne ran a private laboratory at his estate in Berlin Lichterfelde that surpassed American uranium enrichment efficiency by as much a four times. His lab was funded not by the military but by the postal service, whose national leader was a physicist.

Before working for Ardenne, Professor Fritz Houtermans was released from Soviet prison to the Gestapo as part of a prisoner exchange. Despite orders he never work for the German government, with Ardenne he was a key contributor to Nazi efforts to enrich uranium.

Wilhelm Ohnesorge, center, in civilian attire holding hat, the physicist and mathematician who was Postal Minister.

Chapter Four

The Hidden Bomb

Ardenne worked on some sort of atomic project approved at the highest level, his villa was visited on several occasions by Hitler during the latter's periods of residence in Berlin.[1]
— Dr. Henry Picker, author, *Hitler's Tabletalk*

Ardenne's technology "had clear similarities to the tracks at Oak Ridge,"[2]
— David Irving, author, *The German Atomic Bomb*

On 13 March 1945, one and a half months before the end of the war in Europe, Adolf Hitler addressed the officers and generals of the German 9th Army. The English and Americans were closing in from the west and south; the Wehrmacht was in shreds in front of them and falling back. The Russians were just outside Berlin and closing from the east; in three weeks the German capital would be surrounded. The Luftwaffe was decimated; it could barely get aircraft off the ground. Germany had all but lost. Yet Hitler stood before his soldiers and announced, "we still have things that need to be finished, and when they are finished, they will turn the tide."[3] Their Fuehrer was intent on buying time until he could thrust his newest, most secret weapon into battle.[4] Scores of later observers and historians would attribute his "miracle weapon" rhetoric to dementia that had set in under the influence of drugs and duress. Or they suggested it was a vain and empty promise meant to buttress German military resolve to buy time while the Fuehrer tried to negotiate with the Allies – or break them up, depending on to whom one listens. But Hitler's visionary prediction now appears to have been more than war-wearied wishing or drug-induced hallucinations.

In fact, according to Albert Speer, Martin Bormann, Hitler's top lieutenant and in all things the consummate realist, believed Hitler was neither wishing nor hallucinating. With the end of the war closing in, Bormann had told Gauleiter Hellmuth that a secret weapon soon would be forthcoming even as Germany was being defeated.[5] Party bigwigs were being told by Bormann, and firmly believed, that a most-secret miracle weapon was about to be unleashed,[6] wrote Bormann biographer Jochen von Lang. SS General Karl Wolff, who at the end of the war secretly negotiated with the OSS to surrender all of Italy – Germany's southern front – revealed in postwar interrogations that he had spoken privately with Hitler about the secret weapon.[7] Wolff was in position to know about such a weapon. Before his assignment in Italy he had been a key link between Himmler and his SS, I.G. Farben and its plant at Auschwitz, and Bormann and the rest of Hitler's headquarters (see Chapter Sixteen). According to Lang, Bormann focused "all his energy" on making sure the miracle weapon would happen.[8]

Bormann insisted a miracle weapon was coming because, in all likelihood, Bormann had seen it – or at least he had seen its most integral and difficult-to-obtain component – while touring with Hitler the laboratory where it was created. As was his fashion, Bormann followed Hitler almost everywhere and had his clerks write down on small white cards almost every word that fell from Hitler's lips and nearly all his comings and goings. From these references Dr. Henry Picker wrote his book *Hitler's Tabletalk*. Using those references, Picker confirms that Hitler (probably accompanied by Bormann – author's note) made a habit of visiting the private laboratory of nuclear physicist Manfred von Ardenne.

"Ardenne worked on some sort of atomic project approved at the highest level," wrote Picker, "his villa was visited *on several occasions* by Hitler during the latter's periods of residence in Berlin"[9] (emphasis the author's).

Such singular attention by the leader of the Third Reich, whose time was in great demand and who during this period thought and worked only on important issues relating to the war, bespeaks a man fully supporting a program he perceived was integral to Germany's ability to win the war. Hitler by these repeated visits, despite later assertions otherwise, appears to have understood the importance of the Ardenne nuclear program in the worldwide military/political arena. His visits to the laboratory show Hitler was aware, knowledgeable, involved and supportive of nuclear weapons and that, interpreting the reason for his successive visits, the program must have been progressing. So if Hitler believed a miracle weapon was forthcoming when he addressed the Ninth Army, having been an eyewitness to Ardenne's developments, he probably had good reason for that conviction, as did Bormann.

Bormann, in fact, had already focused a considerable amount of energy on making the "miracle weapon" happen.[10] He had actively resisted

Armaments Minister Albert Speer's attempts to induct almost 15,000 scientists and technicians into the military – 5,000 who had already been
inducted were released – so they could continue their research efforts on
weapons development. Among them were several atomic scientists saved
from conscription into Bormann's own Volksturm Army. Bormann then
issued a decree that protected all scientific personnel from any future
combat assignments other than as required for defensive operations in
the regions of their own homes.

But evidence exists Martin Bormann had a more direct connection to
nuclear development than establishing and enforcing broad policies about
scientific personnel and their relationships with the military, and irregular
tours through nuclear laboratories. In his book *Inside the Third Reich*, Albert Speer related how Hitler received an update about the development
of nuclear weapons from Bormann's old friend Dr. Wilhelm Ohnesorge.
Speer gives a brief accounting of Ohnesorge and his chief physicist, the
young Manfred von Ardenne.[11] Bormann, under Hitler's instructions, had
assigned Dr. Ohnesorge, the mathematician and physicist who was Minister of Posts, the task of capturing and deciphering intercepted messages
between Franklin Roosevelt and Winston Churchill. And Bormann had
arbitrated a deal between the postal ministry and Hitler on the usage of
Hitler's likeness. Now Ohnesorge was in Hitler's presence again reporting
on the nuclear program, an achievement not possible without Bormann's
approval at the very least, and most probable only with his wholehearted
support.

Given the many references to Bormann's efforts to move forward the
miracle weapon, the possibility seems worth considering that Bormann
was not only Ohnesorge's champion and intermediary in Hitler's court,
based on their previous success decrypting the Roosevelt-Churchill hotline and the fact that almost nothing was presented to Hitler without
having received Bormann's support first, but Bormann appears to have
been involved with a nuclear program at a high level, as well. Whenever
the miracle weapon was mentioned, Bormann's name was always tied to
it, as Speer reported in the case of Gauleiter Hellmuth having been told
about the weapon by Bormann, and Jochen von Lang documented that
"Bormann's commissars" revealed the existence of the weapon.

The miracle weapon program, in its entirety, also follows what had long
been the pattern and had all the earmarks of a Bormann program: it was
a shadow program composed of people strongly aligned with Bormann –
ill-regarded, perceived quasi-experts who performed their tasks outside

of the orthodox structures one would expect people in those functions to work within. Ohnesorge and Ardenne, and to a degree Houtermans, were looked down upon by their more orthodox counterparts as less-than-capable scientists: Ohnesorge because he was bastardizing his training in mathematics and physics by mixing the beloved disciplines with the bureaucracy of the postal ministry, Ardenne because he was not a credentialed physicist, and Houtermans because he had a somewhat checkered past.

Speer, without mentioning Bormann by name, complained about amateurism and "Sunday-supplement" reporting of the program, a complaint he often threw at Bormann but at few others. The production component of the project also appears to have been assembled and controlled by a web of close Bormann cronies at I.G. Farben, Auschwitz and in the Gestapo and SS.

Many of Hitler's most powerful leaders, led by Speer, resisted the miracle weapon enterprise but were unable to bring it down, a strong sign Bormann was involved and used his power with Hitler and elsewhere to maintain the program. Speer usually had his way with the Fuehrer on most subjects; except those Bormann contested.

Hitler's personal and long-standing interest in the project supports widely documentable evidence that Ardenne, with Ohnesorge's backing, was working with commitment and aggressiveness equal to those who were striving to develop an atomic bomb in America. Additional evidence suggests Ardenne achieved far more success in nuclear energy development than any other German research team – including and far surpassing Dr. Werner Heisenberg and his followers, who in the traditional history are summarily held forth as Germany's nuclear leaders.

At his villa in Berlin Lichterfelde, supported by Ohnesorge's massive funding, Ardenne had built his own first-rate underground laboratory safe from the intermittent bombings delivered by Allied airplanes.[12] Contrary to efforts during and after the war to minimize Ardenne's achievements, he actually succeeded by mid-war in developing an isotope separation technology that "had clear similarities to the tracks at Oak Ridge."[13] "The Ardenne Source," as the most critical component of the device came to be known, was, in fact, better than those at Oak Ridge and continued to be the ion plasma source of choice globally for decades after the war.[14] Ardenne conceived the idea of his magnetic isotope separator in early 1940, at the beginning of the war, well early enough to be of service to the German program, despite the opposite often being inferred in the traditional history. In fact, development of Ardenne's technology occurred at the same time Ernest Lawrence first began toying with the idea of converting his cyclotron into a similar type of device. By then, Ardenne had already drawn up plans for his own isotope separator.[15] Before the year was out, Ohnesorge had underwritten Ardenne's effort and the equipment for the great laboratory had been purchased.

By mid-1942, at the same time the modification of Lawrence's experimental cyclotron in America[16] was completed, Ardenne's isotope separator had been completed as well,[17] construction having begun in 1941.[18] Ardenne's work during the war has been described as "the most far-reaching work on isotope separation."[19] Based on reports of the success of the Ardenne ion source, his isotope separator appears to have been superior – possibly substantially superior – to the American calutrons. In essence, at mid-war, Ardenne was at least neck and neck with America's leading electro-magnetic isotope separation bomb program and probably ahead of it – especially when considering the superiority of his technology – something neither Heisenberg nor any of his German cohorts could ever claim. It is a fact never openly admitted by the United States at that time or during the years since.

When calutrons technology had been proven in the American uranium enrichment program, it was handed over to the big industrial combines for transitioning into production models and methods. Subsequently, uranium enrichment production on an industrial scale was begun. Because such a course of research and development and then production was, and still is, the normal and expected paradigm of technology development; and because the two nation's programs so closely resembled each other in so many other facets, it seems probable the Germans went into development of a production phase of Ardenne's technology no later than the United States' program started building commercial calutrons, and probably earlier. There is no indication that German separation technology development was disrupted at this stage. Once the technology had been created by Ardenne and his team, it would not have been necessary for the developers to participate in its adaptation to industrial production processes, although it certainly would have been helpful.

The assumption that Ardenne's device went into production development can be made because there remains the important questions of what were Ardenne and his program doing between mid-1942, when his machine was successfully enriching uranium, and the end of the war; and what happened to any enriched uranium that may have been produced at

Ardenne's laboratory or in any resulting production facilities? At the end of the war, nearly 2,400 tons of German uranium was missing and unaccounted for (see Chapter Three), almost certainly because it had been enriched down to the 1/140th of its mass that was U235. At that ratio, seventeen tons of enriched uranium could have been produced from the missing material. This does not include the massive discovery of approximately 375,000 tons of spent uranium discovered in salt mines in 2011, covered earlier in this text, documented as having come from the Nazi atomic bomb program.

The point is, Ardenne's program had been moving with great momentum right up until the moment no more was heard about it, by then the program had already been advanced enough to go into production development and a facility most likely was under construction at Auschwitz. Besides, Ardenne repaired the isotope separator quickly after the bombing and it appears reasonable to believe additional experimentation could have continued right up to the end of the war. The questions of whether Ardenne continued to operate and improve his enrichment process, and what happened to any enriched product created as a result of it, are unanswered by the traditional history. *Critical Mass* proposes answers to these mysteries.

Despite his achievements, significant effort has been made to discredit Ardenne's wartime work[21] and, in fact, to hide it whenever possible, including by Ardenne himself. Ardenne, who was essentially self-taught in physics and mathematics but whose zeal for the subject matter and his personal connections allowed him to make great strides with his unconventional projects, was belittled personally and professionally by many of his counterparts for not being a true academic, most especially by Heisenberg and another leading German theoretician Carl-Friedrich von Weizsacker. Because of them the animosity from Albert Speer grew.[22]

Ardenne was supposedly, in turn, misled by Heisenberg and Weizsacker into thinking a bomb was not possible for technical reasons, even though Heisenberg, along with Hahn, was one of the theoreticians who had revealed to Ardenne the estimated critical mass of an atom bomb.[23] Despite Heisenberg's later alleged disinformation to Ardenne regarding the technical unfeasibility of a weapon, Ardenne, using Heisenberg's previous argument for a bomb, secured Ohnesorge's funding for his project, who in turn used the argument to gain his audience with Hitler – almost certainly through Bormann.

Ardenne's practical application of physics was not without the direction of a strong theoretical mind that kept him current and gave him guidance in his quest to unleash the atom. He had hired Professor Fritz Houtermans, a fascinating and brilliant Austrian who while still a student in Germany, like Oppenheimer in America, had worked out the thermo-nuclear theory of solar energy; that is, what fuels the stars – and hydrogen bombs. In fact it is Houtermans and astronomer Robert Atkinson who, together, are given credit for first deciphering and articulating the ther-monuclear theory; so named because of the immense heat inside the stars that is released when hydrogen atoms collide and fuse together to form helium.[24] Thus an atom bomb is the result of energy released by the fis-sion, or splitting of atoms, while a thermonuclear warhead – or hydrogen bomb – is the result of energy released by the combining, or fusion, of atoms.

Houtermans' genius was not limited to astrophysics. As early as 1932, the same year the nucleus of the atom was discovered and six years be-fore the atom was first split, Houtermans was the first to recognize and champion the potential for nuclear power from atomic chain reactions.[25] When Houtermans' hated Austrian compatriot Hitler came to power in Germany, Houtermans immigrated to the Soviet Union. While there he advanced the theory, in 1937,[26] of neutron absorption, which would even-tually be used to create plutonium, another first. Before the war had even begun, Houtermans' powerful and imaginative mind in Stalin's hands could have placed the Soviet Union as front-runner among the nations in the race for an atomic bomb, had the Russian leaders paid more attention to the unusual physicist. By not doing so they committed a serious error.

For having thus jumped out of Hitler's frying pan and into Stalin's fire, Houtermans was arrested in one of Stalin's paranoia-driven purges in 1937. Houtermans' wife and children escaped to the United States but Houtermans was imprisoned for two years, constantly at threat of death, and tortured in an effort to gain a confession of having been a saboteur. In one 72-hour session all of his teeth were knocked out. In a following interrogation, Houtermans falsely admitted to having spied for Germany by ascertaining Russian aircraft speeds using a device he had "invented."[27] The torture stopped while his "invention" was reviewed. The contrivance turned out to be wholly invalid on scientific grounds, as Houtermans planned it would, and higher officials correctly deduced his confession had been coerced from him by "unscientific" means, all according to Houtermans' plan.

While Houtermans awaited review of his case in 1939, Hitler nego-tiated what would be a short-lived peace with Stalin and Houtermans was turned over to the Gestapo as part of a general prisoner exchange. Heinrich Mueller's police force locked him up again for a short time, then freed him on request of Nobel laureate Dr. Max von Laue, with the proviso Houtermans was under Gestapo supervision and the understanding he would not be allowed to work for any state agencies or universities.[28] Soon he was employed in the private laboratory, funded by the Postal Service, a state agency, of the unorthodox Baron Manfred von Ardenne.

The renowned theoretical and experimental mastery of Houtermans – who despite his proven theoretical leadership was actually degreed as an experimental physicist – certainly provided significant contributions to the unappreciated but substantial enthusiasm and experimental genius of Ardenne. For example, Ardenne had been told by Heisenberg and Hahn the required critical mass of a uranium bomb was estimated to be "only a few kilos."[29] Houtermans actually performed the exact calculations for critical mass, however, while working for Ardenne in 1941,[30] thus providing a crucial piece of information. For comparison, the United States' program did not deduce its final figure for the amount of enriched uranium needed for a bomb until four years later, in April 1945. Houtermans also had calculated not only the cross sections of a fast, or exploding, chain reaction, but the cost of various isotope separation methods, as well. In addition, while in Ardenne's employ Houtermans performed serious research on development of a nuclear reactor.

Much has been made in previous histories of Houtermans' covert resistance to the Nazis waging war using the fruits of his mind and the infinite powers of the universe it discovered; and undoubtedly much, if not all, of what is reported about his opposition to Hitler is true. For Houtermans appears to have been a man of quality conscience. This fact and his contributions to a German bomb as listed above, notwithstanding, history suggests his main obstruction to the Nazis co-opting his marvelous mind came in the form of steering Ardenne and others away from a bomb and toward the development of nuclear reactors for creating energy for industrial purposes.

Besides Houtermans' research into reactors, there is evidence Ardenne's laboratory was, in fact, actually building a reactor as well as the magnetic isotope separator.[31] The fact, however, both Ardenne and Ohnesorge understood and promoted the development of a bomb before Houtermans arrived on the scene and that they continued to pursue one after his employment indicates Houtermans' politics had little effect upon the purposes of the laboratory or upon its achieving those objectives. In addition, given the Gestapo's close control of Houtermans, it can hardly be expected he would have effectively tried to thwart Ardenne's, and by extension Ohnesorge's, efforts toward a bomb. Although all German scientists were watched closely, because of his history none had his actions so carefully scrutinized as Fritz Houtermans.

In fact, it is entirely possible Houtermans' working at Ardenne's laboratory was the result of Gestapo Mueller having informed his mentor, Bormann, that the eminent physicist was in Gestapo hands following the prisoner exchange with the Soviets. Upon hearing this, Bormann, in an effort to expand his own nuclear program, may have manipulated his bureaucratic strings, steering Houtermans into "his" program run by Ohnesorge and Ardenne, knowing they could use Houtermans' substantial capabilities. Considering the Gestapo's order for Houtermans not to work at any state program, and then Houtermans ultimately working for a state agency that was under Bormann's control, such a course seems likely. For working at Ardenne's facility, which though private was funded by a major government branch that performed important war research on the most secret weapon of all, would almost certainly have been considered a breach of the Gestapo directive. Only with the Gestapo's blessing, and by extension Bormann's, is it likely Houtermans would have been allowed to work on the Postal Ministry's nuclear bomb project.

The Gestapo's directive ordering Houtermans to stay out of government nuclear programs may have been a device to keep the scientist out of the control of competitors to Bormann's nuclear bomb program, as well. Fritz Houtermans had been the "guest" of one too many state police organizations not to know what was expected of him if he wanted to survive. But he was a physicist at heart – to not pursue his work was the same as not breathing.

On the smoky, ash-covered banks of the Vistula River hulked the miserable Polish town of Oswiecim. The cause of its wretchedness surrounded it: To the southwest one kilometer stood a concentration camp established by the occupying Germans who had overrun Poland. To the west two kilometers stood another, much larger camp with an even more nefarious purpose. From its smokestacks the constant snow of human ash settled upon the town. Between the stacks and the town stood the train station through which humans, like ignoble beasts of the field, were trundled to these abject camps. To the east six or seven kilometers was a third camp, reserved for prisoners of conscience who dared defy the

Nazi regime, as compared to most of those in the other camps who just happened to be unlucky and were born across the wrong boundary line or of arbitrary parentage. A few kilometers north of that stood yet another camp, where the "lucky" prisoners were starved more slowly on slightly higher rations while their military masters in the SS sold their 18-hours-a-day labor for a pittance but kept all of the earnings for themselves. The town had even been severed from using its own Polish appellation and was forced to use the Teutonic version of its name: Auschwitz.

To slap wicked insult on cutting injury, 12,000 residents of the town had been thrown out of their homes and German scientists, technicians and factory workers, all employees or contract workers of the world's largest chemical cartel, Hermann Schmitz's I.G. Farben, moved in.[32] From then until the end of the war nothing would be held back in the effort to erect and put into operation what would be one of the most, if not *the* most, technically advanced processing plants in the world, according to authors Peter Hayes and Richard Sasuly, who wrote *Industry and Ideology* and *I.G. Farben*, respectively.

The site had been carefully selected for its purpose: it was outside of Germany and far from Allied bombing and the watchful eyes of American and English reconnaissance operations; it was next to a major railroad center allowing easy access for moving equipment and materials from around Europe to and from the site; it had a nearly inexhaustible supply of manual labor from the death camps for building the fences, barracks, offices and other non-technical structures required and for operating equipment that might otherwise be deemed too dangerous for individuals whose lives were valued; and it had quick and ready access to vast stores of coal from the Brzeszcze-Jawiszowice coal mine.[33]

The purpose of the plant appears to have been hidden behind an illusory wall carefully crafted to camouflage the truth from the world. So much of what went into building and operating the plant, and the paucity of product it was reportedly constructed to produce – no product whatsoever was made – is not congruent with the history of the company that owned and operated it or its alleged purpose: the making of synthetic gasoline and synthetic rubber, known as buna.

First, and most telling, the plant consumed more electricity than the entire city of Berlin.[34] Considering the installation never made a pound of buna, never even went into production, and is alleged to be the biggest failure in the history of I.G. Farben because of that fact, such electrical consumption is incredible if not entirely unbelievable. That such huge quantities of power were required to build the facility – not run it – is inconceivable. Certainly Berlin, the eighth-largest city in the world at the time, constantly bombed by the Allies and continually rebuilt to keep the war machine going, had many construction and re-construction projects within its boundaries that individually matched or exceeded the electrical demands of a single buna plant under construction, not to mention the total consumption of all of Berlin's construction projects combined. Add to these the electrical consumption of the hundreds of thousands of businesses and residences throughout the sizable city, and the electrical consumption of the buna installation is massive and unexplainable in comparison.

Even had the plant been completed and making buna, but it was kept secret after the war for some unexplained reason, the electrical consumption described still would have been astronomical given the buna manufacturing process, far exceeding any power usage that could have been expected for the facility. The only explanation had the plant been making buna that could begin to explain such a high level of electrical consumption, although this even stretches the bounds of plausibility, is the plant was designed to be powered totally by electricity, including heating the buna directly with electrical power.

But such an approach would have been extremely inefficient because burning coal created electricity at Auschwitz. Normally, the burning of coal heats water to create steam, which would then efficiently be used for the buna heating processes. To burn coal to create steam to create electricity, which was and is the conventional way to create electricity, which would then be used to heat buna, adds an extra step – and cost, time and vulnerability – to the heating process while losing considerable energy, which is fundamentally very inefficient. There is no conceivable reason to have done such a thing.

Ed Landry, President and General Manager of Keystone Polymers, Inc. of Houston, Texas at the time he was interviewed, and an expert on synthetic rubber production, when told about the electrical consumption of the buna plant, responded, "that was not a rubber plant – you can bet your bottom dollar on that." Based on other information provided, as well, Landry believes it is hardly conceivable that the so-called "buna" plant at Auschwitz was primarily designed to make synthetic rubber.

When the author contacted George Ladzun, another leading expert on buna production and a senior manager with over 35 years of experience building and operating at least three buna plants from the 1960s through the 1990s, Mr. Ladzun supported Mr. Landry's assessment completely. Mr. Ladzun and Mr. Landry furnished considerable additional details about the construction, costs of construction, and development of buna facilities that have been confirmed by the author using current trade journals, and the author has used this information to substantiate other evidence.

Second, the plant had cost over 900 million reichsmarks, over 250 million[35] 1945 United States dollars based on the initial currency exchange of marks for dollars following the war. The value of the mark, however, had already begun spiraling before the end of the war. Using the conservative $250 million figure adjusted for inflation to today's dollars, the buna plant would have cost $2 billion.[36]

"That's a hell of a lot of money for a buna plant," asserts Mr. Landry, again questioning the assertion that buna was, in fact, what the facility was built to produce.

The average buna plant that produces 150,000 tons of buna annually cost approximately $80 million to build in 1999 dollars, the timeframe in which these calculations were made. That is $10.5 million adjusted down to 1945 dollars. The expenditure of $250 million 1945 dollars reported to build the buna plant at Auschwitz is not just twice the amount expected, or even three or four times the sum one would anticipate the plant would have cost, but twenty-five times that of the average buna plant of the day. And today's costs are greatly inflated in comparison with 1945, in order to meet the higher costs of environmental protections that doing business in the 21st Century entails.

In addition, a plant that produces 150,000 tons of buna per year is producing the same amount of buna the Auschwitz plant *and* the two other existing plants of the time at Schopau and Huens, Germany[37] *and* one additional plant equal to that at Auschwitz were all intended to produce, *combined*.[38] This being the case, in essence the alleged buna plant at Auschwitz should have cost only half as much to build as the 1945 $10.5 million estimate; in other words, $5.25 million – about one-fiftieth the cost of the Auschwitz construction. Upon careful consideration, it is hard to conceive of a buna plant being built that cost 50 times more than existing buna facilities.

Some critics have decried a comparison of calculations between marks and dollars adjusted for inflation rates, saying the extensions can allow no fair correlation between the German and American spending. While admittedly such calculations are not exact, they nonetheless reflect spending differences showing many times the order of magnitude of spending that would be expected in such comparisons. These differences – at ratios of anywhere from 1:25 to 1:50 – are so large as to suggest there is no correlation between the values of dollars and marks at all, if the critics are correct, which obviously is inconceivable.

But for the sake of argument, we will use another comparison that may be more satisfactory to these critics. Instead let's compare the buna construction costs with something of near-equal uniqueness, technical requirements and importance within Germany's own sphere of economic influence, the V-2 rocket program. Germany spent $120 million to build the entire V-2 complex, according to Brian Ford in *Germany's Secret Weapons*.[39] At a 1945 exchange rate of three marks to the dollar, the V-2 plant cost 360 million marks, compared to 900 million marks for the buna plant. In other words, the famed pioneering, high tech rocket facilities at Peenemunde cost only one-third the cost of the buna plant at Auschwitz. This despite the fact buna technology had already been around for two decades and its processes were already well established and at least two other plants had already been built, while the V-2 represented the development of technology that no other nation at the time even considered possible.

Third, the suggestion that I.G. Farben in the four years between the beginning of construction of the plant in early 1941 and the plant's shutdown at the end of 1944, completed only one installation in the buna plant and was still unable to produce buna[40] runs counter to the commission given to Farben regarding the construction of the plant; counter to the priority given to the building of the plant by both the Nazis and Farben; and counter to the history of the company and its experience building buna facilities and its proven capabilities as the largest chemical concern on earth. Considering its great investment,[41] the 25,000 inmates and 12,000 German employees and contractors who worked on the project;[42] and the intense interest and pressure put on the project by Hitler and his SS, it seems doubtful if not inconceivable I.G. Farben would have come up empty on such an important venture. Especially since buna product was allegedly needed so badly and buna technology had already been developed two decades earlier and therefore required little, if any, experimentation or research and development time.

I.G. Farben's reputation had been made on technological achievements. The forerunner to the I.G. Farben company was BASF, whose founder, Carl Bosch, had been one half of the two-man team that first developed synthetic nitrates for fertilizers and explosives[43] – called the Haber-Bosch process. The new process, with another Farben-developed technology, the Bergius synthetic oil and rubber process with which buna is made, was the first production-level technology that required extremely high pressures;[44] a challenge Bosch met with great success and for which he was the first engineer to earn the Nobel Prize.

In 1921, BASF's synthetic nitrate plant in Oppau, Germany exploded, killing 600 and wounding 2,000.[45] BASF needed to rebuild the facility fast but required 10,000 skilled workers to do so. The problem was solved by hiring entire companies, paying them so well that they dropped all other business and went to work concentrating only on Oppau's reconstruction and operation. As a result the plant, previously estimated to require a year in reconstruction, was rebuilt and operating within three months; a testament to the acumen and boldness of Carl Krauch, who had been assigned by BASF's Hermann Schmitz, to rebuild Oppau.

Twenty years later, Krauch, now wearing the duel hats of chief operations officer of I.G. Farben and plenipotentiary general for special chemical production for the Third Reich,[46] had been assigned, once again by Martin Bormann's old business buddy Hermann Schmitz, to a task that required similar handling, the buna plant at Auschwitz.

"In the new arrangement of priority stages ordered by Field Marshal Keitel, your building project [the buna plant] has first priority," wrote Krauch to Otto Ambros, who headed the day-to-day building of the buna facility.[47] General Keitel, with whom Krauch liaised, was Hitler's chief military advisor, and eventual co-chairman with Martin Bormann on what would come to be known as the highest seat of Nazi power, subservient only to Adolf Hitler himself – the Committee of Three.

Ambros was an interesting choice for the assignment. He was considered the leader in the field of high-pressure and synthetic rubber technology and he was the man who oversaw the construction and operation of BASF's first large-scale buna plant at Schopau in 1935.[48] He was also, oddly, Farben's leading expert on poison gas. Ambros dabbled in physics as well, having pioneered theory in magnetic tape technology in 1932; and he studied under Nobel Prize-winning organic biologist Richard Willstaeter. On all fronts, Ambros had special qualities for the special project at Auschwitz.

Given Farben's experience with the Oppau reconstruction and the priority placed by the highest powers in the land – political, commercial and military – on building the buna plant, there appears to be little reason for the installation's construction having taken four years and yet to not have been completed at all. The buna process had been invented two decades earlier[49] and was, by now, old hat so to speak; two large production plants were already built and operating successfully. Manpower, both skilled and unskilled, was massively available at Auschwitz. Even though efforts were supposedly being made to update buna technology, there seems to have been little to hinder Farben from repeating Krauch's success at Oppau when constructing the buna plant at Auschwitz.

Given the directive of high priority and the quick results the directive demanded,[50] certainly the buna plant would have come to fruition within four years had buna been the project's true objective. But after four years in construction, at the end of 1944 when it was dismantled and carried away in the face of the approaching Soviet Army, the buna plant at Auschwitz still had not produced a drop of buna.

Certainly there is something wrong with this picture.

A compilation of the three central and readily known facts just outlined – electrical consumption, construction costs, and I.G. Farben's previous record – does not easily form a picture that a buna processing plant was the type of project being constructed at Auschwitz. Such a compilation does sketch a picture, however, of another important wartime production process being developed, though very secret at the time. The process is uranium enrichment.

First, while buna requires almost no electrical consumption to produce, electro-magnetic isotope separation requires staggering amounts of electricity to power the immense magnets used to separate the ionized uranium particles. As already documented, the buna plant at Auschwitz devoured as much electricity as the entire city of Berlin, the eighth-largest city in the world in the 1940s. Few things, even today, consume as much energy as the buna plant did given its small relative size. The fact I.G. Farben had built an electrical plant next to the buna operation, a very rare occurrence in those days of inexpensive electricity,[51] especially given buna production used little electricity, is stark testimony of the plant's voracious appetite for voltage.

Second, although at least 25 times the construction cost of a buna installation, the cost of construction of the Auschwitz plant is strikingly in line with what one would expect to see for an isotope separation plant. For comparison, the United States calutrons program at Oak Ridge spent $20 million on research, $6 million on engineering, $204 million on construction, and $10 million on operations, for a total of $240 million, according to General Groves' own figures.[52] This compares to $250 million for the "buna" plant at Auschwitz. The harmony of the German and American figures is striking if not compelling. If that correlation is ignored, one must believe the buna plant actually cost more to construct than the Oak Ridge calutrons plant.

Third, while Farben had a strong reputation for quick construction of its priority projects, the delays of the "buna" facility and the problems that caused those delays mirrored to a significant degree the chief difficulties experienced at Oak Ridge. The buna plant enjoyed top-priority status over all other projects in the Reich, "even at the expense of other important building projects or plans which are essential to the war economy,"[53] Krauch had declared. Thus, priority-wise, the "buna" plant held a position roughly equal to that enjoyed by the Manhattan Project in the United States. But even early in the war, before shortages became prevalent, the "buna" plant at Auschwitz suffered continual delays caused by malfunctioning equipment and material shortages.[54] Such setbacks were totally out of character for the technically advanced and highly efficient (even for a German company) I.G. Farben that was *supposed* to be installing buna technology already well-developed; and that was supposedly being led by managers who were the leading experts in their fields, and, personally, had already successfully overseen similar projects. In addition, due to the vast numbers of people required for the installation, there were difficulties providing housing and transportation as well as the other essentials of daily life, again paralleling similar challenges within the Manhattan Project.[55]

The most obvious characteristics and history of the facility strongly indicate an isotope separation installation rather than a buna processing plant. Add to this the requirement for absolute secrecy about uranium enrichment during wartime and the fact isotope separation was such a unique and costly process at the time, unlike any other, and it becomes hard to imagine the so-called buna installation being anything but a cover for a uranium enrichment facility.

Other clues, while not conclusive individually, dovetail so strikingly with the premise that the alleged buna plant was actually a uranium enrichment facility as to place a collective exclamation mark after the conclusion. A few examples: First, despite reported drastic and ongoing setbacks, the I.G. Farben leaders, Nazi bigwigs and the SS command at Auschwitz appear not only to have worked amicably hand-in-hand throughout construction to resolve the problems, but they even cordially wined and dined one another throughout the duration of the project, without allowing their supposedly dismal failures to interfere with their personal relationships.[56] Such relaxed accord could not have been expected within the Nazi regime – nor, indeed, within many other regimes – with this project as high-priority, essential, beyond schedule and over-budget while at the same time being such a familiar construction process as the alleged buna plant. If the challenge, however, was pioneering unknown science with the hope of creating a decisive miracle weapon, certainly an atmosphere of teamwork and esprit d'corps would have prevailed, as was the case within the Manhattan Project, and as seems to have been the case within the leadership at Auschwitz.

Second, I.G. Farben, traditionally known as a chemical concern, on the heels of developing synthetic nitrates shortly after the turn of the century actually had built an explosives empire unequaled in Europe by gaining controlling interests of the other major munitions manufacturers on the continent. Farben then aligned the operations to create Europe's largest broad-based vertical explosives manufacturing empire, causing Joseph Borkin to write that Farben "had focused a portion of its strategy on the waging of war."[57] Wouldn't it have been the natural next step in that strategy to be the manufacturer of the next generation of weapons – nuclear weapons? And wouldn't it have been the Nazi's most logical next step to ask the leading munitions provider to undertake this endeavor; especially when the relationship was as close as that of I.G. Farben and the Nazi Party?

Once again, the German and American nuclear programs appear to have followed similar paths on this front; chemical companies led the key industrial concerns that produced the American atomic bombs, DuPont and Tennessee Eastman among the largest. Such institutions were the only organizations that worked with the high-pressure and high temperature technologies that most closely resembled nuclear technologies.

Even combined, however, America's largest four chemical companies did not equal the size, stature, capabilities or expertise of I.G. Farben. And at least one such American company, DuPont, participated only for the sake of patriotism and to ensure the conservation of democracy. The leaders of DuPont not only intentionally precluded nuclear weapons development from its business strategy, but they accepted only one dollar over and above its expenses for the entire wartime project, as a showing of their refusal to profit from war. After the war, DuPont withdrew from nuclear weapons development altogether. Presumably, a company like Farben that had integrated war production into its business plan as a basic strategy of its growth would be the first to jump on such a potentially profitable market as nuclear weapons.

Third, even the backgrounds of the chief men involved at Farben all appear, to one degree or another, to lend themselves to atomic involvement. Ambros' association and apparent willingness to lead the development of weapons of indiscriminate mass-destruction, as illustrated by his expertise with poisonous gases; combined with his interest and knowledge of theoretical and experimental physics, as shown by his pioneering work with magnetic tape technology; and his ultimate vocation as the chief high-pressure expert and construction project manager at Farben, combine to provide a man singularly prepared to be the chief architect of a uranium enrichment mass-production facility.

Martin Bormann's relationship with Schmitz, and through Schmitz control of Krauch and Ambros, as well as Bormann's relationship with General Keitel, who has already been connected militarily with the alleged buna plant, and Auschwitz commandant Hoess, SS Reichsleiter Heinrich Himmler, and even Himmler's adjutant and liaison with both Hitler and I.G. Farben, SS General Karl Wolff, can all be connected – through Bormann – to the German atomic bomb. Wolff, in fact, had been Himmler's liaison to the buna plant as well.

Fourth, construction on the plant had started some time during or shortly after February 1941. The time frame is salient because it was a year after Ohnesorge's first atomic conference with Hitler and about the same time Ardenne started building his isotope separation machine. German and United States atomic programs often paralleled each other. Both the Oak Ridge and the Hanford facilities' construction were begun while the technologies for each were still in developmental stages. Probably Germany mirrored the United States' policy of starting to build facilities for technologies that were still on the drawing board, in order to gain an advantage in time. As in America, time constraints were a chief issue, and with the risks of failure being geo-political and economic inferiority and military oblivion, it seems reasonable Germany would also begin construction before the technology had been proven. To fail to initiate concurrent design, engineering and construction would have consumed additional months, or even years, when time was of the essence. Alternatively, perhaps the installation was in fact originally intended to be a buna facility and its design was modified to enrich uranium only after the project was begun.

Albert Speer in his recounting of history, perhaps believing a nuclear program under Bormann never could have succeeded, holds the post-war party line that Germany had not pursued a nuclear initiative with any conviction; a claim employed to hide the fact Germany had indeed vigorously pursued an atomic bomb but hid that effort, with help from the Americans after the war. Speer castigates Ohnesorge's and Ardenne's efforts and minimizes Hitler's conviction to nuclear weapons and even his ability to comprehend their usefulness.

Speer stated in a bizarre sort of argument, that Hitler resisted development of a bomb out of a moral sense. He then falls back on the work of Dr. Heisenberg as the unquestioned leader in German nuclear physics to substantiate his position that Germany never gave atomic weapons serious consideration. But in his diatribe Speer totally fails to address the idea that it was common political wisdom by then – even if the potential for a bomb was known only within very high national leadership and scientific circles – that whoever obtained the bomb first would control the world, which was the essence of Adolf Hitler's life, the Nazi cause, and the reason Hitler had begun the war. Roosevelt, Churchill, Hirohito and Stalin all understood this precept. To think Hitler did not is ludicrous, causing one to question whether Speer was stating his actual perceptions or intentionally disseminating disinformation.

If an enemy achieved nuclear weapons before Germany, Hitler would have lost his life's task by default whether he liked the idea of having or using a bomb or not. His "moral sense" did not stop him from committing a plethora of the most heinous atrocities experienced in this world. Hitler could do nothing to prevent his enemies from creating a nuclear weapon. Therefore, the only possibility he had to counter their efforts was to develop and use an atomic bomb first. Can anyone really believe that, based on moral grounds, the man who introduced to the world Blitzkrieg, terror bombing, Auschwitz and the "scorched earth" policy, so gallantly rejected a nuclear weapon at the cost of his own life's work and his nation's final fulfillment of what he believed to be its supreme purpose?

Speer's argument that the Fuehrer was too dull to understand the abstract physics of a nuclear bomb seems most strained, too. Hitler had been capable of understanding and visualizing the benefits of such cutting edge technology as jet propulsion and rocketry, both of which Germany first introduced to the battlefield, not to mention some of the politically ingenious advances he executed in his rise to power and European domination. It sounds a hollow claim that Hitler did not have the intellect to "grasp the revolutionary nature of nuclear physics," as Speer suggested.

The time frame of Speer's reference to the Ohnesorge report is mid-1942, the middle of the war. Ohnesorge had first approached the Fuehrer eighteen months earlier, at the end of 1940, with his nuclear proposal.[58] Hitler is said to have scoffed at the suggestion at that time, and joked that while his other leaders "were worrying about how to win the war, it was his Minister of Posts who had to bring him the solution."[59]

One must ask if following the first meeting and Hitler's reputed rejection, Ohnesorge would have gone forward with nuclear weapons research in the face of Hitler's supposed jeering? Possibly. But if he had, he probably would not have done it openly and with disregard for the Fuehrer's feelings about it. One did not expect to be smiled upon by Hitler if one were openly questioning, by his own actions, the Fuehrer's judgment. So why would Ohnesorge expose himself to Hitler's reproach, as Speer's later account suggests, by giving him an update on the project, especially if it showed the lack of promise Speer insinuates, which would have confirmed Hitler's supposed reservations about nuclear weapons? The fact Ohnesorge was discussing nuclear arms with the Fuehrer again and Hitler was intermittently visiting Ardenne's laboratory, probably means either Ohnesorge and Ardenne had in fact achieved a significant level of success that validated the program, or at the very least Hitler was not actually averse to the program in the first place, as so many interpretations of history, including and often based on Speer's assertions, have tried to make us believe.

Endnotes: Chapter Four – The Hidden Bomb

1   Dr. Henry Picker, *Hitler's Tabletalk*, (as quoted by Brooks, Hirschfeld in *Hirschfeld: The Story of a U-Boat NCO, 1940-1946* p. 230

2   David Irving, *The German Atomic Bomb*, p. 235

3   Jochen von Lang, *The Secretary*, p. 316

4   Paul Manning, *Martin Bormann: Nazi In Exile*, p. 113

5   Albert Speer, *Inside The Third Reich*, p. 531

6   Jochen von Lang, *The Secretary*, p. 313

7   U.S. National Archives II, War Crimes Records, *Interrogation Summary #4476*, 1 December, 1947, RG 238 M1019, Roll 80; also *Interrogation Summary #4453, 16 December, 1947*, RG 238 M1019, Roll 80

8   Jochen von Lang, *The Secretary*, p. 315

9   Dr. Henry Picker, *Hitler's Tabletalk*, (as quoted by Brooks, Hirschfeld in *Hirschfeld: The Story of a U-Boat NCO, 1940-1946* p. 230

10   David Irving, *The German Atomic Bomb*, pp. 256, 258

11   Albert Speer, *Inside The Third Reich*, p. 271

12   David Irving, *The German Atomic Bomb*, pp. 78, 290

13   David Irving, *The German Atomic Bomb*, p. 235

14   David Irving, *The German Atomic Bomb*, p. 235

15   David Irving, *The German Atomic Bomb*, pp. 77, 116; Richard Rhodes, *The Making Of The Atomic Bomb*, p. 360

16   Richard Rhodes, *The Making Of The Atomic Bomb*, p. 488

17   David Irving, *The German Atomic Bomb*, p. 76 - 78, 116; Richard Rhodes, *The Making Of The Atomic Bomb*, p. 487

18   Richard Rhodes, *The Making Of The Atomic Bomb*, p. 371

19   David Irving, *The German Atomic Bomb*, p. 235

20   David Irving, *The German Atomic Bomb*, p. 290

21   Brooks and Wolfgang Hirschfeld in *Hirschfeld: The Story of a U-Boat NCO, 1940-1946* p. 230

22   David Irving, *The German Atomic Bomb*, p. 78

23   David Irving, *The German Atomic Bomb*, p. 77

24   Robert Jungk, *Brighter Than A Thousand Suns*, pp. 27, 28, 93; Richard Rhodes, *The Making Of The Atomic Bomb*, p. 370

25   Robert Jungk, *Brighter Than A Thousand Suns*, pp. 48, 94

26   Robert Jungk, *Brighter Than A Thousand Suns*, p. 94

27   Robert Jungk, *Brighter Than A Thousand Suns*, p. 93, 94

28   Robert Jungk, *Brighter Than A Thousand Suns*, p. 94

29   David Irving, *The German Atomic Bomb*, p. 77

30   David Irving, *The German Atomic Bomb*, p. 92

31   Brooks and Wolfgang Hirschfeld in *Hirschfeld: The Story of a U-Boat NCO, 1940-1946* p.

32   Peter Hayes, *Industry and Ideology*, p. 349

33   Peter Hayes, *Industry and Ideology*, p. 349; Yisrael Gutman and Michael Berenbaum, *Anatomy of the Auschwitz Death Camp*, p. 38

34   Paul Manning, *Martin Bormann: Nazi In Exile*, p. 153; Joseph Borkin, *The Crime and Punishment of I.G. Farben*, pp. 3, 116;

35   Joseph Borkin, *The Crime and Punishment of I.G. Farben*, p. 116

36   *Forbes, For Your Information*, 25 November 1991 (the exchange rate was US$1.00 to Deutschmark 7.56)

37   Joseph Borkin, *The Crime and Punishment of I.G. Farben*, p. 63, 114

38   Joseph Borkin, *The Crime and Punishment of I.G. Farben*, p. 114

39   Brian Ford, *Germany's Secret Weapons*, p. 22

40   Peter Hayes, *Quest For Economic Empire*, p. 63; Joseph Borkin, *The Crime and Punishment of I.G. Farben*, p. 127

41   Joseph Borkin, *The Crime and Punishment of I.G. Farben*, p. 116

42   Joseph Borkin, *The Crime and Punishment of I.G. Farben*, p. 3; Paul Manning, *Martin Bormann: Nazi In Exile*, p. 153; Peter Hayes, *Industry and Ideology*, p. 349

43   Joseph Borkin, *The Crime and Punishment of I.G. Farben*, p. 8,9

44   Joseph Borkin, *The Crime and Punishment of I.G. Farben*, pp. 45,46

45   Joseph Borkin, *The Crime and Punishment of I.G. Farben*, p. 37

46   Joseph Borkin, *The Crime and Punishment of I.G. Farben*, p. 115

47   Joseph Borkin, *The Crime and Punishment of I.G. Farben*, p. 116

48   Joseph Borkin, *The Crime and Punishment of I.G. Farben*, p. 115

49   Joseph Borkin, *The Crime and Punishment of I.G. Farben*, p. 50

50   Joseph Borkin, *The Crime and Punishment of I.G. Farben*, p. 114

51   Primo Levi, *Survival In Auschwitz: The Nazi Assault On Humanity*, p. 123

52   Leslie Groves, *Now It Can Be Told*, p. 97

53   Joseph Borkin, *The Crime and Punishment of I.G. Farben*, p. 116

54   Joseph Borkin, *The Crime and Punishment of I.G. Farben*, p. 118

55   Joseph Borkin, *The Crime and Punishment of I.G. Farben*, p. 119

56   Joseph Borkin, *The Crime and Punishment of I.G. Farben*, p. 119

57   Joseph Borkin, *The Crime and Punishment of I.G. Farben*, p. 43

58   David Irving, *The German Atomic Bomb*, p. 77

59   David Irving, *The German Atomic Bomb*, p. 77

Chapter Five

Oak Ridge

"To separate 100 grams (3.5 ounce – author's note) ... of U²³⁵ per day ... 2,000 4-foot ... calutrons could enrich material enough for one bomb core every 300 days."[1]

Richard Rhodes, author, *The Making of the Atomic Bomb*

The mad scramble that marked the beginning of the Manhattan Project under General Groves' administration would have seemed like a carnival to the outside viewer. The inertia that marked Colonel Marshall's administration was quickly replaced by frenetic activity, but very little of it seemingly tied to a master plan. Much of the Manhattan Project would be operated this way throughout the war. Often parallel programs were being developed and implemented that depended on one another for success, even though neither may have been proven, and none of the interdependent, very sophisticated, highly technical and extremely demanding parts of the project had been demonstrated successful before the next component was begun. From the very beginning, the demands of time did not allow for this. Huge investments were made of time, money and effort, sometimes speciously, justified only on a tremulous belief technologies could and would be created, the need for which had not yet been conceived much less the technologies themselves envisaged. Groves had faith required answers would be found before they were needed, and that all risks were justified by the global imperatives reflected in the war itself, even if it was all lost in the end.

In Germany in April 1942, Baron Manfred von Ardenne already had completed an operational magnetic isotope separator in his laboratory in Berlin Lichterfelde, his associate Fritz Houtermans having correctly calculated the critical mass of U²³⁵ the previous year. Manhattan Project scientists, too, had tried to calculate the critical mass of enriched uranium but came up with a surprising range of values. The MAUD Committee, the British group that liaised with the Manhattan Project providing technical support and personnel, calculated the critical mass of U²³⁵ to be 25 pounds. Physicists Frisch and Peierls had at different times predicted the baneful number to be either eight kilograms (17.6 pounds) or five kilograms (11 pounds).[2] Robert Oppenheimer himself, before joining the project, had estimated critical mass at about 100 kilograms (220 pounds). His theoretical group at Berkeley quickly upped that by three times to 300 kilograms (660 pounds).[3] As late as August 1943, when theorists at Los Alamos provided a critical mass estimate of 40 kilograms (88 pounds), the United States' target number for enriched uranium production was still unknown.

And while Ardenne was already proving the effectiveness of his isotope separator, the United States program at the end of April 1942 was still trying to complete development of its calutrons.[4] Despite this and other setbacks, in the United States electro-magnetic separation was unrealistically anticipated to begin enriching uranium by the summer of 1943.[5] Suffering from choking fits and starts, the program would not actually begin any kind of serious production until over a full year later, in the summer of 1944. Even then, production would be in such small quantities as to be valuable only for experimentation.[6] Arthur Compton, in a report he wrote for the Uranium Committee in 1941, had already stated "atomic bombs can hardly be anticipated before 1945."[7] Now, in 1944, it appeared bombs would not be available until November or December of 1945, possibly not until 1946.

The challenges and uncertainty being clear, Groves supported other forms of uranium enrichment – as had the Germans – along with electro-magnetic separation, including gaseous diffusion, a method of separating the lighter isotope of uranium from the heavier ones by vaporizing the uranium and forcing its atoms through a series of "filter" barriers. The Lewis Committee, the select scientific/political review board responsible for oversight of nuclear development at the time, during the winter of 1942 had approved gaseous diffusion as the most likely method to achieve success.[8] The committee made this prediction despite the fact gaseous diffusion was calculated to require one hundred thousand barriers and the vessels in which to contain them, and several months of processing, to enrich enough uranium for a bomb.[9] In fact, at the time, the technology was totally unproved.

Groves also had supported an effort initiated at the University of Virginia to study isotope separation by centrifuge.[10] Outside of the Manhattan Project, the United States Navy was developing a liquid thermal diffusion process for producing enriched uranium to power its warships. Roosevelt had given strict instructions to the Manhattan Project not to co-mingle its efforts with the Navy technology,[11] which pumped liquid uranium through concentric tubes of differing temperatures to separate the lighter from heavier isotopes. Liquid diffusion allowed one more avenue of success for making the bomb, and was alleged to eventually have been utilized despite the President's orders.

In addition to building up these technologies, Groves had to establish a laboratory to study the fundamentals of making and detonating a bomb itself, and to maintain a center for studying and implementing the metallurgical processing of uranium that would be required.

The development of the atomic bomb depended on the disparate programs moving along parallel and often unrelated tracks. As the laboratories hopefully started providing accurate answers and technologies began yielding usable results, General Groves would then try to cobble them together to produce a viable program from this eclectic assemblage of science.

Realizing it was critical each step was accomplished at the earliest possible moment, during his first week as chief Groves signed the approval to purchase 59,000 acres in Tennessee to house uranium enrichment efforts. He also quickly approved a $100 million expenditure to begin construction of the gaseous diffusion plant – the plant was code named K-25 – before the technology had even been proven.[12] K-25 would eventually consume half a billion dollars[13] of the Manhattan Project's $2 billion war-time outlay but, despite the traditional history if recently declassified records are believed, the program would contribute nothing to the atomic bombs dropped on Hiroshima and Nagasaki. Gaseous diffusion did, however, produce enriched uranium that was important in the design of the post-war nuclear generation of weapons – the hydrogen bombs.

K-25 would stand for many years as the world's largest totally enclosed single building, and was twice the size of the calutrons' electro-magnetic isotope separation facility,[14] which became known as Y-12. But Y-12 would be the one and only United States plant through which, Groves claimed, every gram of American-made bomb-grade enriched uranium was processed.[15] Despite the Lewis Committee's recommendation for gaseous diffusion, the calutrons was still the only technology that had successfully enriched uranium and it remained the method of choice for the Manhattan Project.[16]

The original calculations of calutrons requirements were nearly as dizzying as those for K-25. Experiments indicated that 2,000 uranium ionization sources and collectors were required to yield 100 grams – only three and one-half ounces – of enriched uranium per day.[17] Such a rate would require 455 days, one and one-quarter years, to produce enough enriched product for the bomb that was dropped on Hiroshima.

Rhodes calculated 300 days per bomb, based on a smaller bomb estimate than the bomb actually dropped.

> "To separate 100 grams (3.5 ounces – author's note) ... of U235 per day ... 2,000 4-foot ... calutrons could enrich material enough for one bomb core every 300 days."[18]

The above calculations assumed the calutrons worked reliably. In fact, the calutrons throughout their wartime lives proved to be models of inefficiency and poor operation.

Another humbling fact was, the first production calutrons contained only two,[19] not two thousand, sources and collectors. Ground had been broken on 18 February 1943 for the first "track," as it was called – because its oval shape resembled that of a racetrack – and it went into operation for the first time 1 August 1943. But the device ran so poorly due to mechanical and technical shortcomings that it still had not been tested by September. Stone and Webster, the engineering contractor hired to run the operation, finished repairs and final installation in October and "powered up" the machine again in November[20] – to experience a similar failure.

An engineering flaw required the apparatus be completely disassembled and shipped back to its manufacturer in Milwaukee to be cleaned and rebuilt. The entire year passed with so little uranium enriched it could not be counted as production material but only as experimental stock.[21] Also, the enrichment rate was nominal, having raised the level of U^{235} from .7 percent to 10 percent. True, this was over a 1,000 percent improvement, but bomb-grade enrichment needed to be 80 to 90 percent. A 1,000 percent increase to just 10 percent was disheartening. There was still a long way to go.

Despite the setbacks, Groves not only approved a second track to be built, with more sources and collectors, but he added plans for a third, with a new wrinkle. The newest track would be used to run already-enriched product, the end result being bomb-grade enrichment resulting from using the "seeded" slightly enriched feedstock. The two different types of track were designated "Alpha" and "Beta" tracks. Alpha tracks consumed natural uranium and produced 10 percent-enriched uranium. Beta tracks consumed the 10 percent-enriched feedstocks and produced bomb-grade uranium. At about the same time, the General decided to reduce the number of planned tracks from 2000 to 500, trusting that technology would become more efficient and bridge the wide chasm between political/military requirements and uranium's realities.[22]

The second Alpha track was in operation by mid-January 1944[23] and the first, dismantled, Alpha track was returned to Oak Ridge, reinstalled and placed back in operation by March.[24] But problems persisted. To make matters worse, spare parts were non-existent and operators ruefully inexperienced. There was no track record to guide them; no experience base upon which to rely. And the Beta tracks were suffering much the same setbacks.

In the spring of 1944, with only a year left to achieve success – and estimates requiring a year of enrichment to produce just one bomb – no enriched uranium had been produced in anything close to production quantities.[25] Still, Groves and his gargantuan and dubious enterprise pressed on. By the end of June, five Alpha tracks and two Beta tracks were operating, but with very poor performance levels. The ionization sources in the calutrons that converted the uranium feedstocks to be "sublimed" in order to separate the atoms, only converted up to 75 percent of the feedstocks. The condition remained prevalent to the end of the war, leaving anywhere from 25 to 40 percent of the material with enrichment potential sitting useless in the "feed can" at the beginning of the process.[26] All of this had to be cleaned out, reclaimed chemically, and reprocessed before it could be rerun – to the same result.

In addition, losses of sublimed material accumulated throughout the mechanical system of the calutrons,[27] leaving partially-enriched uranium embedded in equipment surfaces and linings within the device. The enriched material, even in microgram quantities, was so valuable that following each run the tracks were disassembled and thoroughly cleaned using Geiger-counters to locate each small particle, and appropriate chemicals and special technologies in order to reclaim every scintilla of enriched uranium. Calutrons operators' clothing was specially laundered each day as well, to ensure every microgram that may have been captured in their clothing was recovered. The sum total of all of these efforts, on all of these machines that were operated and supported by tens of thousands of people, was still an enriched uranium average daily yield of 11.5 grams per day throughout the month of July[28] – not even half an ounce.

More calutrons were installed to increase production levels. By November, nine Alpha tracks and three Beta tracks – most of them with a full contingent of 96 sources and collectors each – were operating to only slightly better efficiency performance than their predecessors, but the additional machines put production on the rise. Daily output in the first week of November averaged 45 grams (1.6 ounces). In the second week it was 57 grams (2 ounces). The third week production dropped slightly to 50 grams, but in the fourth week of November production climbed to 81 grams

(2.84 ounces) per day.[29] The increase was significant but it was still a rate that would require 620 days, a full one and two-thirds years, to accumulate enough enriched uranium to fuel the bomb that must be used within eight short months. Otherwise the United States would forever lose the politico/military nuclear initiative it stood on the cusp of grasping.

Despite the fact output had not yet reached required levels, the enriched material that had been produced was many times more valuable than any other commodity present on earth at the time – if a price could be set on it at all. The potential this material held to change the world made it, in many ways, valuable beyond the mere computation of cost-to-produce versus volume-of-grams. To protect his enormous investment from what could be an immeasurable loss Groves prepared a secure location for its storage before transit. Inconspicuous among the indigenous households of east Tennessee, a lone farmhouse stood at the end of a dusty dirt road. Lumbering farmhands in the peaceful pastures were actually a patrol of security guards. The innocuous silo next to the barn hid a machine-gun nest. The embankment that framed the picturesque homestead harbored a submerged bunker made of reinforced concrete buried under the escarpment's leafy foliage. Inside the invincible bunker a vault stood, surrounded on all sides by yet another cadre of watchful guards.[30] Inside the vault were a few ounces of enriched uranium waiting for the weekly courier to whisk it away to a mountaintop in New Mexico.

Gaseous diffusion at K-25 was just beginning to receive working barriers for filtering the atoms, but only a few at a time.[31] As they arrived, the barriers were installed in the headmost possible stage and tested for integrity, the plan being to allow the first-stage of "converter" vessels to begin the long diffusion process before the numerous ensuing stages had even been completed. In other words, K-25 was making excruciatingly slow headway and had still not produced an atom of enriched uranium.

To crank overall production to a higher level, General Groves either ignored, circumvented or had otherwise persuaded President Roosevelt to belay his presidential order refusing Navy involvement in the atomic bomb project, and Groves adopted as his own the liquid thermal diffusion technology the Navy had been devising. Within K-25, besides the massive gaseous diffusion apparatus, stood a 100-column liquid thermal diffusion pilot plant that Groves had ordered constructed in January 1944 and that had been completed in July of the same year.[32] The plant was soon expanded and began production, according to Groves, at the end of October, and, he claims, it reached peak production in June 1945.[33]

Herbert Childs, author of *An American Genius,* differs from Groves in his account of events, stating liquid thermal diffusion production output actually began on 1 March 1945, four months after Grove's assertion. Careful review of charts based on daily Beta calutrons production output, however, shows no hint whatsoever of an increase of production in March or any other time between the beginning of the year and the middle of June. In fact, charted average daily output is so consistent throughout the first six months of the year as to form an almost straight line, with the lone exception being a small dip in production during the third week of January.

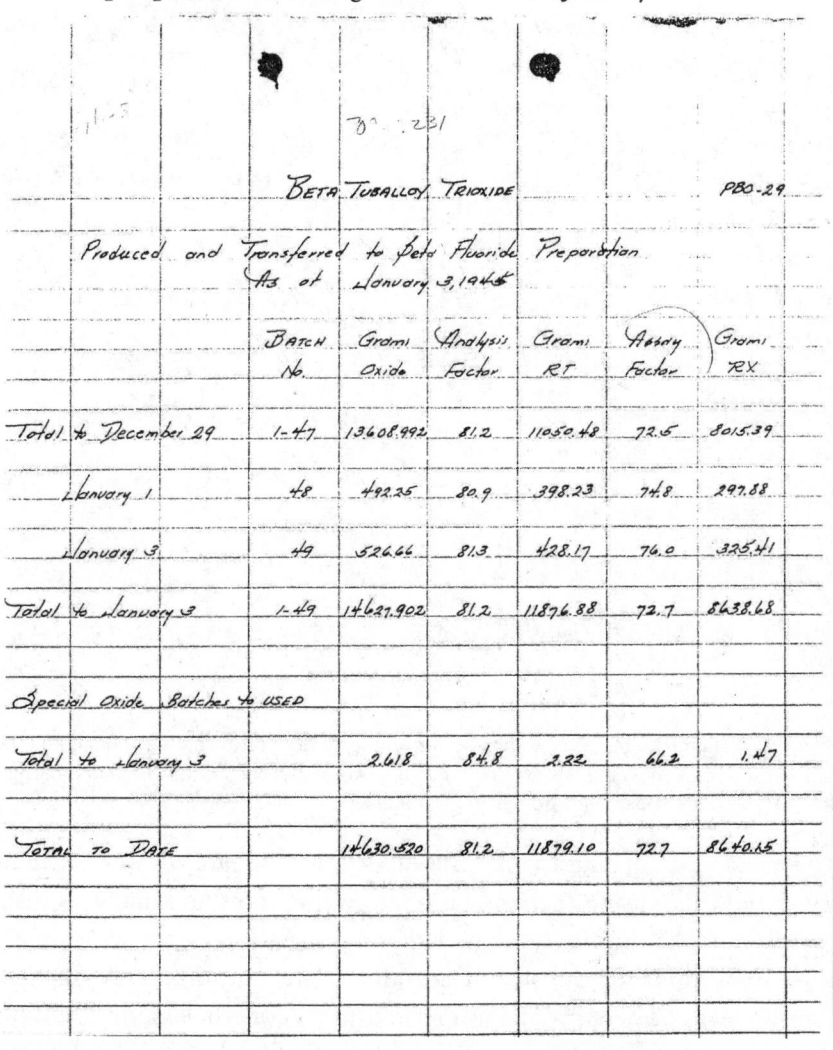

	Batch No.	Grams Oxide	Analysis Factor	Grams RT	Assay Factor	Grams RX
Total to December 29	1-47	13,608.992	81.2	11050.48	72.5	8015.39
January 1	48	492.25	80.9	398.23	74.8	297.88
January 3	49	526.66	81.3	428.17	76.0	325.41
Total to January 3	1-49	14627.902	81.2	11876.88	72.7	8638.68
Special Oxide Batches to USED						
Total to January 3		2.618	84.8	2.22	66.2	1.47
TOTAL TO DATE		14630.520	81.2	11879.10	72.7	8640.15

BETA TUBALLOY TRIOXIDE — PBO-29

Produced and Transferred to Beta Fluoride Preparation As of January 3, 1945

Figure 1: The first page of the "Beta Tuballoy Trioxide" report for 1945. "Tuballoy" was the code name for uranium. Trioxide was the chemical form the enriched uranium was converted to after coming from the calutrons in a tetraflouride state.

This fact not only draws into question Mr. Childs' statement regarding when the thermal diffusion plant began operating, but General Groves' implication, as well, that liquid thermal diffusion continuously rose in productivity until it peaked in June. The record shows a significant upward production curve from mid-November 1944 to the end of the year, corroborating Groves' version of when thermal diffusion production began and the remaining calutrons were put into operation. But by the beginning of the new year, daily production had reached a plateau, consistently producing about 240 grams (8.4 ounces) per day, with no further increase in daily productivity until late June 1945.[34]

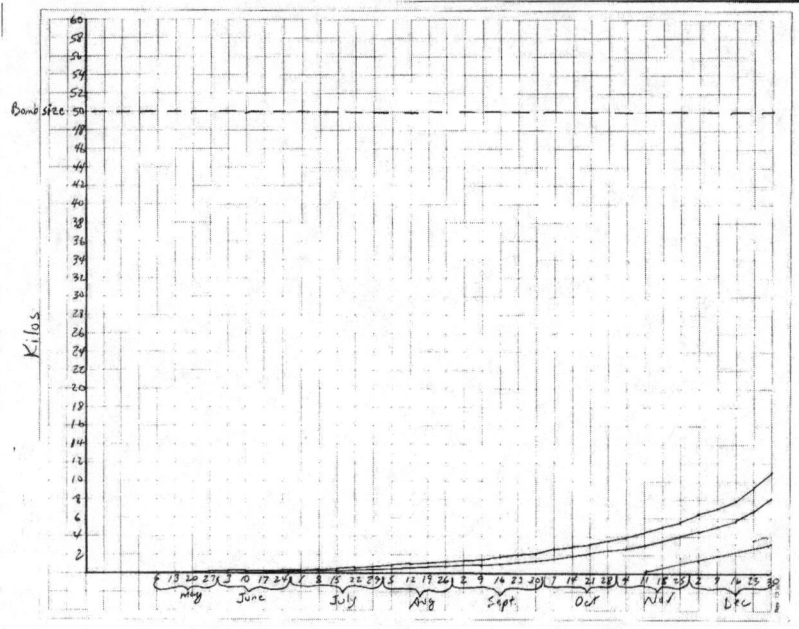

Figure 2: Using the "Beta Tuballoy Trioxide" report the author plotted enriched uranium output. These graphs show the rate enriched uranium was produced. Contrary to traditional history, output was a straight line from late December to late June.

Production did take a significant upward turn the fourth week of June 1945, but this advance certainly is related more to other influences than to continuously improving liquid thermal diffusion processes. Groves tried to account for the significant, immediate, and otherwise anomalous increase of Beta production in mid-June, nonetheless, by crediting the increase to the thermal diffusion process. Later, he credited the gaseous diffusion process with the same production expansion,[35] although gaseous diffusion is reported to have first gone into operation less than three months after thermal diffusion did, on 20 January 1945.[36] While the process would not have borne an immediate impact on output, since it is a cumulative process that requires several weeks for end-product to be available, it undoubtedly would have begun producing enriched material long before the six-month time period reflected between the January start-up and the June surge. And this does not take into account the fact product was already being partially processed as the gaseous diffusion tubes were being added, effectively reducing time-to-product after the

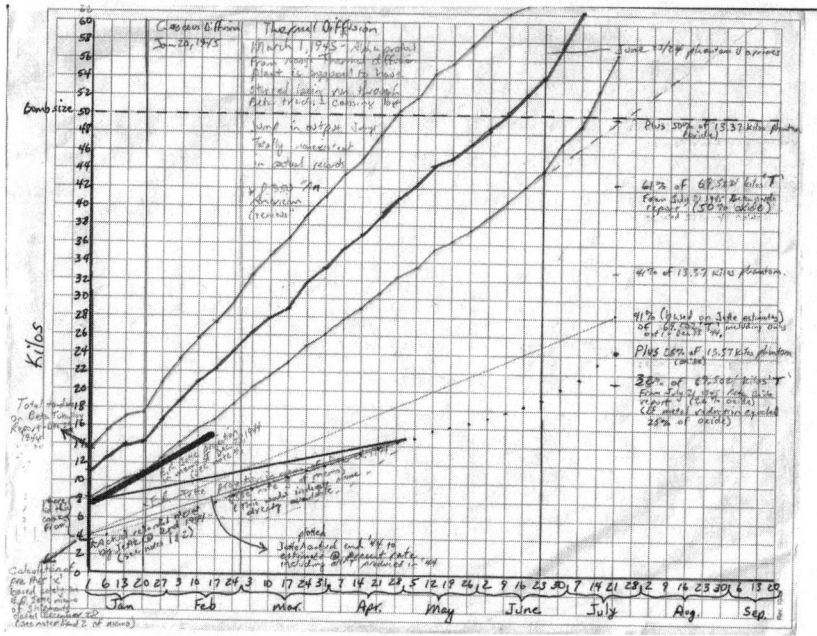

last stage was installed.

Figure 3: The three angled lines at the top represent, from upper to lower, 1) complete output including contaminants, 2) all uranium, 3) level of U235. Horizontal dotted line at top is actual bomb size. Angled dotted line at bottom is Jette's prediction, extended throughout. Vertical line 20 January marks when gaseous diffusion alpha feedstock was supposed to have begun increasing beta output. Line at March 1 marks when liquid thermal diffusion product was supposed to have been added. Neither exhibited any increase in beta output. Line June 23 marks first jump in output, two weeks after U-234 uranium is known to have entered the Manhattan Project.

The Beta output records themselves show that neither gaseous diffusion nor thermal diffusion caused a late-spring dramatic production upturn, as the traditional history asserts. In fact, liquid thermal diffusion was shut down for good just a few weeks after the material for the bomb was accumulated in June,[37] suggesting its ineffectiveness. The mid-June 1945 timing of the increase in enrichment shown in the Beta output records fits perfectly, however, with events of the offloading of the enriched uranium captured from U-234.

With or without the June 1945 spike in enriched uranium output, Manhattan Project isotope separation appeared to be successful. Enough calutrons were working well enough that output was finally on track to produce 50 kilograms of enriched uranium by early August – the amount used in the Hiroshima bomb.[38] And even better, the critical mass for U^{235} finally had been determined to be only 15 kilograms, only one-third the amount of enriched uranium that would be available when it came time to start fabricating bombs. If desired, at the present rate of production, the United States would be able to assemble as many as three uranium bombs in early August. Despite all the difficult challenges, success appeared to be just around the corner.

Endnotes Chapter Five – Oak Ridge

1   Richard Rhodes, *The Making of the Atomic Bomb*, p. 488

2   David Irving, *The German Atomic Bomb*, p. 69, 95

3   Herbert Childs, *An American Genius*, pp. 321, 344

4   Robert Serber, *The Los Alamos Primer*, p. xxix

5   Richard Rhodes, *The Making of the Atomic Bomb*, p. 406

6   Leslie Groves, *Now It Can Be Told*, p. 96

7   Richard Rhodes, *The Making of the Atomic Bomb*, p. 365

8   Richard Rhodes, *The Making of the Atomic Bomb*, p. 489

9   Leona Libby, *The Uranium People*, p.53

10   Richard Rhodes, *The Making of the Atomic Bomb*, p. 550

11   Leslie Groves, *Now It Can Be Told*, p. 22; Herbert Childs, *An American Genius*, p. 350

12   Richard Rhodes, *The Making of the Atomic Bomb*, p. 493

13   Richard Rhodes, *The Making of the Atomic Bomb*, p. 496

14   Richard Rhodes, *The Making of the Atomic Bomb*, pp. 494, 495

15   Richard Rhodes, *The Making of the Atomic Bomb*, p. 601

16   Herbert Childs, *An American Genius*, p. 333

17   Herbert Childs, *An American Genius*, p. 341; Richard Rhodes, *The Making of the Atomic Bomb*, p. 488

18   Richard Rhodes, *The Making of the Atomic Bomb*, p. 488

19   Herbert Childs, *An American Genius*, p. 341

20   Herbert Childs, *An American Genius*, p. 346; Richard Rhodes, *The Making of the Atomic Bomb*, p. 491

21   Leslie Groves, *Now It Can Be Told*, p. 110

22 Richard Rhodes, *The Making of the Atomic Bomb*, p. 489

23 Richard Rhodes, *The Making of the Atomic Bomb*, p. 492

24 Herbert Childs, *An American Genius*, p. 349

25 Stephen Groueff, *The Manhattan Project*, p. 312

26 Jerry Rice, Y-12 Beta shift supervisor, interview via telephone with the author (date not recorded)

27 Herbert Childs, *An American Genius*, p. 350

28 US National Archives, Southeast Region, East Point, Georgia, P.B.O. Report (daily Beta output report), RG 194 – 69 A 406 section 326 box 17

29 US National Archives, Southeast Region, East Point, Georgia, P.B.O. Report (daily Beta output report), RG 194 – 69 A 406 section 326 box 17; also compare to Richard Rhodes, *The Making of the Atomic Bomb*, pp. 600, 601

30 Richard Rhodes, *The Making of the Atomic Bomb*, p. 602

31 Richard Rhodes, *The Making of the Atomic Bomb*, p. 602

32 Richard Rhodes, *The Making of the Atomic Bomb*, p. 551

33 Leslie Groves, *Now It Can Be Told*, p. 122

34 compare to Richard Rhodes, *The Making of the Atomic Bomb*, p. 601

35 Leslie Groves, *Now It Can Be Told*, p. 69

36 Richard Rhodes, *The Making of the Atomic Bomb*, p. 602

37 Leslie Groves, *Now It Can Be Told*, p. 379

38 US National Archives, Southeast Region, East Point, Georgia, Beta Oxide Transfer Report, RG 194 – 69 A 406 section 326 box 17; also compare to Richard Rhodes, *The Making of the Atomic Bomb*, p. 601

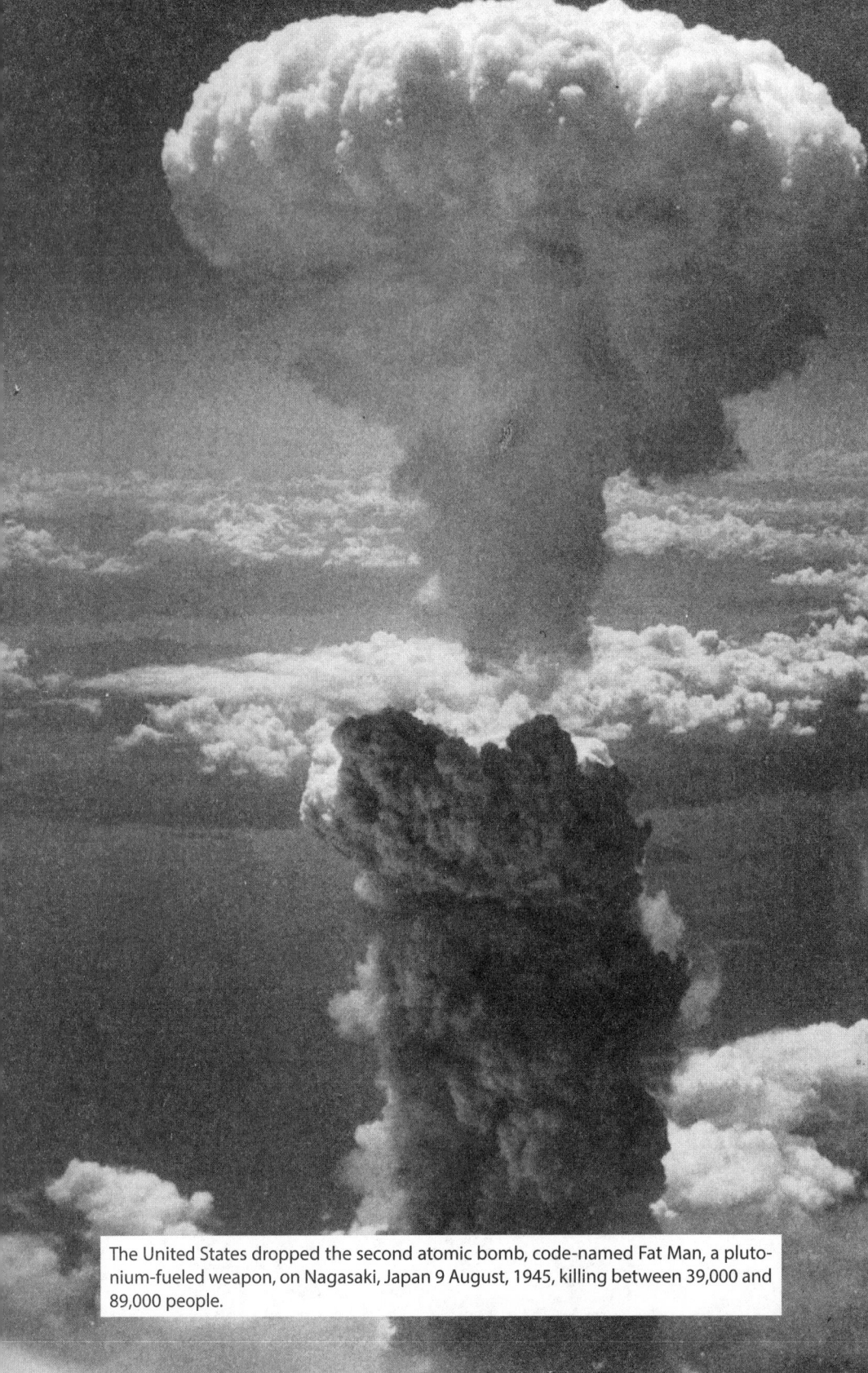

The United States dropped the second atomic bomb, code-named Fat Man, a pluto-nium-fueled weapon, on Nagasaki, Japan 9 August, 1945, killing between 39,000 and 89,000 people.

PART TWO

The Plutonium Bomb

J. Robert Oppenheimer, scientific lead of the Manhattan Project, and General Groves after the first nuclear test explosion, a plutonium weapon called "the Trinity Shot," proved the capacity to detonate 64 triggers all within 1/50,000th of a second. The Manhattan Project was far from achieving this accomplishment before fast-operating energy transfer technology arrived on American shores aboard U-234.

Chapter Six

Timing

Lt. (JG) H E Morgan, Lt. (JG) F M Abbott, Ens F L Granger with Dr. Schlicke POW in custody leaving Anacostia noon Friday via plane. This party expert in bomb disposal and proximity fuses and being sent to assist in securing certain infra red proximity fuses important BUORD [Navy Bureau of Ordnance – author's note] and in cargo U-234. Fuses when secured to be returned Washington custody above party.[1]
– Dispatch from Commander Naval Operations to Portsmouth Naval Yard, 25 May 1945

After Dr. Schlicke completes his lecture he will be available for questions that people ask. But we will kindly ask you not to ask any questions during the lecture and after the lecture Mr. Alvarez will sit at the table and the person who wishes to ask a question is asked to come forward so that we can get in the microphone and keep a record of all the questions and answers.[2]
– From the transcript of an introduction to a lecture given by Dr. Heinz Schlicke to the Navy Department. "Mr. Alvarez" appears to be Dr. Schlicke's handler. Manhattan Project physicist Luis Alvarez was credited with, at the last minute, solving the plutonium bomb's multiple fusses firing simultaneously.

Uranium does not appear to be the only component aboard U-234 capable of being used in an atomic bomb. There were the steel drums and wooden barrels full of fluids noted in Chapter One, which Manhattan Project personnel tested, apparently to see if the materials had been, or could be, part of a plutonium breeder reactor.[3] And there were tons of lead, possibly for gamma radiation shielding; mercury, possibly for very fast mercury switches; and infrared proximity fuses.

The infrared fuses were discovered within five days of U-234's landing at Portsmouth, apparently as the result of Dr. Heinz Schlicke's interrogation. A memorandum written by Jack H. Alberti dated 24 May 1945[4] stated, "Dr. Schlicke knows about the infrared proximity fuses which are contained in some of these packages.... Dr. Schlicke knows how to handle them and is willing to do so." According to the following transmission, at noon the very next day Schlicke was placed on an airplane with a three-man escort and flown back to Portsmouth for the sole purpose of retrieving the proximity fuses.

> Lt. (JG) H E Morgan, Lt. (JG) F M Abbott, Ens F L Granger with Dr. Schlicke POW in custody leaving Anacostia noon Friday via plane. This party expert in bomb disposal and proximity fuses and being sent to assist in securing certain infra red proximity fuses important BUORD [Navy Bureau of Ordnance – author's note] and in cargo U-234. Fuses when secured to be returned Washington custody above party.[5]

```
  )
  (        'P )                            94                25 MAY 1945

FROM:            CNO

TO:              NY PORTS

INFO:            BUORD
                 COMONE

SUBJECT:         POW AND FUSES FROM U-234
                 LT (JG) H E MORGAN, LT (JG) F M ABBOTT, ENS F L
                 GRANGER WITH DR SCHLICKE POW IN CUSTODY LEAVING
                 ANACOSTIA NOON FRIDAY VIA PLANE.  THIS PARTY
                 EXPERT IN BOMB DISPOSAL AND PROXIMITY FUSES AND
                 BEING SENT TO ASSIST IN SECURING CERTAIN INFRA
                 RED PROXIMITY FUSES IMPORTANT BUORD AND IN
                 CARGO U-234.  FUSES WHEN SECURED TO BE RETURNED
                 WASHINGTON CUSTODY ABOVE PARTY.
```

Figure 1: Of all of the valuable technology on board U-234, special and immediate attention was given to retrieving the infrared proximity fuses of Dr. Schlicke.

The dossier on the technology portfolio Schlicke was accompanying to Japan was extensive. Beyond fusing and explosives expertise, he was either referenced by other prisoners of U-234, listed in documents onboard U-234, or admitted to being knowledgeable in or responsible for: very high technology radar and radio systems,[6] guided missile development, and V-2 rockets.[7] While still in Germany, he also had met with a long list of scientists. He noted in his interrogation the intent of many of these meetings was for him to receive the transfer of their technologies and to later disseminate them in Japan, and to serve as the listed scientists' liaison and advisor with Japan.[8] Much of the technology accompanying Schlicke to his destination was the product of this group of 54 obviously very high-level scientists. Among the scientists with whom he had coordinated, which he listed for American interrogators, were Professor Dr. Esau and Professor Gerlach,[9] both of whom, at one time or another, were important members of Germany's atomic research programs.[10] Dr. Esau had served as head of the Kaiser Wilhelm Institute and was a member of the Reich Research Council.

That Schlicke was personally and almost immediately flown back to U-234 specifically to retrieve the infrared fuses, from among all the technology for which he was responsible and all the other technology on the U-boat, seems revealing. The infrared fuses were of immediate interest to the United States, apparently, not just as the booty of war, as were all of the other technologies on the boat, but expediting retrieval of the fuses seems to have been driven by a need to have them immediately available for some purpose. We may see a hint of that purpose revealed in the notes of a meeting at the Office of Naval Intelligence held on 19 July 1945. A portion of the transcribed introduction of Dr. Schlicke bears an innocuous clue to the possible purpose of the infrared fuses.

> "After Dr. Schlicke completes his lecture he will be available for questions that people ask. But we will kindly ask you not to ask any questions during the lecture and after the lecture *Mr. Alvarez* [italics added] will sit at the table and the person who wishes to ask a question is asked to come forward so that we can get in the microphone and keep a record of all the questions and answers."[11]

```
     After Dr Schlicke completes his lecture he will be available for
questions that people ask.  But we will kindly ask you not to ask any
questions during the lecture and after the lecture Mr Alvarez will sit
at the table and the person who wishes to ask a question is asked to com
forward so that we can get in the microphone and keep a record of all
the questions and answers.  Thank you.
```

Figure 2: Navy Intelligence had a "Mr. Alvarez" working closely with Dr. Schlicke.

The presence of a "Mr. Alvarez" as Dr. Schlicke's apparent host or "handler" may be a singular indicator regarding the importance of the infrared fuses. The reference to Mr. Alvarez was not the first to be made from among U-234's passengers and crew. Three weeks earlier, General Kessler had written a letter regarding missing personal items in which he identified a "Commander Alvarez" as having seen some of these items.[12] The identification that Alvarez held the rank of commander appears on the face to indicate he was a Navy Officer; no other United States services maintain a rank of Commander except the Coast Guard, which is very unlikely to have been involved with the U-234 intelligence operation.

```
        I should be very much obliged if they were handed over to the Provost
Marshal General for safekeeping on my behalf.

        I trust that a pocket book containing 1500 Swiss francs,
100 Danish kroner, and 200 Norwegian kroner has been kept for me;
Commander Alvarez has told me that he had seen it and put it in safe
keeping; the same applies to a large pair of Weiss binoculars (7x50)
which I had bought from a Navy submarine store at Angers in 1943
(use of this type of binocular had been discontinued in the submarine
service). This last item was also seen by Commander Alvarez.




                        ULRICH O.E. KESSLER
                        General der Flieger
                        ISN 3 NG-1269
```

Figure 3: General Kessler met and identified Alvarez as a commander, but rosters show no officer by the name of "Alvarez" in the Navy at that time.

U-234's skipper, Lieutenant Captain Johann Heinrich Fehler, also identified Alvarez in a letter written decades after the war, but he identified Alvarez as a Lieutenant Commander.[13] The distinction between whether Alvarez was a full Commander or a Lieutenant Commander would be minimal, except that it may be a moot point altogether. Alvarez may not have been a Navy officer at all. In parenthesis in his letter, Fehler, following his identification of Alvarez, noted that Alvarez is "probably not his real name."

Fehler seems to have sensed there was something disingenuous about Alvarez but assumed it was his name, not his rank, that was dubious. The name, in fact, may have been the counterfeit. There is no listing of any officer surnamed Alvarez in the *Register of Commissioned and Warrant Officers of the United States Navy and Marine Corps* for either July 1, 1943 or its publication two years later on July 1, 1945.

When discussing this enigma with Dr. Alan Bath, a specialist in naval intelligence, Dr. Bath indicated ONI interrogators were often drawn from the naval reserves, which would not be listed in the rosters of full-time officers. This author twice tried to locate reserve rosters for the time period, but none were available. On the other hand, Dr. Bath admitted it was highly doubtful a first-level interrogator of POWs would be the handler of Schlicke at the high-level lecture given. Dr. Bath supposed there were two Alvarez's, one an interrogator from naval reserves (who, by the way, was never mentioned in any interrogation reports) and the handler at the lecture, who Dr. Bath conceded may well have been Luis Alvarez. To Bath it was a mere coincidence they were both named Alvarez.

The name Alvarez may have been real, however, and the rank of commander – in fact, the entire military persona – may have been a fraud, and that was the ill-defined deception Fehler was sensing.

At the time U-234 was escorted into Portsmouth Harbor, the Manhattan Project was near desperation. Because Groves appears to have decided to use some of the already enriched uranium to fuel the plutonium reactors at Hanford, he was short of enriched uranium for the uranium bomb. On the other hand, the Manhattan Project scientists had not figured out a way to efficiently trigger the plutonium bomb. Therefore, at that point in time, neither bomb was viable. And the mid-August requirement for any kind of bomb was fast approaching.

The plutonium bomb consisted of a solid sphere of plutonium the size of an orange. The key requirement to make the bomb explode – besides the creation of the requisite amount of plutonium – was to compress the plutonium sphere so it would reach critical mass. To achieve this compression, 32 redundant detonators – 64 in all – needed to be fired within $1/50,000^{th}$ of a second or less, or the bomb would likely fail.

The challenge was daunting. For a year-and-a-half, the Los Alamos scientists tried to develop a simultaneously firing detonation system. Just a month before U-234 landed, there was "more than a bare possibility that the detonators will be unsatisfactory"[14] wrote Norris Bradbury, who headed the team responsible for triggering the explosion. Indeed, into late June and early July, just two weeks before the first atomic bomb test at Alamogordo, New Mexico, the detonator timing problem was still not resolved.[15]

> The schedule of tests on informers and related equipment was predicated upon 300 detonators having been tested on the ground and the results shown to be satisfactory. Two difficulties may crop up judging from present observations— (a) a much smaller number of tests than 300 will have been carried out; (b) there is more than a bare possibility that the detonators will be unsatisfactory.
>
> Particularly in the latter event it may be necessary to postpone final Raytheon tests until the detonator difficulty is unscrambled.

N. L. Bradbury

Figure 4: Norris Bradbury, in charge of testing the plutonium bomb, identified continued failure of the bomb detonators three months before the Trinity Test.

The experts at Los Alamos had been working on the timing problem since the fall of 1943,[16] but had failed to solve it when, in October 1944, Robert Oppenheimer created a committee to tackle the detonator problem. The first name on the three-man team was Luis Alvarez.[17]

Page 2
Memo from Bainbridge dtd. 10/25/44
Subj: Minutes of a Meeting on the
Electric Detonator Program.

CLASSIFICATION CANCELLED
PER DOC REVIEW JAN. 1973

d. J. R. Oppenheimer appointed a committee of three, L. W. Alvarez,
K. T. Bainbridge and Lt. Col. Lockridge, to consult on the procurement of detonators to insure that the designs are satisfactory to G-7
and to X-2.

Figure 5: Luis Alvarez was the first member of a three-man team personally assigned by Oppenheimer to resolve the detonator timing problem of the plutonium bomb. Alvarez is credited with getting 32 detonators to fire within 1/50,000th of a second of each other.

Alvarez had begun his wartime work in the Radiation Lab at MIT, then worked on Ground Controlled Approach Radar, which allowed controllers to "talk down" a pilot whose vision was impaired.[18] He then worked on Phased-array Radar, which allows a radar system to track an object electro-magnetically rather than steering the system by manual means. After the war, Alvarez went on to win the Nobel Prize for Physics in 1968 for his work on aeronautical navigation systems. And he, with his son Walter and geologist Frank Asaro, were the first to forward the theory that Earth was struck by a meteorite that caused the extinction of the dinosaurs. They based their theory on findings of high levels of iridium in concentrated locations on earth. At first scorned, the theory has become widely accepted.

Luis Alvarez also became one of the great heroes of the atomic bomb story when he solved the plutonium bomb detonator timing problem in the last days before the Trinity Test.[19] In his own account of his work in the Manhattan Project, he wrote simply that he "cleaned up some loose ends in detonator design."[20] The understatement and lack of detail may be telling – especially if it was meant to hide how he "cleaned up" those details.

Of all the Manhattan Project personnel whose name one would expect to see connected to Heinz Schlicke's and U-234's infrared proximity fuses, if there was a connection, Luis Alvarez's name would be at the top of that list. The two scientists' backgrounds were strikingly similar; both men were leaders in the field of high frequency light waves. When it came to science, they spoke the same language. If the Manhattan Project wanted somebody to debrief Schlicke, or anyone aboard U-234, about any category of atomic bomb development, Alvarez would have been the logical choice. By assignment and as a close confidant of Oppenheimer, he was one of the very few people who had a broad view and understanding of all the aspects of the Manhattan Project. By late spring 1945, when U-234 arrived on American shores with just two months left until the Trinity Test, the detonator problem was still unsolved and its resolution was now paramount to the success of the entire program. Alvarez, as the key man assigned to the problem, was in desperate need of a fusing system that could fire multiple detonators simultaneously. Schlicke had fuses that worked on the principles of light – presumably they worked at the speed of light.

In fact, among the documents Schlicke was accompanying to Japan was a report on "the investigation of the usability of ultraviolet (invisible) light for transmitting messages or commands and particularly for the remote ignition of warhead fuses."[21] The report had been prepared based on research done from 1939 through 1941 by Hans Klumb and Bernard Koch. In suggesting that "the ultraviolet method permits the transmission of much more concentrated energy compared with the infrared method," the inference is made that infrared was also being studied and was usable for similar purposes, though lower concentrations of energy made it problematic. Ultraviolet light, on the other hand, according to the same report, appears to have presented its own challenges to the task because it had a "stronger absorption rate."

Certainly nothing is proven regarding Schlicke's fuses from this independent report, but the document appears to show the technology could be used for controlling the type of warhead detonation Luis Alvarez required for the plutonium bomb. The fact somebody named "Alvarez" was in contact with Schlicke and apparently involved in his and others of U-234's passengers' interrogations, seems to be more than a coincidence.

And the fact "Commander Alvarez" was not actually perceived by Captain Fehler as being who he claimed to be, provides an interesting, if subjective, observation regarding Commander Alvarez. Fehler mentioned in his letter that Alvarez, who was his interrogating officer, "has always been correct, even when sometimes trying to press some knowledge out of me and to threaten me in a rather primitive way." (sic) The statement that Alvarez was "always been correct, even when threatening in a primitive way" seems on the face of it to be incongruent. But if Alvarez, whoever he was, was not used to interrogating people – as Luis Alvarez surely would not have been – if he was doing his best without the interrogation skills required, would that not qualify as a primitive interrogation, too? Especially if the language in which you were describing the event – English – is your second language, as it was Fehler's?

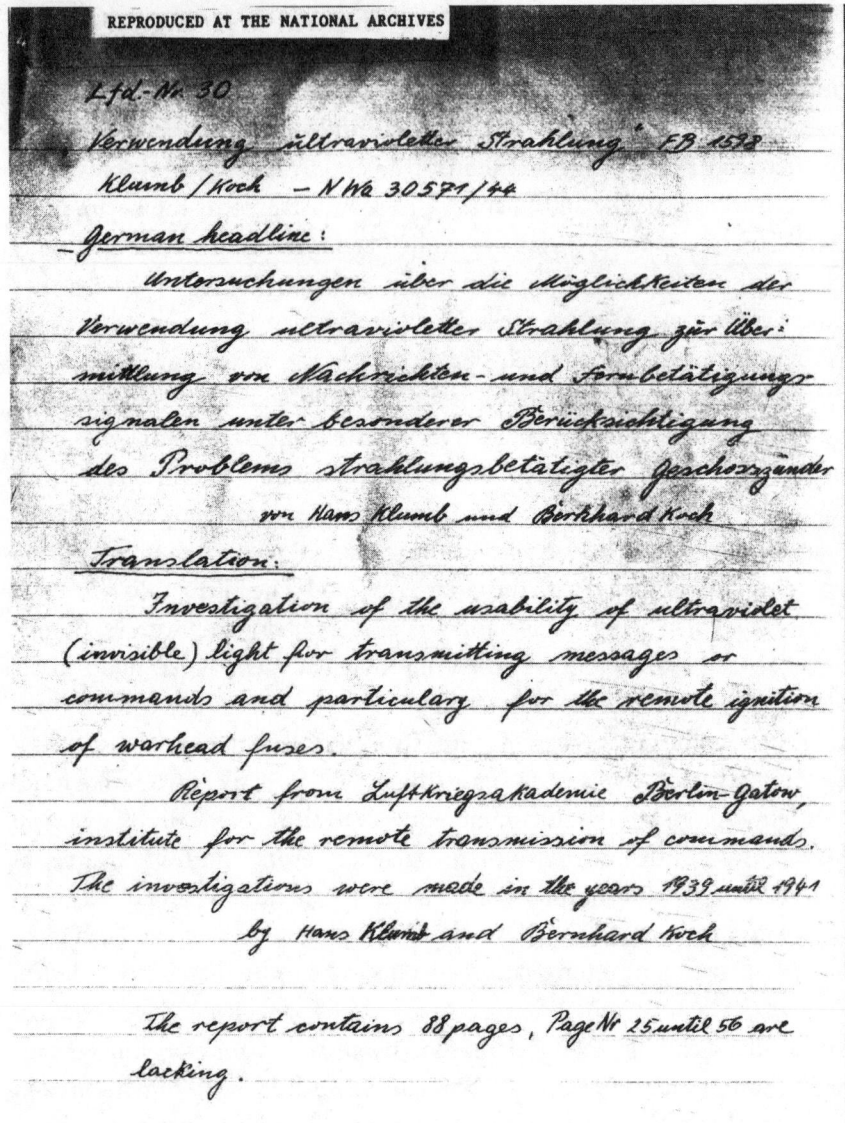

Figure 6: Handwritten translation of documents captured onboard U-234 showed that for two years the Germans had been working on using light pulses to trigger warheads. Later in this document infrared is discussed.

But what about the identification of Alvarez as a Commander in the Navy? General Groves supplied military identities – uniforms, ranks and papers – to scientists Robert Furman and James Nolan, so they could escort the enriched uranium bomb cores to Tinian on board the USS Indianapolis without raising suspicion.[22] Harlow Russ also recounted in his writings how a Major Vanna, an intelligence officer responsible for the technical crew of the plutonium bomb, always carried with him a cigar box full of rank insignias from every military service. He passed one to each of the team of civilian technicians to wear on their uniform-looking coveralls, so military personnel would not hinder them as they concluded their secret project.[23] General Groves, himself, corroborated this story in his book *Now It Can Be Told*, when he recounted how each civilian in the 37-man team of the First Technical Service was required to wear a uniform with a simulated Army rank.[24]

That Schlicke was returned to U-234 specifically to pick up the proximity fuses further seems to substantiate that Commander Alvarez, Schlicke's handler, and Luis Alvarez, who solved the plutonium bomb fusing problem, are one and the same.

This suggestion is also strongly supported by two factors. First, according to Harlow Russ, who wrote in his book *Project Alberta* about his work on the team that assembled the plutonium bomb, two significant changes were made to the bomb design at the last minute. One was the development and inclusion in the plutonium bomb of "detonator chimneys"[25] that were developed so late in the process they were not included in the first four shipments of equipment to Tinian, the Pacific airfield from which the bombs were dropped on Japan. The second design addition was a series of small-diameter stainless steel tubes called "hypodermics" that vented radiation from the plutonium core, according to Russ's explanation, to allow the technicians to monitor activity at the core.[26]

Russ makes a point of stating both additions were new and just in time for the Trinity Test. These modifications suggest that very late before the plutonium bomb's use, passages were being built into the bomb that, presumably, would allow the free flow of light waves throughout the device. Theoretically then, with these passages in place, once any one of Schlicke's 64 repurposed infrared detonators was ignited in the bomb, the new system allowed waves – including infrared waves – to race at the speed of light through the "detonator chimneys" and "hypodermics" to the other infrared fuses to simultaneously ignite them all.

Given the timing of the developments, from Alvarez's arrival on the U-234 scene, to Schlicke's special trip to retrieve the fuses, to Alvarez's solving the timing problem so late in the process, and Russ receiving last-minute design changes apparently initiated to provide paths for the free movement of light waves within the bomb, such a scenario certainly seems viable.

In an effort to substantiate or eliminate this theory, I tried to call Harlow Russ on the telephone at his home in Los Alamos to ask him about the detonator chimneys, venting tubes, and if in general, there were any significant changes to the actual detonators themselves. Unfortunately my call came too late; I was informed Mr. Russ had died in the few months between when I received from him his book and when I had developed the above scenario.

Dr. Delmar Bergen, retired director of the Nuclear Weapons Program at Los Alamos, is unconvinced of the idea the chimneys and hypodermics were used for the simultaneous flow of infrared light waves. His interpretation is they were used to vent heat generated by the radioactive plutonium core which, according to him, had the potential to heat and possibly sensitize the high-explosive, making it somewhat unstable and dangerous.

This, of course, may be the case. It should be noted, however, that when I presented my findings during each of the four presentations I gave at Los Alamos and Oak Ridge about the infrared fuses and how I believed they were used, I was never challenged by any of the scientists or technicians that my interpretation was incorrect. I admit, however, that it may be.

Even if the infrared fuses were not used in the bomb, though, Dr. Bergen still considers the probability high that fusing technology from Heinz Schlicke and U-234 were used to contribute to the solution of the detonation problem.

"The rapid and consistent release of electrical energy was a key part of the problem the (Los Alamos) scientists were experiencing triggering the detonators with the simultaneity necessary to achieve a clean spherical implosion," explained Dr. Bergen. "The scientist Heinz Schlicke had knowledge of fast operating energy transfer systems .... There apparently is no written unclassified record available to provide us with what may have come from the debriefing of Heinz Schlicke but this we do know, over the summer months since his capture and the surrender of U-234 the confidence in the detonation system greatly improved...."[27]

The second factor suggesting the detonators used to fire the plutonium bomb came from or were assisted by Dr. Schlicke's technology is indeed the striking success of the Trinity Test itself. Trinity was "successful beyond the most optimistic expectations of anyone," wrote General Groves.[28] "Nearly everyone was surprised,"[29] Robert Serber recorded. In his quintessential tome on the subject, *The Making of the Atomic Bomb*, Richard Rhodes wrote that Trinity was four times its expected yield.[30]

What could have caused such a remarkable under-estimation by the experts? Those who knew the problems the system was experiencing in firing all of the detonators at once by mechanical means, but were unaware that fast-operating energy transfer systems from U-234 had become available, certainly would not have expected the profoundly superior results. The scientists were expecting a much less dramatic event.

In fact expectations were so low, Dr. Bergen explained, a large cast iron containment vessel called a "pig" was created, inside of which they planned to detonate the bomb. The scientists were so unsure of the success of Trinity, they planned to capture the plutonium inside the pig if the test did not result in a nuclear explosion; which would have devoured the pig if it occurred, but the pig would have contained the plutonium if it did not. Just days before the event, it suddenly was decided the test was certain enough the pig would not be needed.

Instead, the vast majority of those who witnessed the test were surprised by the power and efficiency of the explosion. That so many who knew what the outcome of the detonation should have been were so surprised by how efficient it actually was, tends to indicate Schlicke's technology was used to compress the plutonium core at sufficient simultaneity to create a very powerful explosion.

Endnotes: Chapter Six – Timing

1   U.S. National Archives, Northeast Region, Waltham, Massachusetts, *secret dispatch from Commander Naval Operations to Portsmouth Naval Yard, POW and Fuses From U-234, 25 May 1945*, RG 181, box 531

2   U.S. National Archives II, transcription of a lecture given by Dr. Heinz Schlicke to the Navy Department, 19 July 1945, RG 38 – 370 15/09/04 box 13, also RG 165 – 390 39/34/06 box 93 file 413.68

3   U.S. National Archives II, *Manifest of Cargo For Tokio On Board U-234*, translated from German, 23 May, 1945; original German loading manifest, RG 38 – 370 15/05/07 box 3

4   U.S. National Archives II, memorandum written by Jack H. Alberti to Captain John L. Rihaldaffer, 24 May, 1945, RG 38 – 370 15/09/01 box 2

5   U.S. National Archives, Northeast Region, Waltham, Massachusetts, *secret dispatch from Commander Naval Operations to Portsmouth Naval Yard, POW and Fuses From U-234, 25 May 1945*, RG 181, box 531

6   U.S. National Archives II, Report of Interrogation, U-234 POW Kay Nieschling, 24 May 1945, RG 38 – 370 15/09/01 box 2

7   Geoffrey Brooks and Wolfgang Hirschfeld, *Hirschfeld: The Story of a U-boat NCO 1940-1946*, pp. 212, 213

8   U.S. National Archives II, Report of Interrogation, U-234 POW Heinz Schlicke, Appendix V and VI, RG 165 – 390 35/11/05 box 540

9   U.S. National Archives II, Report of Interrogation, U-234 POW Heinz Schlicke, Appendix V and VI, RG 165 – 390 35/11/05 box 540

10   Richard Rhodes, *The Making of the Atomic Bomb*, pp. 402, 513; David Irving, *The German Atomic Bomb*, pp. 172, 173, 230-237, 305 and elsewhere throughout

11   U.S. National Archives II, transcription of a lecture given by Dr. Heinz Schlicke to the Navy Department, 19 July 1945, RG 38 – 370 15/09/04 box 13, also RG 165 – 390 39/34/06 box 93 file 413.68

12   U.S. National Archives II, letter written by Luftwaffe General Ulrich Kessler, *Subject: Personal Belongings – Through: Channels*, 28 August 1945, RG 38 – 370 15/09/04 box 13

13   Heinrich Fehler, in an undated letter to Sharkhunters, p.2

14   U.S. National Archives, Southeast Region, East Point, Ga, memorandum from N.E. Bradbury to N. Ramsey, 18 April 1945, A-84-019-82-16

15   U.S. National Archives, Southeast Region, East Point, Ga, memorandum from G.B. Kistiakowsky to L. Fussell, *X Units for Trinity*, 6 June 1945, A-84-019-55-9; memorandum from N.F. Ramsey to J.R. Oppenheimer, W.S. Parsons and Norris Bradbury, *Unsatisfactory Features of Weapons Program*, 23 June 1945, A-84-019-82-25; memorandum from F. Oppenheimer to K. Greisen, D.F. Hornig, E.J. Lofgren, *Rehearsals at TR*, 26 June 1945; memorandum from D.F Hornig to N.E. Bradbury, *Schedule of Firing Team at TR*, 28 June 1945, A-84-019-55-9; memorandum from D.P. Irons to W.S. Parsons, *July Kingman Schedule Revision I*, 3 July 1945, A-84-019-67-7

16   Robert Serber, *The Los Alamos Primer*, p. 60

17   U.S. National Archives, Southeast Region, East Point, Ga, *Minutes of Meeting on the Electric Detonator Program*, p.2, 25 October 1945, A-84-019-41-11

18   Glenn Seaborg, *The Plutonium Story: The Journals of Professor Glenn T. Seaborg*, pp. 862, 863 note

19   Robert Serber, *The Los Alamos Primer*, p. xvii note

20   Luis Alvarez, *Alvarez*, p. 137

21   U.S. National Archives II, document surrendered with U-234 titled, *Verwendung ultraviolette Strahlung*, FB 1598, by Hans Klumb and Bernhard Koch – N Wa 30571/44, RG 38 – 370-15-09-04 box 6

22   Max Morgan Witts and Gordon Thomas, *Enola Gay*, pp. 169, 170

23   Harlow Russ, *Project Alberta*, pp. 18, 58

24   Leslie R. Groves, *Now It Can Be Told*, p. 282

25   Harlow Russ, *Project Alberta*, p. 55

26   Harlow Russ, *Project Alberta*, pp. 55, 56,

27   Delmar Bergen, Forward to *Critical Mass* 3rd Edition

28   Leslie R. Groves, *Now It Can Be Told*, p.433

29   Robert Server, *The Los Alamos Primer*, p. 60

30   Richard Rhodes, *The Making of the Atomic Bomb*, p. 677

Chapter Seven

Hanford

"Irradiated enriched sample intended for you being removed from Clinton (Oak Ridge) pile today... "[1]
– Samuel Allison: From a cable to J. Robert Oppenheimer, 17 March 1944. (The traditional history insists plutonium was bred in reactor piles fueled with natural uranium, not enriched uranium)

B ecause the long road to a valid uranium enrichment program from the beginning was thought to be a long shot, the discovery of plutonium in December 1940 was a godsend to the bomb makers. More than a year after Glenn T. Seaborg, Joseph W. Kennedy and Arthur C. Wahl confirmed they had re-created an element[2] heavier than uranium that had long ago disappeared from earth, Seaborg and his team, along with Italian physicist Emilio Segré, proved that the new substance would fission. The cleaving of this first man-made element allowed the great American nuclear braintrust a second, more sensible option than trying to pluck a small minority of nearly identical atoms from an otherwise homogenous body of matter, as was the requirement for enriching uranium.

Plutonium was an element unto itself, with characteristics all its own.[3] The difference was significant. Whereas separating uranium isotopes required devising methods to differentiate and take advantage of infinitesimal weight discrepancies between sub-microscopic atomic particles, plutonium was created by bombarding natural uranium with neutrons, morphing the uranium to plutonium as the neutron count increased. The plutonium then could be separated from the uranium simply by dissolving the mass and rinsing the solution with a chemical found to bind with plutonium but not with uranium. As the "binder" later was separated away, the plutonium would be exposed for the taking. Such an explanation is a vast oversimplification but suitable for a basic understanding.

The process was substantially simpler than enriching uranium. There still existed significant barriers to overcome, however; like, how could uranium be bombarded with enough neutrons to achieve production levels of plutonium? The cyclotron that Seaborg's team used to create plutonium was far too small and neutron-anemic to produce anything but microscopic amounts of plutonium. And once the irradiated, plutonium-carrying slugs of uranium were ready to be dissolved, how could the task be accomplished without radiation poisoning the people assigned to work with the highly radioactive material? Plutonium, in theory, was a great solution, but its practical application would prove to be a prickly challenge in and of itself.

The chemical differences, however, were not the only advantages plutonium held over enriched uranium. With U^{238} being 139 times more common in natural uranium than U^{235}, and plutonium being a product of neutron bombardment of U^{238}, it was possible to create much more plutonium out of an equal amount of uranium than would ever be possible to separate U^{235} from the mother substance.[4] And conversely, even while more plutonium fissile material could be made faster and cheaper than enriched uranium, only one-third as much plutonium was needed for a bomb than enriched uranium because plutonium more readily fissions when exposed to thermal neutrons.[5] More nuclear fuel, at higher quality, for less time and money – the advantages were obvious. Despite all of the time and effort and money being poured into uranium enrichment, pursuit of plutonium quickly became the primary objective of the Manhattan Project.

The Manhattan Project's scientific community rallied around the proposal. In fact, Ernest O. Lawrence, the father of the calutrons, plutonium's "competitor," led the charge in favor of plutonium with Oppenheimer's blessing.[6] In 1941, before isotope separation had been proven in the United States, Arthur Compton, Nobel laureate in physics and one of the original movers and shakers that made the bomb project possible, thought the plutonium alternative saved American bomb research altogether.[7] Compton's committee, in fact, recommended the creation of a central lab just to handle the development of a plutonium bomb.[8] Jewish-German war refugee Hans Bethe, whom one would have thought would jump at the slightest chance of developing a successful bomb to be used against the Nazis – who had driven him from his home – had refused to join an atomic bomb research group. Bethe considered the creation of a bomb from enriched uranium impossible and therefore a waste of time. But when the plutonium option became available, he jumped into the project with both feet.[9] General Groves, who received his assignment to lead the Manhattan Project in the midst of the plutonium option development, believed his best hope was to create a plutonium-fueled bomb[10] and made it the number one priority of the entire program.

All of this was well and good but plutonium research, though an excellent prospect, was "getting out of the blocks" late. Assessing a plutonium bomb's legitimacy took time. An answer for the weak neutron bombardment problem caused by the cyclotron's limitations was not found until almost the end of 1942. On 2 December of that year, Enrico Fermi's research group successfully sustained the first man-made nuclear chain reaction during their famous experiment in a squash court under the bleachers of the University of Chicago's football stadium. The astounding success meant neutrons could be released in unimaginable numbers, to be absorbed by U^{238} and thus transmute the uranium into plutonium.

The success of Fermi's plutonium breeding pile resulted in a major change of plans. While the original purchase of the property at Oak Ridge included plans to house plutonium development facilities, General Groves soon realized the risks of building production-size breeder reactors were too great for a highly populated area like Knoxville, which was close to Oak Ridge. A new reservation had to be found, far from large population centers and prying eyes.

A site team was dispatched to locate such a property, visiting areas in California, Oregon, Idaho and Washington, and eventually returning to Groves with a recommendation – Hanford, on the barren, eastern plains of the state of Washington.[11] Groves flew out to Washington and approved the site. But in February 1943, with barely two and one-half years left to successfully fulfill the future time objective (as yet unknown, since Russia was not showing any signs of declaring war on Japan) the property at Hanford was still in the process of being purchased.[12]

Construction on the site was officially begun 22 March, with a multitude of development, construction and research projects running concurrently, not only at Hanford, but at Oak Ridge, Los Alamos, Chicago, and elsewhere. By the end of 1943, however, the construction of the first reactor pile – so named because a reactor was simply a sophisticated pile of graphite blocks with uranium slugs and control devises inserted in holes drilled through the graphite – had not been started. Only eighteen months until the – then unknown – time objective, and still no production reactors were being built.

Which is not to say no work was being accomplished. A small pilot reactor at Oak Ridge had been assembled and was beginning to provide milligram quantities of plutonium for experimentation and metallurgical research.[13]

Progress in the chemical process of plutonium separation was being made, with the proposal and eventual validation of bismuth phosphate as a plutonium binder to separate plutonium from uranium. Innovative methods in miniaturization and robotics, and to some degree television, which would lay the groundwork for the future high-tech industry and its resulting economy that would burst forth a quarter-century later, were being developed to perform the dirty, dangerous work of separating plutonium from its mother, natural uranium, without irradiating the people performing the tasks.

And at Hanford, although reactor piles had not been started, great strides were already being made toward the construction of the mechanical aspects of the chemical separation facilities.[14] The separation team had devised a semi-automated system where irradiated slugs would be dropped mechanically into a huge "trough" that contained the equipment and substances required to run the slugs through the series of steps necessary to dissolve the slugs and then to separate the different elements according to requirements. The trough was buried almost completely in the ground and was lined with huge cement walls and 20,000 tons of steel plate and cellulose, as well as 7,500,000 square feet of Masonite,[15] all forms of biological shielding to protect operators from the dangers of radioactivity.

At its peak, 42,400 construction workers plied their trades building the Hanford reservation.[16] Even more than in the uranium enrichment program, everything was being thrown into the endeavor to make the plutonium bomb succeed. Still, the chance of producing more than just a few grams of plutonium in 1943, and not much more in 1944, even under the best of circumstances, was all they could hope for, according to General Groves.[17] Groves did not expect production levels of plutonium until 1945, and there were many doubts about that.

The doubts were well-founded. A year earlier, in the beginning of 1942, Seaborg had written that bombs were planned to be in production around the beginning of 1944.[18] Obviously, that had not occurred. No plutonium was produced in 1943 at all at either Hanford or at the scaled-down experimental pilot reactor at Oak Ridge, which had been built as a working model to develop the Hanford technology. The Oak Ridge plant had been loaded with uranium fuel in early November, however, and went critical soon afterward.

As a result, the first day of 1944 saw the inaugural delivery of milligram quantities of plutonium sent to Chicago for experimentation.[19] The Oak Ridge reactor continued to send experimental quantities of plutonium to the metallurgical laboratory in Chicago and to the nuclear laboratory at

Los Alamos. But bomb-production quantities from Hanford would not be produced for almost another full year, beginning on 24 November 1944 (B reactor, the first to be fueled at Hanford, went critical 26 September 1944).[20] Only eight months were left on the countdown to August 1945, at the time the first small quantity of production plutonium was created.

Like the uranium enrichment effort, continual dilemmas and delays had slowed the plutonium program. A most serious problem, realized before production even started, was the low concentration of plutonium the initial pile design would produce.[21] The difficulty, simply put, was that natural uranium contains so few U^{235} atoms, only one out of every 140 uranium atoms. These U^{235} atoms fission and release neutrons that in turn either fission more U^{235} – continuing the chain reaction – or are absorbed into U^{238} atoms and thus transmute the uranium to plutonium, which is the desired end-product. But even after the maximum amount of fission occurred, after long weeks in the reactor when the U^{235} was finally spent, much more U^{238} remained that could have been transmuted to plutonium. Plutonium production, while better than enriched uranium output, was still woefully lean. Available records of the time appear to indicate the plutonium content of the initial Hanford discharge was so low the chemical separation process had to be further refined to optimize the product yield to an acceptable level.[22]

As early as 1941, however, Philip Abelson, a physicist for the United States Navy, had realized that using enriched uranium to fuel a reactor would make the reactor rich in free neutrons. The reactor would not only be more powerful, with a greatly reduced size requirement,[23] but, most importantly, the modification would produce significantly more plutonium.

From the beginning and throughout the Manhattan Project, all avenues to improve success were pursued. So it appears to be with efforts to increase plutonium yield. Plutonium was the top priority for a bomb; and with a growing arsenal of newly developed technologies from which to draw, Groves and his advisory board appear to have made a logical and, one would think, obvious but very fateful, decision. Unknown to historians up until discovered by this author, they appear to have used the invaluable enriched uranium from Oak Ridge – which was fat in U^{235} that would provide the neutron flood needed to create significantly more plutonium per production run – to fuel the reactors at Hanford. The decision was not without risk and potential political fallout, however, and so its secrecy was vigilantly guarded at the time; and following later dubious developments, it appears to have remained carefully buried afterward, until now.

The traditional history simply tells us that in early 1943, the Hanford reactors' design was modified from helium-cooled piles to water-cooled piles.[24] This was done to eliminate the need for building compressors and very technically demanding containment vessels for producing and using helium as a coolant.

According to Dr. Bernard Wehring, Director of the J.J. Pickle Research Center for Nuclear Engineering at the University of Texas in Austin, [25] and Dr. Delmar Bergen, water cooling a pile would be used only to cool a reactor fueled by enriched uranium, not one fueled by natural uranium. Both scientists agree that water, while a great moderator to slow neutrons, also absorbs neutrons – as opposed to helium, which has no neutron absorption cross-section – and the water therefore is in competition for neutrons with both U^{235} and U^{238}. Therefore a natural uranium reactor cooled by water would produce less plutonium than would a helium-cooled pile, not more. The neutron-hungry water in the pile would consume the very neutrons needed to make plutonium.

Fueling the reactor pile with uranium enriched in U^{235}, on the other hand, would increase the amount of free neutrons in the pile to a level that supported a high rate of fission, resulting in more neutrons for transmuting much greater quantities of U^{238} to plutonium, all the time feeding the cooling water's appetite for neutrons as well. The water would be required to cool the more powerful enriched reactor, for which helium would be insufficient. The end result, depending on the level of uranium enrichment utilized, would be more plutonium per production run, and thus plutonium would be produced at a faster rate.

Drs. Wehring and Bergen both admit to not being historians of nuclear physics and that without knowing the full background of the Hanford reactors they could not declare with certainty the reactors were fueled by enriched uranium. But on theoretical grounds alone, neither of them could conceive of a case in which a natural uranium reactor would be cooled by water.

The historical documentation does indicate, however, that natural uranium *was* used in the first run of a water-cooled Hanford pile. B-Reactor was energized on 26 September 1944 with 901 of its 2004 fuel tubes loaded with natural uranium metal slugs.[26] Eventually, 2002 of its tubes would be loaded with fuel. Los Alamos scientist Eugene Wigner had calculated that a water-cooled, natural uranium-fueled reactor was viable if both the water and the graphite moderator were pure enough.[27] Water cooling would alleviate the problem of constructing the very demanding helium-cooled system. Apparently the savings in time realized by building the water-cooled system would offset the more desirable non-neutron absorbing characteristics of the helium-cooled pile.

But when the reactor first reached critical mass and was operated, within a day the chain reaction died.[28] Reactivity did not cease directly because of the cooling fluid chosen, but because the reaction inside the pile had unexpectedly built up excess levels of xenon-133 and other fission poisoning elements that starved the reaction of neutrons, no doubt helped by the cooling water's neutron appetite. As these daughter products degraded through the chain reaction process, the reaction began again in the pile; but as the technicians continued to load the reactor the xenon poisoning increased, continuing to reduce reactivity.

The designers of the reactor realized the pile now needed higher levels of reactivity to overcome the fission poisoning. The original 1,500 fuel channels in the pile had been increased to 2004 in case of just such an emergency.[29] Additional water-cooling channels now also had to be drilled to cool the extra fuel channels,[30] adding again a higher neutron absorption factor. The process was becoming a careful balancing act between neutron absorbing elements and low-reactivity-yielding natural uranium. Would it not have been preferable to use a type of uranium that could overcome easily all of these reaction-stifling influences? Would it not be preferable to use U^{235}? There is significant evidence that this is exactly what occurred.

The first, very convincing piece of evidence that enriched uranium went into the plutonium program, is a memo written by J. Robert Oppenheimer on 28 July 1945. In the document, Oppenheimer reports to the key leaders of the plutonium bomb project, Samuel K. Allison, Joseph W. Kennedy, and Robert F. Bacher, the delivery schedule of multi-kilogram-per-week – production level – quantities of 86.5 percent enriched uranium.[31] In it, Oppenheimer wrote:

> I have a schedule of 25 ("25" was one code word for U^{235} – author's note, emphasis the author's) in the next few weeks. These do not include the material now on hand.
> Up to August 2nd   2.0 kg of X (another code for U^{235} – author's note)
> August 2nd to 9th   3.5 kg of X
> August 9th to 16th   3.5 kg of X
> The concentration for the above will be 86.5%

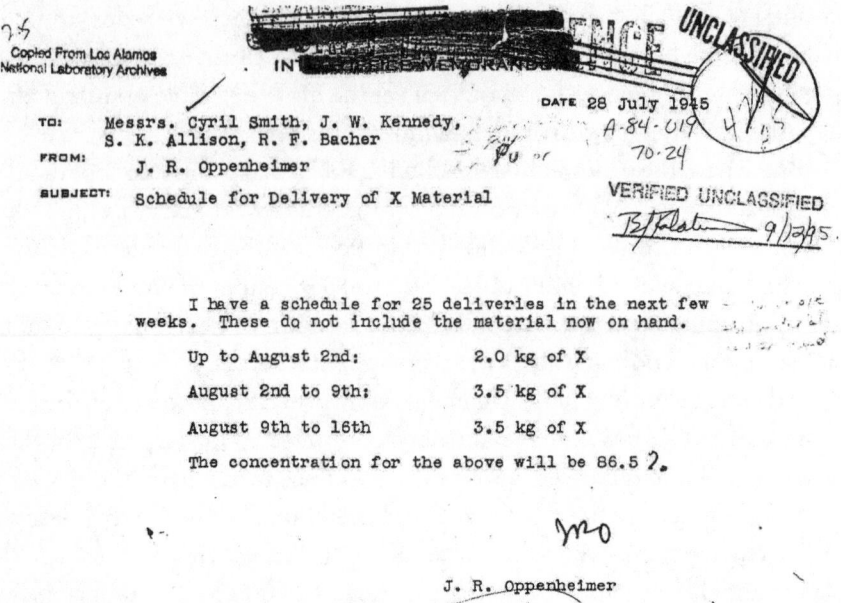

Figure 1: Oppenheimer sent production quantities of enriched uranium to the plutonium bomb team, suggesting that a serious shortage of enriched uranium was caused by using it as fuel to produce more-favored plutonium.

Why would Oppenheimer make such a report to the plutonium bomb's leaders instead of to the heads of the uranium gun bomb project, who would be the obvious recipients, if the information was not important to – and the shipments not intended for – the plutonium bomb program?

While the date of the report is too late for the shipments to have influenced the plutonium bomb dropped on Nagasaki, the fact that it proves enriched uranium was going into the plutonium bomb program in production quantities is evidence enriched uranium was used in reactors during the war. The huge quantities, from two to three-and-one-half kilograms per week, represent production-level amounts, not just experimental levels of enriched uranium. Additional documentation proves the experimentation required had been done long ago, but whether in experimental or production level quantities, either instance contradicts the traditional history. The tone of the communication also infers these deliveries were an expected and probably common event, and therefore had occurred in the past, as well.

In this vein, a second compelling indicator that the Hanford reactor piles were fueled by enriched uranium lies in the uses of, and changes made to, their forbearer and model, the pilot reactor at Oak Ridge. Communications beginning in March 1944 between Samuel K. Allison,[32] once again, a co-discoverer of plutonium and one of the men listed in Oppenheimer's memo above, and Oppenheimer clearly show that in the spring of 1944 the Oak Ridge reactor was being used to test enriched uranium as a reactor fuel. Apparently Phillip Abelson's recommendations three years earlier were being explored.

"*Irradiated enriched* sample intended for you being removed from Clinton (Oak Ridge – author's note, emphasis the author's) pile today...,"[33] states the first communiqué from Allison to Oppenheimer matter-of-factly.

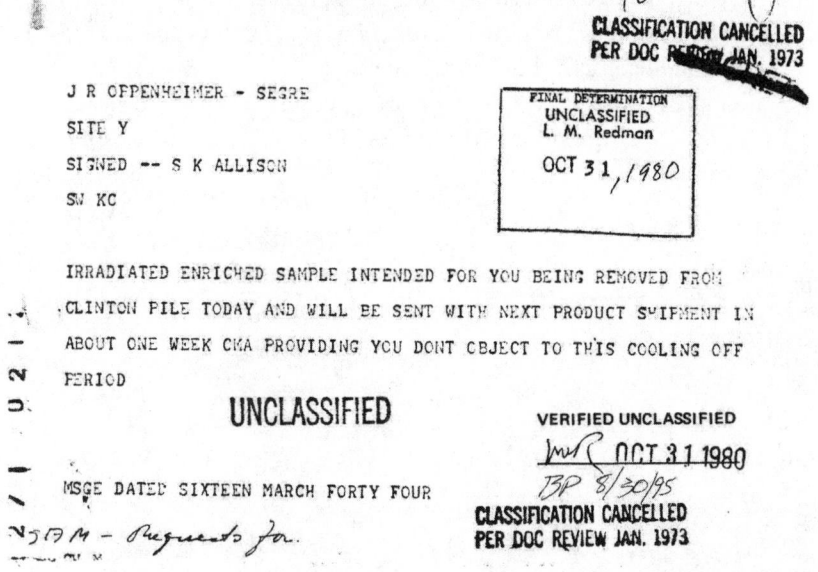

Figure 2: In March of 1944, according to a cable to Oppenheimer, the Manhattan Project was already experimenting with enriched uranium to increase plutonium output at the Clinton experimental reactor in Oak Ridge.

A portion of a letter sent from Allison to Oppy the following day to provide more details said,

> I am sending you in a separate package 57 milligrams of *enriched* T_3O_8 ['T' stood for 'Tubealloy,' a code name for uranium, 'O' for oxide; thus the material was enriched uranium oxide]. This is part of the sample which was exposed at X ['X' was a code name for Oak Ridge].

> You should receive the *irradiated* material directly from X in the next shipment of product within about a week, and material I am sending you will serve as a control. (emphasis the author's)

Allison's plainly written communications reveal with certainty that experimentation using enriched uranium as a reactor fuel at the Oak Ridge pile was underway in March of 1944, over a year before the plutonium bomb was created and six months before the first Hanford reactor was loaded. By May, experiments had progressed to the point Oppenheimer, upon Allison's request, asked General Groves to provide enriched uranium from production stocks for testing the build up of U^{236} in a pile using enriched uranium in the fuel.[34]

> Have had request from Allison for one half gram of X of Y *Twelve Alpha* [Alpha was low-enriched uranium] product for experiment to be conducted at X [the Oak Ridge reactor].... We cannot supply this material from experimental stocks (period) We should be glad to provide it from production stocks but feel that we need your approval on this.[35]

```
YC 663     27 MAY 44      2039 Z      JRO/AJR
```

```
MAJOR GENERAL L R GROVES
WASHINGTON D C
```

VERIFIED UNCLASSIFIED

BP 8/30/95

```
SIGNED... J R OPPENHEIMER
CLEAR CREEK
```

```
SW KC
HAVE HAD REQUEST FROM ALLISON FOR ONE HALF GRAM OF X OF Y TWELVE ALPHA
PRODUCT FOR EXPERIMENT TO BE CONDUCTED AT X WHICH SEEMS TO US VERY
MUCH WORTH DOING PERIOD REFERENCE YC DASH SIX SIX THREE PERIOD WE CANNOT
SUPPLY THIS MATERIAL FROM EXPERIMENTAL STOCKS PERIOD WE SHOULD BE GLAD
TO SUPPLY IT FROM PRODUCTION STOCKS BUT FEEL THAT WE NEED YOUR APPROVAL
ON THIS PERIOD PURPOSE OF EXPERIMENT IS DETECTION OF TWENTY
SIX
    TK AMC       UNCLASSIFIED    CLASSIFICATION CANCELLED
                                 PER DOC REVIEW
```

Figure 3: Barely two months after experimenting with small quantities, presumably resulting in success, Oppenheimer requested permission from Groves to experiment with production quantities of enriched uranium in the Oak Ridge experimental reactor.

It is difficult to believe the plutonium-multiplying enhancement of using enriched uranium to fuel the reactors, which history demonstrates proved completely successful, was ignored at such a critical moment in history when it was being carefully studied and was so greatly needed. Virtually all later reactors were fueled by enriched uranium or plutonium. Indeed, it is even documented that Hanford's B Reactor was transitioned to enriched uranium fuel after the war, as the others certainly were, proving the modification could be done easily to the piles.[36] With the need so great and the probability of success undoubtedly proven shortly after these early 1944 experiments, why would the decision to use enriched fuel be postponed until after the war, as the traditional history suggests?

While Allison's multiple references to irradiating enriched uranium in the Oak Ridge pile are the only direct documentation the author has found that expressly states enriched fuel was used in reactors during the war, the implication of a cover-up is seen in how this modification was later recorded for official history. H.D. Smyth, who wrote the first official history of the Manhattan Project, *Atomic Energy For Military Purposes*,[37] wrote that in the spring of 1944, "a change was made in the distribution of uranium" within the Oak Ridge reactor. Without mentioning *enriched* uranium, he goes on to describe how the uranium fuel channels were reconfigured with fewer uranium slugs in the middle so power could be increased without overheating the reactor pile.

The result was reactor performance in June 1944 "considerably exceeded expectations." To produce *more* plutonium with *less* uranium would have been highly unlikely, if not impossible, unless the uranium was enriched. And thus, it appears, is purposely hidden the real reason for the increased output – enriched uranium had apparently worked its magic.

The timing of the reconfiguration of the pile in the spring not only coincided with Allison's enriched uranium experiments, but the description coincides with a statement made by Dr. Wehring when such a reconfiguration was described to him. Dr. Wehring theorized that the core realignment would have been required to increase the size and/or number of cooling passages in order to control the additional heat created by the introduction of at least some enriched-uranium to the pile. In addition, although Smyth stated flatly the pile was run at higher power levels as a result of the reconfiguration, he never suggested the pile was expanded in size to achieve that increase. In fact, research shows no increase in the size of the reactor, leaving enrichment the only option.

A third evidence that enriched uranium was used to fuel the Hanford reactors is the fact the fuel storage room and other fuel handling areas were marked with "Danger" signs.[38] Natural uranium metal is not dangerous. Indeed, at the pilot reactor at Oak Ridge technicians worked very closely with the natural uranium fuel slugs, pushing the slugs into the fuel channels protected by nothing more than a cloth jumpsuit.[39]

Even enriched uranium metal is only radioactive to the extent that it is no more powerful than to burn the dead skin off a person's hands. Hardly dangerous. But unlike the metal slugs made of natural uranium, if enough enriched slugs were assembled, the increased density of, and additional neutrons generated from, the U^{235} could instigate a chain reaction spontaneously. Since the word "radiation" was not allowed used openly at Hanford, "Danger Zone – Keep Out" was the prescribed signage that heralded radioactive threats.[40] Placing a "Danger" sign, therefore, over the reactor fuel storage area certainly suggests more than natural uranium was being stockpiled as the reactor fuel.

All of this information, the knowledge that the Hanford reactors were water-cooled; the revelation that six months before the first Hanford reactor was loaded, Oak Ridge was experimenting with enriched uranium as a reactor fuel; the memo from Oppenheimer to the heads of the plutonium bomb project apparently reporting multi-kilogram shipments of highly enriched uranium into the plutonium bomb project; and the documentation the fuel storage rooms at Hanford were protected with danger signs, all form a picture of enriched uranium being used as a fuel in the Hanford reactors.

Other, less obvious, details of the Hanford reactors' development support the probability the reactors were fueled by enriched uranium, as well. For example, the management of Hanford went to great expense and effort after the redesign of the reactor piles from helium- to water-cooled to remove an existing system designed to store radioactive waste by-products of the pile. In its place a system for extracting uranium from the effluent was installed.[41] The reason given was to reduce the waste of uranium. But the cost of the modification probably far exceeded the value of the spent uranium salvaged – unless the uranium still contained residual amounts enriched in U^{235}. In that case, the inestimable value of the reclaimed uranium would have justified the expense of the reclamation project many times over. A similar reclamation process was already being used in the calutrons for the same reason.

And shortly after the war ended, the system was discontinued, suggesting the successful plutonium bomb negated the need for the expense of reclaiming the entrained enriched uranium. Presumably, the reclamation system would not have been discontinued if the true reason was to reclaim spent uranium because of its value, as the traditional history insists. The spent uranium still would have been just as valuable. With a viable plutonium bomb, however, minute quantities of enriched uranium would not. Perhaps the reclamation system is yet another evidence the Hanford piles were fueled by enriched uranium.

In April 1944 – again, note the time frame relative to Allison's experiments in March of the same year – a purification system for Hanford's reactors also was redesigned when "criticality requirements" were eased.[42] The essence of using enriched uranium instead of natural uranium in the pile was to increase neutron activity to well above the level required to achieve criticality, for which natural uranium was just barely sufficient. Might a modification to the system allowing critical requirements to be eased herald a change regarding the type of fuel used?

And lastly, just as when the Oak Ridge pile garnered an unexpected increase in plutonium output when apparently secretly using enriched uranium as a fuel, the chemical laboratory at Los Alamos experienced an unexpected increase in the output of plutonium received from Hanford also, straining the department's resources.[43] Estimates of plutonium product coming from Hanford should have been easy to calculate and reliable not to have drastically increased, unless something drastic had been done to increase plutonium output – something drastic, like fueling the reactors with enriched uranium. This unexpected upswing is particularly telling when one remembers the change to water cooling and the xenon-133 poisoning should have had an effect of *reducing* plutonium output of a natural uranium-fueled reactor, not increasing it.

With so much evidence suggesting at least part of the calutrons-enriched uranium was sent to Hanford, it seemed logical the next step was to look for records of calutrons shipments of enriched uranium to the Hanford site. Reviewing the 'documents findings lists' I had copied during previous visits to the Southeast Regional Branch of the National Archives, where the records of Oak Ridge are maintained, I found the entry for a document or series of documents titled "Reactor Shipments." The documents were listed to be in 67A1545, Record Group 326, Box 1. I wrote a letter to the Southeast Branch of the Archives requesting copies of the documents. I received a reply in July 1997 from archivist Charles Reeves, who said the referenced documents had disappeared from their file with no apparent explanation.

This was the first of only two incidents during ten years of research within the National Archives and Library of Congress – reviewing thousands of formerly classified documents within a broad range of files and subjects – in which I came across records that were absent from their appointed place. I was stunned. How could such records be missing? Suspicious, but convincing myself to give the archive managers the benefit of the doubt, I chalked it up to an honest mistake of misplacing the documents, and thought no more about it. Until two years later, when another batch of missing records – to be summarized in a later chapter – were also missing. A batch that, like the "Reactor Shipments" file, suggest certain events and facts accepted by the traditional history may not be true. But the records required to prove it are now mysteriously gone.

Not to be discouraged, and calculating that enriched uranium would have gone to Los Alamos for reduction into the reactor slugs before being sent to Hanford, I requested records from Los Alamos that would possibly identify the material reduced and transited. I was informed by Los Alamos archivist Roger Meade, that these records are still classified and clearance requires a Freedom of Information Act request and approval. Having spent almost one thousand dollars previously on a Freedom of Information Act request and receiving in return a large stack of papers with most of the information I was looking for still censored, by way of huge white blanks where strips of empty paper had been photocopied over the sections whose headings indicated they included the information for which I was looking, I chose not to pursue further. The next step would have been legal proceedings to force the archives to declassify the documents. The cost of such exercises is expensive and the results not guaranteed. I chose to save my money and my sanity. Perhaps someday, somebody with deeper pockets than I will pursue this trail. The idea that these shipments are still somehow important to national security, half a century after the event, seems suspect to me.

Despite the missing records – perhaps strengthened by the knowledge these important documents have mysteriously disappeared or are being hidden from view – the circumstantial evidence seems powerful if not incontrovertible that enriched uranium was used as a fuel in the plutonium breeding reactor pile at Oak Ridge, as well as in the production piles at Hanford. The enriched uranium could have come from no other source than the hard-earned but negligibly growing cache of U^{235} produced by the calutrons at Y-12.

Endnotes: Chapter Seven – Hanford

1   U.S. National Archives, Southeast Region, East Point, Georgia, Samuel Allison classified cable to J.R. Oppenheimer, 17 March 1944, A-84-019-16-8

2   Harry Thayer, *Management of the Hanford Engineer Works In World War II*, p. 133

3   Leona Libby, *The Uranium People*, p. 77

4   David Irving, *The German Atomic Bomb*, p. 93

5   Robert Serber, *The Los Alamos Primer*, pp. xv, xvi

6   Herbert Childs, *An American Genius*, p. 325

7   Richard Rhodes, *The Making of the Atomic Bomb*, p. 368

8   Leona Libby, *The Uranium People*, p. 79; Richard Rhodes, *The Making of the Atomic Bomb*, pp. 388,389

9   Richard Rhodes, *The Making of the Atomic Bomb*, p. 416

10   Richard Rhodes, *The Making of the Atomic Bomb*, p. 431

11   Harry Thayer, *Management of the Hanford Engineer Works In World War II*, p. 136

12   Leslie Groves, *Now It Can Be Told*, p. 76

13   Harry Thayer, *Management of the Hanford Engineer Works In World War II*, p. 139

14   Harry Thayer, *Management of the Hanford Engineer Works In World War II*, pp. 138, 139

15   Harry Thayer, *Management of the Hanford Engineer Works In World War II*, pp. 138, 139

16   Richard Rhodes, *The Making of the Atomic Bomb*, p. 557

17   Leslie Groves, *Now It Can Be Told*, p. 51

18   Richard Rhodes, *The Making of the Atomic Bomb*, p. 412

19   Harry Thayer, *Management of the Hanford Engineer Works In World War II*, p. 139

20   Harry Thayer, *Management of the Hanford Engineer Works In World War II*, p. 141

21   Harry Thayer, *Management of the Hanford Engineer Works In World War II*, p. 140

22   Harry Thayer, *Management of the Hanford Engineer Works In World War II*, p. 140

23   Richard Rhodes, *The Making of the Atomic Bomb*, p. 549

24   Richard Rhodes, *The Making of the Atomic Bomb*, pp. 497, 498

25   Dr. Bernard Wehring, personal telephone conversations with author, August 6, 1997 and October 10, 1997

26   Michele S. Gerber, PhD, *B Reactor Museum Association-History of 100-B/C Reactor Operations, Hanford Site*, 1.2.1; Richard Rhodes, *The Making of the Atomic Bomb*, pp. 557, 558

27   Richard Rhodes, *The Making of the Atomic Bomb*, pp. 497, 498

28   Michele S. Gerber, PhD, *B Reactor Museum Association-History of 100-B/C Reactor Operations, Hanford Site*, 1.2.1; Richard Rhodes, *The Making of the Atomic Bomb*, pp. 557, 558

29   Richard Rhodes, *The Making of the Atomic Bomb*, pp. 559, 560

30   Michele S. Gerber, PhD, *B Reactor Museum Association-History of 100-B/C Reactor Operations, Hanford Site*, 1.2.5; Richard Rhodes, *The Making of the Atomic Bomb*, pp. 559, 560

31   U.S. National Archives Southeast Region, East Point, Georgia, J. R. Oppenheimer memo to Messrs. Cyril Smith, J. W. Kennedy, S.K. Allison, R.F. Bacher, 28 July 1945, A-84-019-70-24

32   U.S. National Archives Southeast Region, East Point, Georgia, Samuel K. Allison, classified cables to Robert J. Oppenheimer of March 17 and 18, 1944, and May 22, 26 and 27, 1944; also cable from Robert J. Oppenheimer to General Leslie Groves, May 27, 1944, both documents at A-84-019-16-8

33   U.S. National Archives Southeast Region, East Point, Georgia, Samuel Allison classified cable to J.R. Oppenheimer, 17 March 1944, A-84-019-16-8

34   U.S. National Archives Southeast Region, East Point, Georgia, Samuel Allison classified cable to J.R. Oppenheimer, 22 May 1944; Samuel Allison classified cable to J.R. Oppenheimer, 26 May, 1944; cable from Robert J. Oppenheimer to General Leslie Groves, May 27, 1944, all documents at A-84-019-16-8

35   U.S. National Archives Southeast Region, East Point, Georgia, cable from Robert J. Oppenheimer to General Leslie Groves, May 27, 1944, Georgia, A-84-019-16-8

36   Michele S. Gerber, PhD, *B Reactor Museum Association-History of 100-B/C Reactor Operations, Hanford Site*, 1.3.1

37   H.D. Smyth, *Atomic Energy For Military Purposes*, p. 143, 144

38   Michele S. Gerber, PhD, *B Reactor Museum Association-History of 100-B/C Reactor Operations, Hanford Site*, 1.2.4

39   Oak Ridge National Laboratory brochure, *The Graphite Reactor-A Historical Landmark at the Oak Ridge National Laboratory*, p. 1 photo and caption

40   Michele S. Gerber, PhD, *B Reactor Museum Association-History of 100-B/C Reactor Operations, Hanford Site*, 1.2.4

41   John F. Hogerton, *The Atomic Energy Desk Book*, p.220

42   Harry Thayer, *Management of the Hanford Engineer Works In World War II*, p. 140

43   Anthony Cave Brown and Charles B. MacDonald, *Secret History of the Atomic Bomb*, p. 507

Chapter Eight

Simple Math

"By April 1945 Oak Ridge had produced enough U²³⁵ to allow a near critical assembly…."[1]
– Richard Rhodes, author *The Making of the Atomic Bomb* (The uranium bomb prepared for Hiroshima barely three months after this first critical mass was accumulated, which was the result of a year of enriching uranium, contained three critical masses)

The determination to use Y-12's hard-won enriched uranium to power the Hanford reactors, like so many other decisions General Groves had to make in order to advance the program, must have been a finely balanced judgment, which is reflected in how he appears to have chosen to execute it. As has been outlined, only two options were scientifically feasible for creating an atomic weapon before the end of the war: a uranium bomb and a plutonium bomb. Technologies to create the bombs varied widely in some areas and were interchangeable in others but progress on all fronts was going forward. Research had shown that economically and technically, plutonium was the better bet – faster, cheaper, easier.

But the whole program was still a wager. Nothing was guaranteed. Even considering plutonium's appreciable promise, Groves could not afford to put all his eggs in one basket. To cover all eventualities, the imperative still was to achieve independent success with each weapon. He simply must be careful allocating resources along the way – ensuring he had enough of the necessary materials to make at least one of each bomb – and then he could weight any surplus resources in favor of the preferred plutonium prospect.

Such an intermingling of resources, however, could prove to be tricky. Enriching uranium came at a very high price in every form of currency, whether it was money, time, or the energy exerted to achieve the endeavor. It was the most valuable substance on earth. Consuming the hard-earned enriched uranium in an effort that failed, therefore, would be anathema to those powerful men who had invested so much into it and were counting so much upon it. Should the plutonium bomb fall short of success for whatever unknown reason, and the uranium program require more fissile material to produce multiple bombs – which was believed the requirement to achieve victory[2] – the entire project could fail as a result of using the enriched uranium in reactors. The objectives of all involved would be thwarted.

For Groves, the risks of taking enriched product originally intended for the uranium bomb and using it in the plutonium weapon must have seemed great, but worth taking; especially if the intermingling could be camouflaged from the eyes of those who had a vested interest in the success of the overall program. If no one knew that one program had been put at risk for the other, if either program failed it would be considered to have failed on its own deficiencies.

While the brilliant brain-trust Groves had hired was ciphering the universe of factors and exponents, of calculus and algorithms, Groves' decisions, though pressure-packed and frequently daunting, were usually solved using simple math and inspired resourcefulness. For example, each calutrons required a great amount of electro-magnetic energy, and there was a given number of calutrons at any one time; therefore, a huge amount of silver was required for the electro-magnet windings of the calutrons. Eventually, over 13,000 tons of silver was required for the calutrons[3] (copper windings were not favored because silver conducts electricity more efficiently). Groves borrowed 13,540 tons of silver – normally measured in troy ounces, in this case 395 million troy ounces worth over $300 million – from the United States Treasury.

Groves also faced the decision of how many Alpha calutrons should be built and how many Betas? Through the course of time and the accumulation of experience he eventually settled on nine Alpha and three Beta calutrons.[4]

And lastly, how should he allocate the enriched uranium between the plutonium bomb option and the uranium bomb alternative?

Probably the simplest, most natural solution was to evenly split the enriched uranium originally dedicated to the uranium bomb, and use half of the concentrated product in the plutonium reactors while saving half for its original purpose in the uranium bomb.

A second, more logical, option took into account critical mass for the uranium bomb finally had been calculated at approximately 15 kilograms.[5] He could have gauged the accumulation of enriched uranium so 15 kilograms, expected to be enough for one uranium bomb, would be on hand at a date early enough before the target drop date in August 1945.[6]

The bomb's enriched uranium would need to be reduced to metal, fabricated into its sub-critical slugs, and the bomb assembled in time for use when required. And, of course, he would still need enough time to transport the weapon to the base of operations, from where it would be flown to its target for its deadly delivery. Once he had made the calculations of how much enriched uranium per day would be required to accrue the 15-kilogram critical mass by, say, 1 May 1945, two months before the drop date objective, any surplus of the precious product could be invested in the favored plutonium-breeding reactors at Hanford. If everything went right, then in August General Groves could simply add the lone uranium bomb to his reserve of two or three plutonium bombs. Once the more prolific plutonium started being produced using the enriched uranium the cache would grow by one bomb every two to three weeks,[7] compared to one uranium bomb every five months.[8] He simply had to choose now to fuel the reactors with the surplus enriched uranium.

Actually, both options fortuitously came to the same result. The rate of enriched uranium production as of the beginning of 1945 was setting a pace to be just at 30 kilograms around the beginning of May – 33 kilograms was actually achieved – according to the Beta Tuballoy Trioxide Transfer Report that documented bomb-grade enriched uranium production.[9] If the plutonium bomb was granted half of the enriched uranium produced, by early May, 15 kilograms (33 pounds) could still be set aside for the uranium bomb – the amount needed for critical mass. Enough surplus enriched uranium would still be available for the plutonium project to receive 15 kilograms for feeding the Hanford reactors, providing a sizable boost to the more promising effort.

Supporting the suggestion this path was chosen is a pair of references by author Richard Rhodes in his book *The Making of the Atomic Bomb* that, combined with two contemporaneous documents, reveals how the enriched uranium appears to have been used.

First, based on assumptions recorded by James Bryant Conant at the beginning of 1945, Rhodes calculated the uranium bomb would eventually need "about 42 kilograms – 92.6 pounds," which Rhodes then stated was approximately 2.8 critical masses.[10] In other words, critical mass can be calculated to be about 15 kilograms. At that time, nobody knew that three times that amount would actually be required for the uranium bomb, probably due to reduced efficiency due to contamination. Robert Serber, author of the quintessential textbook on Los Alamos, *The Los Alamos Primer*, which was the basis of all Los Alamos' scientists' orientation at the time, writes the amount actually used in the bomb was somewhat higher at approximately 50 kilograms.[11] Serber also validates the 15-kilogram critical mass figure in a footnote.[12] Rhodes stated later in his book that enriched uranium production achieved critical mass in mid-April 1945.[13] One can therefore recap that critical mass was 15 kilograms (33 pounds), and that the first 15-pound critical mass became available in mid-April.

But Oak Ridge by then had enriched roughly 30 kilograms of uranium to bomb quality, according to the *Beta Tubealloy Trioxide Transfer Report* – twice the amount Rhodes records was reported as available at the time for the uranium bomb. In other words, half of the enriched material was gone and, outwardly at least, unaccounted for.

Also suggesting Groves was dividing the enriched uranium between the uranium bomb and plutonium bomb projects is a document written at Los Alamos by Eric Jette, the chief metallurgist at the New Mexico laboratory. He wrote in a memo dated 28 December 1944, that "At the present rate we will have 10 kilos about February 7 and 15 kilos about 1 May."[14] Jette's 15 kilograms estimate by 1 May falls almost perfectly in line with when the uranium bomb program reportedly achieved that milestone in mid-April. All documentation and predictions appear to agree on this point: only half of the enriched uranium processed was accounted for in the report stating that enough material for critical mass had been produced.

But these discrepancies do not directly connect Hanford and Y-12. A document found at the National Archives and Records Administration Southeast Region in East Point, Georgia does. Proof that enriched uranium was shared between weapons programs lies in the memorandum (reviewed in Chapter Seven) written by J. Robert Oppenheimer to Cyril Smith, Joseph W. Kennedy, Samuel K. Allison and Robert F. Bacher. This document ties these plutonium program-dedicated personnel to the shipment of production quantities of enriched uranium. The memo, dated 28 July 1945, is four days too late to be discussing product used in the actual uranium bomb dropped on Hiroshima.

The identities of the four men to whom it is addressed, the timing of the document, and its discussion of the amounts of enriched uranium being shipped, however, makes it relevant to this review. With the exception of metallurgist Smith, who helped fabricate both the uranium and the plutonium bombs, each man was in his own right a driving force in the field of plutonium bomb development, with little or no involvement in the uranium bomb. Kennedy was a co-discoverer of plutonium and continued, with Emilio Segré, to research the radioactivity of the newfound element. Allison worked closely with Enrico Fermi on the first chain reacting pile in Chicago that made plutonium production a possibility, and continued the bulk of his research in the field of breeder reactors as it related to making a plutonium bomb. Remember, it was Allison who wrote to Oppenheimer identifying experiments using enriched uranium in the Clinton research pile and who requested additional production-level amounts of enriched uranium for further reactor experimentation.

Although the memo was sent too late to be referencing material in the bombs dropped on Hiroshima or Nagasaki – the last shipment overseas had been sent four days earlier – the fact such large quantities of enriched uranium were being shipped at all is revealing, and begs certain questions. The favorability of the plutonium bomb had been proven two weeks earlier with the successful completion of the Trinity Test, eliminating the need for continuing the cumbersome and expensive uranium enrichment project. The plan almost since Groves took over was to focus all effort on plutonium, once it was proven. Why continue uranium enrichment if it was no longer needed and it was more expensive, inefficient to produce and a less powerful explosive? And what did they do with uranium that had been enriched in the time between when the first bomb's needs were fulfilled and the successful Trinity Test? All that expensive enriched uranium had to be used somehow. Why not in the same way all other uranium was now being focused – in slugs to feed the plutonium breeding reactors, the enriched uranium providing significantly more productive output?

In the memorandum, Oppenheimer reported the delivery schedule from Oak Ridge of uranium enriched to a level of 86.5 percent. There is no apparent reason for Oppenheimer to have reported enriched delivery schedules to these men unless enriched uranium was an important component of the plutonium bomb project. The memo appears a very direct validation of production quantities of highly enriched uranium being used in the plutonium bomb project.

Metallurgical fabrication of uranium for both the uranium bomb itself and the uranium fuel slugs for the reactors was performed first at Chicago and then at Los Alamos. Given Smith's inclusion with the others in the memorandum, for even Smith's involvement was focused more on plutonium development than on the uranium bomb,[15] it appears the enriched material referred to was intended for use in uranium slugs for the Hanford reactors.

131

Reactor fuel is not composed of bomb-grade enriched uranium but normally is enriched to only between one and five percent. The level of possible enrichment of the wartime Hanford reactors is unknown. Neither the Alpha nor the Beta calutrons, however, were so sophisticated at the time as to be able to control with any certainty their capacity to enrich uranium to a preset level. And considering the circumstances, it would have been risky and inefficient to produce anything but the highest enrichment possible. Instead, the managers of the calutrons must have found it most productive to make the highest-enriched product possible and then calculate the content of enriched versus natural uranium required in the reactor fuel to achieve the desired concentration, and mixed the two uranium stocks to suit the reactors. Such a controlled process, presumably, would have produced optimum results for plutonium output.

So in late 1944, General Groves appears to have split the enriched uranium stocks between the weapons programs. Progress appeared to be going well for Brigadier General Leslie R. Groves and the Manhattan Project. Then the first shoe dropped.

As has been noted, sometime during the middle of April – Rhodes places the time between President Roosevelt's death on the twelfth and his funeral on the fifteenth[16] – Otto Frisch reported to Robert Oppenheimer the staggering fact that one critical mass, the amount just becoming available as the bombing target date was fast approaching, would not be enough to fuel a viable uranium weapon. No reason is given for the critical calculation adjustment but the culprit appears to be the purity of the enriched material. At its best the uranium was enriched to 90 percent, leaving ten percent either U^{238} or other, mostly non-fissile, elements, any of which would obstruct the efficiency of the chain reaction. To overpower the neutralizing effects of these contaminants and produce an explosion that justified the expense and reliance that had been placed in its potential, the bomb required significantly more than the 15 kilograms that was the minimum amount of pure U^{235} initially thought to be required. As has been noted, the quantity ultimately used according to Robert Serber and indirectly but roughly validated by Rhodes' own calculations, was 50 kilograms – over three times critical mass.

Rhodes lightly dismisses the shortfall, however, suggesting that it "was now only a matter of time" before the deficit would be overcome. While the statement is obviously true, so it is true that time was a crucial matter – more so than Rhodes seems to have comprehended. Or else the Pulitzer

Prize-winning author, despite his laudable achievement assembling the most comprehensive reference work ever compiled about the Manhattan Project, and for which he justly earned the Pulitzer and many other awards, fell short in one small but important measure. While his historical references are extensively documented, footnoted and cross-referenced, he either fails to apply a similar standard for reconciling mathematical anomalies in his book; or he, like many other authors on the subject, chooses not to question unresolved discrepancies. Possibly he assumes any explanations for disparities are buried beyond the value of their pursuit, and the fact that history suggests they were resolved is explanation enough. Or, also understandably, he may have chosen not to pursue the incongruity as being outside the scope of his already massive work. Whatever the reason, the numbers, as has been demonstrated, do not add up.

Oak Ridge, splitting its enriched uranium stocks between the uranium bomb and the Hanford reactors, had taken almost a year to provide the 15 kilograms available for the uranium bomb by May 1945. In an effort to produce the balance of the 50 kilograms needed, even at its top production capacity in that spring of 1945, the uranium enrichment program could only have produced seven or eight kilograms more between 1 May and 24 July. The twenty-fourth of July is the date General Groves, himself, gave as the last delivery date of enriched uranium to Los Alamos,[17] and this is corroborated elsewhere.[18] Seven or eight kilograms added to the 15 kilograms already stockpiled was still less than half of the 50 kilograms actually used in the bomb.

Even in a most drastic action, if contributions to the plutonium reactors had been discontinued and all enriched uranium produced from that time forward had been committed to the uranium bomb effort, the maximum possible enriched uranium available was just over five kilograms per month. Accumulated over the next three months of May, June and July, the total of another 15 kilograms added to the original 15 kilos would have made the material available for the bomb only 30 kilograms. The 50-kilogram uranium bomb still would have been short its enriched uranium needs by over fifty percent. A serious stumbling block had been dropped in the path of the uranium bomb.

The second shoe dropped just a few days later. On 18 April, Norris Bradbury, the man assigned the responsibility of overseeing assembly and final testing of the plutonium bomb[19] and who went on to become the post-war director of the Los Alamos National Laboratory, reported in a memo that the program was experiencing serious problems with detonators. He then concluded:

> (a) a much smaller number of tests than 300 (the scheduled number) will have been carried out; (b) there is more than a bare possibility that *the detonators will be unsatisfactory* [emphasis added].
>
> Particularly in the latter event it may be necessary to postpone final Raytheon tests until the detonator difficulty is unscrambled.[20]

The detonator problem, the "Raytheon tests" and the timing of the memo itself all are salient factors in a central premise of this book; that components from captured U-boat U-234 were employed to successfully complete both of the Manhattan Project's atomic bombs. The detonator problem had been a long-standing issue but those in charge thought it would have been overcome long before the spring of 1945. By now the faulty detonators as well as the delay of the Raytheon tests, combined with the shortage of enriched uranium, actually represented less the final shoe falling upon the program than one standing on the throat of the entire Manhattan Project.

Test detonations almost never went right. Getting all 32 firing points to discharge at the same instant using clumsy electro-mechanical circuits of cables, wires and connectors proved to be almost impossible. If the timing was off by a mere fraction of a millisecond at any one of the 32 firing locations, the mass of plutonium at the center would be shot out of the undetonated hole, like a bullet out of a gun. The plutonium core would have been thrust out the low-pressure point caused by the failed detonator, driven by the shock wave of all the other detonator explosions that fired on cue. According to Luis Alvarez, even "the best detonators then available" were only achieving detonation waves spaced 10 to 20 feet apart, "rather than the required fraction of a millimeter."[21] This situation, so late in the game, was considered openly a crisis equal to the shortage of enriched uranium for the uranium bomb.[22]

Oppenheimer, back in October, had aggressively pursued a resolution to the problem and had assigned a three-man committee to "consult on the procurement of detonators to insure that the designs are satisfactory...."[23] The first man listed on the committee was Oppenheimer's old Berkeley buddy, physicist Luis Alvarez, who had transferred to the Manhattan Project from working on the development of radar and other high frequency wave applications. Two other key high-ranking members of the

Manhattan Project staff, Kenneth B. Bainbridge and a Lieutenant Colonel Lockridge accompanied Alvarez on the team. Alvarez later would be credited with saving the plutonium bomb by developing a network of thread-like fuses that helped the 32 detonators instantly and evenly trigger the high explosives in near-perfect unison, compressing the core to critical mass and resulting in an atomic explosion.[24] The "thread-like" description seems to parallel the network of "detonator chimneys" Harlow Russ described as being installed at the last minute to solve the fusing problem.

But for now, six months had passed since the creation of the detonator team – and only three months remained before the Trinity test – and clearly the detonators were still a serious obstacle to success. Alvarez and his team had thus far failed. And this, compared to the "Raytheon" dilemma, appeared to be only a secondary problem, although the two look to have been inseparably interconnected.

According to General Groves, the main delay of the plutonium bomb resulted from the fact "the company manufacturing certain essential parts for a non-atomic assembly in the Fat Man (the code-name for the plutonium bomb) had been unable to meet delivery schedules."[25] Groves continued, explaining the delayed part hindered testing of the bomb "until a critically late date."

The unidentified component over which he lamented was the control unit for discharging the simultaneous firing signal for the detonators, known as "the X-unit." Raytheon, the unnamed manufacturer mentioned by Groves but referenced by name in Bradbury's memorandum above, was the maker of the X-unit. The device, which cost the equivalent of a Cadillac,[26] was a sophisticated conglomeration of cables, switches, transformers, wires, condensers, capacitors and relays.[27] The complexity of the instrument made it an engineer's nightmare. It appears that X-unit manufacturing was delayed further and made even more complicated because, as the brains of the detonation system, it required that the type of detonators to be used be integrated into the X-unit's design specifications, as seems is suggested in Bradbury's memo. Surely in as precise a piece of instrumentation as the X-unit, detonator selection would be critical. But as late as three weeks before the Trinity test, the detonator to be used still had not been selected. As a result, modifications to the largely untested X-unit were expected to be made on-site just prior to the Trinity test.[28]

Whatever the case, the scientists struggled with the detonator and X-unit problem throughout the fall, winter and spring and still had not resolved it as the fateful summer of 1945 was unrolling. A raft of reports, memoranda and schedules, with addressees including Alvarez as well as those who received the enriched uranium schedule reviewed earlier, Bacher and Allison, flew from office to office as efforts were made to resolve the detonation problems.

The communiqués show that as late as 9 April 1945, the first in-flight tests for the X-unit were finally scheduled[29] to evaluate X-unit operation on an actual bombing run. The tests proved in early May, however, the detonators were still unsafe,[30] with only two-and-a-half months left until the Trinity test. Finally, one-and-a-half months later, on 20 June, with less than a month until Trinity, X-units were scheduled to be delivered to Los Alamos for the Trinity test. But even these were not prepared for their final use.

Apparently detonators finally had been obtained that could do the job, but for some unexplained reason modifications were still planned for the X-unit, even after delivery.[31] In a memorandum written by George Kistiakowsky dated 6 June 1945, instructions were given that one X-unit was to be "*modified*, inspected, and made shippable to Trinity by 1 July. Two more units for Trinity should be on the Site (sic) by 1 July, and should be *modified*, and made shippable to Trinity by 7 July" (emphasis the author's). Given Bradbury's note regarding the interrelationship between detonators and the X-unit and the ongoing problems with the detonators right up to the end, which appear to be the reason for the X-unit's delay, it is logical to conclude the modifications mentioned by Kistiakowsky were related to the new detonators.

What was done to procure the new detonators is unknown, as are the last-minute modifications made to the X-units. But the timing of these important changes and the activities of Luis Alvarez during this same period may be very telling in regard to how the implosion-timing problem was resolved. Apparently, Luis Alvarez is the same Mr. Alvarez as the false "Commander Alvarez" who received Dr. Heinz Schlicke's proximity fuses from U-234.

As it also appears that the same enriched uranium so desperately needed to complete the uranium bomb was received from the gold encased stocks of uranium labeled "U235" that Major Vance had taken from U-234. The facts appear to demonstrate that without the bomb materials surrendered with U-234, the United States' atomic bomb effort to win the war by mid-August would have failed. The question is, how did those powerful nuclear components fall into American hands?

Endnotes: Chapter Eight – Simple Math

1   Richard Rhodes, *The Making Of The Atomic Bomb*, p. 612

2   Leona Libby, *The Uranium People*, p. 244; Richard Rhodes, *The Making Of The Atomic*

Bomb, p. 691; Harlow Russ, *Project Alberta*, p. 55

3   Richard Rhodes, *The Making Of The Atomic Bomb*, p. 490

4   Herbert Childs, *An American Genius*, p. 350

5   Richard Rhodes, *The Making Of The Atomic Bomb*, p. 614

6   Richard Rhodes, *The Making Of The Atomic Bomb*, p. 601

7   Leslie R. Groves, *Now It Can Be Told*, p. 309

8   Harlow Russ, *Project Alberta*, p. 66

9   U.S. National Archives, Southeast Region, East Point, Georgia, *Beta Oxide Transfer Report*, RG 69 A 406 section 326 box 17

10   Richard Rhodes, *The Making Of The Atomic Bomb*, p. 601

11   Robert Serber, *The Los Alamos Primer*, p. xv

12   Robert Serber, *The Los Alamos Primer*, p. 33

13   Richard Rhodes, *The Making Of The Atomic Bomb*, p. 612

14   Eric Jette, memo, *Production rate of 25*, Los Alamos National Laboratory Archives, A-84-019-70-24

15   Richard Rhodes, *The Making Of The Atomic Bomb*, p. 657

16   Richard Rhodes, *The Making Of The Atomic Bomb*, p. 614

17   Leslie R. Groves, *Now It Can Be Told*, p. 124

18   Richard Rhodes, *The Making Of The Atomic Bomb*, p. 691

19   U.S. National Archives Southeast Region, East Point, Georgia, George B. Kistiakowsky, memo, Organization of the X Division Participation at Trinity, A-84-019-55-9

20   U.S. National Archives Southeast Region, East Point, Georgia, Norris E. Bradbury, memorandum to Norman Ramsey, A-84-019-82-16

21   Luis Alvarez, *Alvarez*, p. 133

22   Robert Serber, *The Los Alamos Primer*, p. xv

23   U.S. National Archives Southeast Region, East Point, Georgia, Kenneth T. Bainbridge, memorandum, Minutes of a Meeting on the Electric Detonator Program, p. 2, A-84-019-14-11

24   Robert Serber, *The Los Alamos Primer*, p. xvii

25   Leslie R. Groves, *Now It Can Be Told*, p. 341

26   Max Morgan Witts, Gordon Thomas, *Enola Gay*, p. 103

27   U.S. National Archives Southeast Region, East Point, Georgia, L. Fussell memorandum: Detonator Circuits Visit to Raytheon Co., A-84-019-41-11

28   U.S. National Archives Southeast Region, East Point, Georgia, F. Oppenheimer memorandum to N.E. Bradbury, June 21, 1945, A-84-019-55-9; also F. Oppenheimer memorandum to K. Greissen, D.E. Horning, E.J. Lofgren, June 26, 1945, A-84-019-55-9

29   U.S. National Archives Southeast Region, East Point, Georgia, Norman F. Ramsey, memorandum, 9 April 1945, Boosters for T-26, A-84-019-67-7

30   U.S. National Archives Southeast Region, East Point, Georgia, William S. Parsons, minutes of 1 May 1945 Meeting On Detonators, A-84-019-82-16

31   U.S. National Archives Southeast Region, East Point, Georgia, George B. Kistiakowsky, memorandum 6 June 1945, X Units for Trinity, A-84-019-55-9

The secretive Martin Bormann was the leader of Hitler's Nazi Party – all government entities reported through him, including the Gestapo, the SS, and every other non-military government function. He was also the Fuerher's closest confidant. As such, Bormann was the second most powerful man in the Third Reich. Some say the most powerful, if Nazi leaders Goering, Guderian and Schellenberg are to be believed.

PART THREE

Martin Bormann

The crew of U-234 stand at attention on her deck. Besides the 63 members of her crew, U-234 carried twelve passengers, including a German air force general, two civilian specialized technicians, two Japanese officers who committed suicide when they realized U-234 would be surrendered, and a mysterious scientist, Dr. Heinz Schlicke .

Chapter Nine

Maiden Voyage

"I think it was about 14 April when I gave the captain a signal which read: 'U-234. Only sail on the orders of the highest level. Fuehrer HQ'"[1]
– Wolfgang Hirschfeld, Chief Radio Operator of U-234

"I directed the radar beam directly on the attacker. At 3,300 yards the aircraft inexplicably pulled off its headlong course and turned away.... After thirty minutes there was another approach from the west, but... it disengaged at 3,300 yards.... The game went on all night; three times it was repeated"[2]
– Wolfgang Hirschfeld, Chief Radio Operator of U-234, Describing a curious event when U-234 was located at sea by enemy aircraft but inexplicably was not attacked.

The fact U-234 arrived on American soil carrying 560 kilograms of uranium that was enriched and went on to be used in the bombs dropped on Japan can scarcely be argued any longer except by those who refuse to acknowledge the evidence. There also now exists a considerable case that infrared fuses – or other fast energy transfer systems – on board the U-boat were used in the bomb dropped on Nagasaki.

But how did U-234 fall into American hands? Is it true the U-boat captain, Johann Heinrich Fehler, allowed two Japanese officers to kill themselves so he could surrender the U-boat, its contents and passengers to the United States rather than continue his mission to Japan? On what basis did he make such a deadly decision? Could U-234, in fact, have been surrendered as part of a higher level secret negotiation in which Captain Fehler, his crew, passengers and submarine were all simply pawns?

A substantial amount of evidence of just such a situation does exist. And although the evidence is not as conclusive as that regarding the cargo of U-234 going into the Manhattan Project, responsibility and prudence direct that the perplexing record of the recalcitrant U-boat be examined.

There are numerous anomalies and inconsistencies in U-234's voyage that suggest more was occurring than what is accepted in the existing history. On the other hand, there are references to escapes in U-boats and Allied tracking of enigmatic submersibles matching U-234's description that deserve investigation. All of the evidence is circumstantial. None of it can prove anything conclusively about how U-234 came to be surrendered. And yet to ignore the evidence that does exist leaves significant questions remaining that ought to be researched and answered.

The following is a serious assessment of all that is known regarding U-234 and a surrender theory based on the evidence available. The conclusions cannot be certain until tied to explicit primary documentation, but the evidence is strong enough to suggest collusion at the highest levels between the United States and Nazi Germany governments – and that collusion extends down to those within U-234, its officers, crew and passengers – and has been maintained by powerful parties with vested interests on both sides of the Atlantic ever since.

Laden with 240 tons of war materials, including, according to eyewitness reports and documentation, enriched uranium and infrared proximity fuses, U-234 was prepared for her maiden – and what would prove to be her only – mission. She had recently been equipped with a "snorkel," Germany's newest submarine device that under normal sailing conditions allowed its user to stealthily sail the Seven Seas without the necessity of ascending for air. The 24 mine-laying tubes on the boat had been remodeled as storage compartments. The outer keel plates had been removed and the keel duct was loaded with a cargo of mercury and optical glass before the plates were welded back into place. Two hundred forty tons of cargo destined for Japan was estimated by U-234's officers to have been loaded onto the boat; and now it stood at the dock in Kiel waiting to make its desperate dash to safety.

The chief officers of the boat, like the boat itself, appear to have been hand picked for the assignment. Indeed, it is hard to imagine a commanding officer who would have been a wiser selection for his mission than Captain Lieutenant Johann Heinrich Fehler.

Fehler, like so many U-boat skippers, had begun his career fresh out of naval school on surface ships. He and his eventual first officer, Richard Bulla, brought a breadth of experience to U-234 that had been gained on one of the most famous war vessels – or infamous, depending on one's point of view – in modern times; that of the German raider *Atlantis*.[3]

In the early days of the war, the *Atlantis*[4] had roamed the Pacific, Indian and Atlantic Oceans disguised as a ship neutral or friendly to Allied countries. Upon locating and approaching a vessel from one of these countries, *Atlantis* would unloose its six 150mm camouflaged guns and attack with torpedoes and its two deck attack planes, one of which was piloted by Bulla.[5] The ploy was usually so bold and unexpected the matter was over in moments and Fehler, who was the munitions officer onboard the ship and who had therefore earned the nickname "Dynamite," would then apply charges that scuttled the captured vessel. By such means *Atlantis* sunk or captured 22 Allied ships.

Atlantis' modus operandi took daring and cunning, knowledge of how to execute deception on the open seas, and an understanding of the fine balance between audacity and idiocy that differentiates the successful stratagem that creates a hero from the clumsy ruse, whose outcome is ruin.

The Allies eventually caught on to *Atlantis'* tactics, however, and, its impact neutralized, the ship was forced to forego its actively belligerent role to be relegated to relieving other front boats with supplies and weapons. Even after *Atlantis* was converted from rogue warrior to surface supply ship, Fehler quietly carried within him all of those lessons of stealth and subterfuge learned in battle, to be used later while commanding U-234.

Atlantis' final foray has become legend. While tied to U-126 in the South Atlantic in a resupply maneuver on 22 November 1941, HMS *Devonshire*, a British cruiser, happened upon the boats. Dead still in the open water and intertwined in fuel lines, the two ships' crews suddenly had to race to clear the umbilicals to have a chance at survival. Once free, U-126 dove to safety. With *Devonshire* bearing down on her, *Atlantis* was a sitting duck. To avoid capture of the ship according to standing orders, munitions officer Fehler, as he had done with so many enemy vessels before, scuttled his own ship, adding *Atlantis*[6] to the list of vessels he had sent to the bottom of the sea.

The 100 crew and officers who went into the life rafts were later found on the open sea by U-216, but there was no room in the U-boat for extra hands. Gross Admiral Karl Doenitz thought so highly of Atlantis and her crew, however, that he ordered two additional U-boats to aid the castaways and bring them home alive.

The rafts were tied to these U-boats and the U-boats, traveling on the surface and moving excruciatingly slowly, sailed for France. The plan, should it be required by enemy attack, was to release the rafts upon approach of a hostile craft and allow them to float away, their occupants to be killed or captured, as the escorting U-boats dove for safety. Fortunately for Fehler and his 99 mates, the plan never had to be carried out. The three U-boats, the survivors of *Atlantis* in tow, successfully traversed thousands of miles of open ocean to ultimately reach France. The recovery of the *Atlantis* survivors now stands in the annals of naval history as one of the greatest maritime rescues of any military service.

Three and a half years after the return of the *Atlantis* survivors to Germany, at 3 p.m. on the afternoon of 25 March 1945, fifty-five days before its dubious surrender and entrance into Portsmouth Naval Yard,[7] U-234 with Captain Johann Heinrich Fehler in command, its devastating cargo and many of its passengers sealed away in its bowels, slipped away from its base in Kiel, Germany.[8] Once the tending tugboat had drawn U-234 away from the dock, Captain Fehler took control of the U-boat and raced "with great speed" down the Kiel Fjord. To reduce the chance of being caught and bombed by enemy anti-submarine aircraft while vulnerable in the narrow, shallow waterway, U-234 sailed surfaced and at near-maximum speed down the narrow channel. Heading toward the entrance of the harbor, the submarine passed the towns of Laboe and Friedrichsort and then raced out into the open Baltic Sea, where a two-U-boat escort joined it.

In Kiel, the loading of the boat had been completed and her massive hull sealed up for the journey. The crew of 63[9] (25 percent larger than the average crew of a U-boat – even of this size) would eventually be joined by twelve passengers, including two Japanese officers, Genzo Shoji and Hideo Tomonaga, and enigmatic engineer Dr. Heinz Schlicke, as well as General Kessler and his retinue, according to the abstract of a secret U.S. intelligence report:[10]

> Additional to our ... [segment unreadable] shipped aboard [U-234]:
> a. Japanese commanders Tomonaga, Shoshi; Civilian engineers Bringewald, Ruf.
> b. From Air Force: General Kessler, Colonel von Sandrath, Lieut. (j.g.) Menzell.
> c. From Naval High Command: Commander Falck, Squadron Director Nirschling (sic), senior construction chief Schlicke, Lieut. Bulla, Lieut. (j.g.) Hellendorf.... Secret message 1590/46, dated (3 or 5 – number unclear – author's note) April, itemized the cargo which was shipped aboard the U-234. General Kessler, the new German Air Attache to Japan, left Berlin about 30 January to board the submarine on which he was to take passage to Japan. It had previously appeared he was probably aboard the first of three German

U-boats to sail early in 1945 for Japan, U-8 [3 or 5 – number unclear] 4, which was believed sunk off Norway during February. Instead, it appears he boarded the second of three U-boats...

This message noting previously sent secret message 1590/46 listed the cargo on board U-234 presents the interesting possibility the United States may have known uranium was part of U-234's cargo. The author was not able to find the referenced message in the archives, however, so it is uncertain whether the uranium had been identified by enemy intercept prior to the U-boat's surrender as part of the cargo known to be on the U-boat. The presence of the important passengers, however, is unarguable.

The message is one of several records, from both before and after U-234's surrender, that details considerable information about the passengers – some of the data conflicting. Dr. Schlicke, for example, was dressed in a Luftwaffe (German Air Force) colonel's uniform[11] by some accounts. He was identified in other reports as both a naval officer[12] (perhaps "honorary" according to Wolfgang Hirschfeld[13]), and as a civilian specialist in high-frequency and radar technology who was being transported to Japan.[14] Although United States Navy records refer to him unambiguously as a member of the German Navy, with significant references to his involvement there, such references do not preclude the possibility that he actually worked for a different authority. According to U-234 head radioman Wolfgang Hirschfeld, Schlicke was aboard U-234 as an advisor/consultant for the U-boat's radar system.[15] Schlicke is documented as having shared his substantial intellectual services with Hirschfeld during the voyage. But he is also documented in many other capacities, as will be seen. Despite not knowing exactly who this man was, from all of the evidence available, his services extended far beyond submarine radar technology.

As noted in the secret intelligence report, also on board was Kay Nieschling, who had been listed in the report as a naval squadron director. He was later discovered to be a Nazi bigwig and Naval Fleet Judge who, even as the Reich was falling down around him, was being sent halfway around the world on the now futile mission of trying traitors in the infamous case of the Richard Sorge spy ring.[16]

Joining Nieschling, Schlicke and the Japanese duo were others: Naval Lt. Hellendorn, civilian airplane engineer Bringewald, Naval Captain Falk, and civilian engineer Ruf.

Richard Bulla, Fehler's old mate during their daring *Atlantis* raids and rescue, was shortly removed from the passenger list and added to the crew as Fehler's second-in-command. When the originally-assigned first watch officer, Alfred Klingenberg, was caught personally by Fehler *in flagrante delictato* with another crewmember, the Captain removed him from duty and assigned Bulla in his place.[17]

How Fehler's old *Atlantis* mate Bulla came to be on the list of high priority passengers destined for Japan can only be speculated upon. Bulla had flown deck planes – short takeoff aircraft – off of *Atlantis* and had a wealth of experience earned on the raider during daring assaults on enemy targets on the high seas, and therefore was a valued and knowledgeable naval officer and flyer for such operations. U-234 was full of jet aircraft and rockets – and nuclear bomb materials and technical experts of all kinds – destined for Japan. Japan was trying to find a way to deliver an offensive with teeth in the Pacific that could be successful turning the tide of the war in its favor, but the distances involved in island hopping to attack Allied bases were too great for round trip flights from its home islands.

Allied air superiority also was keeping Japan's less capable planes from having their way against them. At the same time, Germany was developing plans for its V rockets to carry atomic warheads,[18] and to be launched from surface ships. And, according to General Kessler's and Judge Nieschling's later interrogations, Japan was modifying V-1 bombs with kamikaze pilot cockpits built into them, [19] and was quite possibly thinking of doing the same with V-2 and V4 rockets. Germany also had devised another plan for piggybacking a modified Messerschmidt 262 – the same type of jet that U-234 was transporting to Japan – as a bomber on a long-range Henkel aircraft for long-distance delivery of a bomb.[20]

Finally, plans for an airplane listed as the "stratosphere plane" were also aboard U-234, along with the plans and a fully assembled high-altitude pressurized pilot's cockpit. Might Bulla's naval piloting experience be valuable in devising a platform for launching ship-based German-made but Japanese-deployed atomic weapons toward Allied bases in the Pacific – part of what might have been a last-ditch but potentially unstoppable effort to win the war?

As U-234 raced out of Kiel Fjord into the Baltic, she turned west into the open bay leading to the mouth of Eckern Fjord.[21] There she waited until dark to begin the first leg of her run for freedom.

Shortly after midnight, in the early morning hours of 26 March, U-234 and her two-U-boat escort joined three smaller Type XXIII U-boats and turned its course toward Norway. Her orders were to remain in the company of the three smaller boats until they reached the Norwegian coastal town of Kristiansand. The small flotilla traveled East below the island archipelago of Eastern Denmark, then North up the narrow neck of water between Denmark and Sweden. They passed Copenhagen while it was still dark and entered the wider body of water between upper Denmark and Sweden known as The Kattegat. Here the two-U-boat escort broke off and U-234 and her three smaller shadows crept up the Swedish coast, U-234 slowed by the 10 statute-mile-per-hour top speed of the Type XXIIIs.

At 3:00 p.m., chief radioman Hirschfeld requested permission from the bridge to discontinue radar operations momentarily in order to change out a malfunctioning component. The bridge, after reconnoitering the surrounding sea and sky for enemy aircraft or warships, gave the all clear. The radar had been out of service barely 10 minutes when sirens screamed throughout the boat that enemy aircraft were approaching. Aware the newly installed component had a recommended 15 minute warm-up time, and not knowing whether the aircraft had spotted them yet, Hirschfeld turned to Dr. Schlicke, who had been observing the radioman's installation procedure, and asked if it would be permissible to power up the radar system. Schlicke simply nodded.

By the time the system was activated, the aircraft were within 5000 meters, and by the time Hirschfeld sent word to the bridge, they had closed to 3000 meters. Fehler, who had already ordered the anti-aircraft guns manned, now gave the order to fire at will. Nobody responded. In the din of battle preparations the crew had not been able to hear the Captain's command.

As the air armada flew overhead 2000 meters to starboard, Fehler personally went to take control of the anti-aircraft guns for the return engagement. But the airplanes never came about; presumably, according to Hirschfeld, having never seen U-234 and its triple tail (which is doubtful since the radar of Allied aircraft flying at 10,000 feet could spot a normal sized – much less triple sized – surfaced U-boat as far away as 80 miles[22]). The enemy air patrol may have been on a dedicated mission elsewhere and simply was not interested in the mini-armada. Or the aircraft may have been ordered only to reconnoiter the U-boats but not attack, an odd but plausible scenario given ensuing events. Whatever the case, the U-boats continued their course toward Norway.

Just before midnight of the same day, the U-boat brigade passed behind a southbound convoy of German torpedo boats. Shortly afterward, those on the bridge of U-234 saw the convoy attacked by enemy aircraft and the resulting firefight was quite a spectacle. The screen of U-234's radar glowed with swarming enemy aircraft attacking the small armada of surface ships. Fearful the planes would turn on them, and unable to dive because of the shallow, thickly mined waters, the crew of U-234 would have liked to race away; but obedience to the order to remain with the smaller, slower U-boats kept her at their sides.

Very soon the enemy airplanes did, indeed, spot the U-boat convoy – again with curious results. Flying very low to avoid radar, but according to Hirschfeld not succeeding, an enemy aircraft headed directly for U-234 and her diminutive detail. Hirschfeld recorded the event:

> "I directed the radar beam directly on the attacker. At 3,300 yards the aircraft inexplicably pulled off its headlong course and turned away..... After thirty minutes there was another approach from the west, but ... it disengaged at 3,300 yards.... The game went on all night; three times it was repeated "[23]

What could have caused the apparently willing and able assault aircraft to approach the small group of vulnerable U-boats but not attack? Under normal conditions any U-boat, but most particularly a group of U-boats, could expect a full confrontation in such circumstances. In addition, if Allied intelligence knew about the important passengers and cargo scheduled to travel on board U-234 – as the above-mentioned intelligence report and other intercepted radio transmissions make clear it did,[24] – one would think every effort would have been made to sink U-234.

There are examples. According to an article titled *The Last Dive* in the October 1999 issue of *National Geographic Magazine,* great efforts and risks had earlier been taken to sink a U-boat Allied intelligence had learned was traveling from Germany to Japan with a boatload of technology.[25] Why not, then, also sink U-234, when its cargo and passengers were so important to Japan's ability to continue to prosecute the war, and the British and Americans knew it?

And when American forces in the Pacific intercepted a report of Japanese Admiral Isoroku Yamamoto traveling by aircraft near Bougainvilla, a squadron of fighters was sent to find his plane and shoot it down at all costs, which they successfully accomplished. Certainly if the purpose of U-234's voyage – especially the transporting of General Kessler to Japan – was known by those controlling the attacking planes in Hirschfeld's account, a similar attack to that on Yamamoto's plane would have been made on U-234.

That no effort was made to sink U-234, but several attacks appear, in fact, to have been suppressed, suggests that, instead of wanting the vessel and its contents and passengers destroyed, the U-boat was being monitored and its passage protected, for some unknown reason, at a high level within Allied command. Obviously, if U-234's progress was being tracked and protected by the Allies, probably the crews of the attacking aircraft would not have known it. But if there were those in higher Allied circles who knew of and expected to obtain U-234's uranium cargo, certainly they would have kept a close eye on its whereabouts and the conditions under which it was traveling – and had channels to the proper authorities necessary to divert disaster if required.

Without further information, one can only guess what those conditions were that caused the planes to approach at least three times – and apparently as many as five – and then cancel the golden opportunity to eliminate four enemy vessels at once. Fortunately, further information is available and will be reviewed later within these chapters. At any rate, U-234 was allowed to proceed, and the tiny armada slipped safely into Oslo Fjord just before sunrise of 27 March, and anchored at Horten, Norway.

At Horten U-234 began trials of its newly installed snorkel device. Two days after arrival, during one of these trials the U-boat was accidentally rammed by another U-boat that was also undergoing testing. Both boats were slightly damaged. A dive tank and a fuel oil tank of U-234 were punctured but the boat was able to continue its testing for four more days, at the end of which Fehler steered his charge to Kristiansand in hopes of making repairs. A problem arose when it was realized placing the boat in dry-dock while it was so full of cargo may stress the heavily laden hull to the point of further damage. A resourceful solution was found. Since the damage was to the aft of U-234, the forward diving tanks were flooded, forcing the nose of the boat to submerge and the stern to rise out of the water. The innovative idea worked wonderfully and the necessary welding was completed without further problems.

In the meantime, more passengers arrived in Kristiansand,[26] including, according to Hirschfeld, General Kessler and his retinue, Colonel Fritz von Sandrart; Lieutenant Menzel, an airplane torpedo expert; and an engineer Klug. Including the two Japanese officers and other previously boarded guests, U-234 now contained 12 passengers and a crew of 63, a total of 75 people – almost 50 percent more than the average personnel load of the U-boat.

Chief radio operator Wolfgang Hirschfeld reported that during the repair time in Kristiansand he personally traveled each day to pick up radio messages intended for U-234. During one of these visits he claimed to have received the following transmission: "U-234. Only sail on the orders of the highest level. Fuehrer HQ."[27] What occurred before and after receiving this cryptic correspondence, and what went through Hirschfeld's mind as a result, he does not say, but certainly such a communication from the Fuehrer's bunker directly to a specific U-boat, if true, is startling.

Hirschfeld went on to write that when he returned to U-234 with the note and handed it to Fehler, the Captain's immediate response, understandably, was to call for Kessler. The General perused the puzzling order and calmly predicted someone was coming from Berlin.[28] "Probably the Fat One," he lamented, immediately remarking that, if so, he (Kessler) would have to leave the boat. Hirschfeld, whether having heard it from Kessler's lips or otherwise, suggested in his writing the allusion was probably to Hermann Goering, at that time Hitler's heir apparent – though not for long.

Kessler's comment about disembarking U-234 if Goering was going to be along for the ride appears to validate the authenticity of Hirschfeld's account of U-234, since it is a true, if little known, fact that Kessler and Goering disliked one another intensely.[29] To put it bluntly, Goering was "out for" Kessler and, in fact, had demoted him five years previously from a diplomatic position to commander of an air wing during the attack on Poland. Traveling with Goering would have been a very unsatisfactory condition for the General, indeed, and one can be certain Kessler did not look forward to a voyage halfway around the world that would take months, stuck in tight submarine quarters with "The Fat One." Still, Hirschfeld makes the unlikely but accurate statement regarding enmity between Kessler and his superior that validates what he has written.

On that same afternoon, Hirschfeld and Second Watch Officer Karl Pfaff, who was responsible for loading the U-boat but otherwise seems an unlikely candidate to be requested to attend this meeting, were ordered to appear before the Flotilla Chief. After scavenging a pair of acceptable uniforms to wear before the Flotilla Commander, they made their appearance. The Commander placed a *code green* – top leadership frequency – transmission on the table before them and asked what the transmission was and how did it get on a tightly-controlled frequency.

The message read: "To head radio chief Hirschfeld on U-234, for your last trip, much luck and healthy return home. Your Bubbi."

"Who is Bubbi?" asked the Flotilla Commander. Hirschfeld told the commander that Bubbi was "the head radio man of 10th Flotilla in Lorient, Bernhard Geissmann." Apparently this was a lie intended to protect the identity of Bubbi. For Hirschfeld then explained in his narrative that the U-boat base in Lorient, France had been captured by the Allies, and it would be impossible, he went on to explain, for the commander to verify who had actually sent the transmission and therefore determine who Bubbi really was.

Such an explanation was strange and unnecessary if Geissmann truly was Bubbi. If Geissmann had been captured at Lorient, it would be a safe bet the occupying forces were not allowing German radiomen to send unscreened and/or personal messages on high-priority enemy frequencies. Considered in combination with Hirschfeld's admittedly disingenuous explanation of the "Bubbi" message, however, and the day's previous transmission received, as well as future transmissions yet to come, it seems more likely the message from Bubbi had a different purpose. Given the previous message from the Fuehrer's headquarters and Hirschfeld's odd explanation of the "Bubbi" message, it seems the two may have been connected.

What was the origin of the cryptic communiqué? Arrangements may have been made prior to the U-boat leaving Kiel for any high-priority transmissions between U-234 and the Fuehrer bunker to be sent to a communications center at Kristiansand specially equipped to receive such high-level messages. Apparently these transmissions were sent to a specialized communications center the frequencies of which U-234 was incapable of receiving, or to keep the confidential communiqués from the knowledge of the regular U-boat command. Once the initial contact had been made, per plan, then Hirschfeld could inform the sender in the Fuehrer bunker of U-234's location and provide contact information for future correspondence. In response, the mysterious messenger in the Fuehrer bunker could then define a plan for further confidential communications on more open channels, using a code name – Bubbi? – for verification without revealing the sender's actual identity.

Shortly thereafter, apparently still on the same day, Hirschfeld was called to return to the radio station for yet another message. This one read: "U-234 is to leave under my command only. After you have made your calculations, leave. BdU."[30] BdU was the personal command designation of none other than Grand Admiral Karl Doenitz, commander of the German U-boat navy. This transmission is documented not only by Hirschfeld, but in the United States National Archives by OSS records of intercepted German transmissions.[31] Doenitz's message alone makes clear a struggle for control of U-234 was taking place, for, as the supreme commander of the German Navy, why else would the Admiral have to remind one of his captains to depart only on his command? In fact, Hirschfeld identified this struggle directly, commenting that Fehler remarked when he heard of the order, that Doenitz "doesn't let himself be submitted to the top leaders' orders."

Apparently, Doenitz by this time had become aware of the plan to use U-234 as an escape vehicle for very high-ranking Party officials at the Fuehrer's headquarters. Possibly the communications center commander had seen through the inconsistencies in Hirschfeld's story about the mysterious 'Bubbi' message and informed Doenitz. Whether Doenitz's determination to keep control of the boat was an effort simply to maintain proper chain of command while still helping to implement an escape plan, or whether his efforts to control the U-boat were to obstruct the plan, is unknown. The latter is doubtful given later history.

Ultimately, history records Martin Bormann from the besieged bunker in Berlin spent considerable attention on negotiations with Doenitz in order to affect his escape from the strangling city. And it records Grand Admiral Karl Doenitz, without political experience or, indeed, any political following, eventually and very unexpectedly replaced Hermann Goering – whom Bormann had succeeded in bringing down as Hitler's successor – when Bormann notified Doenitz that the Admiral was to succeed Hitler as Chancellor of the Third Reich.

Endnotes: Chapter Nine – Maiden Voyage

1   Geoffrey Brooks and Wolfgang Hirschfeld, *Hirschfeld: The Story of a U-boat NCO 1940-1946*, p. 203; also see Wolfgang Hirschfeld, *Feindfahrten*, p. 357

2   Geoffrey Brooks and Wolfgang Hirschfeld, *Hirschfeld: The Story of a U-boat NCO 1940-1946*, p. 202

3   U.S. National Archives II, *Report On Interrogation Of The Crew Of U-234 Which Surrendered To The USS Sutton On 14 May, 1945 in Position 47°-07'N – 42°-25'W*, RG 38 – 370 15/09/01 box 2

4   Robert K. Wilcox, *Japan's Secret War*, pp. 141-143

5   *Sharkhunters KTB 105*, p. 21

6   *Sharkhunters KTB 106*, p. 9

7   U.S. National Archives II, *Report On Interrogation Of The Crew Of U-234 Which Surrendered To The USS Sutton On 14 May, 1945 in Position 47°-07'N – 42°-25'W*, RG 38 – 370 15/09/01 box 2

8   Wolfgang Hirschfeld, *Feindfahrten*, p. 354

9   U.S. National Archives II, Passenger and Crew List Of U-234, RG 38 – 370 15/05/07 box 3-6; Wolfgang Hirschfeld, *Feindfahrten*, p. 357

10   U.S. National Archives II, Extract of secret U.S. Navy intelligence report dispatch, which begins: *In addition to our… (text unreadable)…shipped aboard (U-234)*, no date given, RG 38 – 370 1/4/7 box 113

11   Peter Hansen, *Sharkhunters KTB 107*, p. 11

12   U.S. National Archives II, General Ulrich Kessler, *Interrogation report #5236, 28 May, 1945*, RG 165 – 390 35/10/05 box 495; U.S. National Archives II, *USS Sutton "Passenger and Crew of U-234"* receipt, RG 38 – 370 15/05/07 box 3-6; *Sharkhunters* KTB 112, p. 31

13   Geoffrey Brooks and Wolfgang Hirschfeld, *Hirschfeld: The Story of a U-boat NCO 1940-1946*, p. 200

14   Peter Hansen, *Sharkhunters KTB 107*, p. 11

15   Wolfgang Hirschfeld, *Feindfahrten*, p.355

16   *Sharkhunters KTB 103*, p. 8

17   Peter Hansen, *Sharkhunters KTB 107*, p. 11

18   David Irving, *The German Atomic Bomb*, p. 185; Geoffrey Brooks and Wolfgang Hirschfeld, *Hirschfeld: The Story of a U-boat NCO 1940-1946*, pp. 212, 213

19   U.S. National Archives II, *Report of Interrogation of Lt. Gen. Ulrich Kessler, 21 May 1945*, RG 165 – 390 35/10/05 box 495 and *Report of Interrogation of Kay Nieschling, 24 May 1945*, RG 165 – 390

20   David Irving, *The German Atomic Bomb*, p. 236; see also U.S. National Archives II, Washington, D.C., *Report of Interrogation # 5399 of Lt. Gen. Ulrich Kessler, 25 June 1945*, RG 165 – 35/10/05 box 495

21   Wolfgang Hirschfeld, *Feindfahrten*, p. 354

22   *Sharkhunters KTB 112*, p. 6

23   Geoffrey Brooks and Wolfgang Hirschfeld, *Hirschfeld: The Story of a U-boat NCO 1940-1946*, p. 202

24   U.S. National Archives II, several radio transmission intercepts regarding General Ulrich Kessler and cargo aboard U-234, including transmissions titled *From: Chief Inspector in Germany #165, 0540/15 April 1945, Anton 1*; *In addition to our… (text unreadable)… shipped aboard (U-234)*, no date given; *From Chief of Bureau of Military Affairs, 2014/9 May 1945*; transmission beginning *"unverified information received from Managing Director of Saudel Aircraft Works…"*, dated 8 May 1945; all references listed are contained in RG 38 – 370 01/04/07 box 113

25   Priit Vesilind, *The Last Dive, National Geographic Magazine*, October 1999

26   Wolfgang Hirschfeld, *Feindfahrten*, p. 357

27   Geoffrey Brooks and Wolfgang Hirschfeld, *Hirschfeld: The Story of a U-boat NCO 1940-1946*, p. 203; see also Wolfgang Hirschfeld, *Feindfahrten*, p. 357

28   Wolfgang Hirschfeld, *Feindfahrten*, p. 35

29   U.S. National Archives II, *Report On Interrogation #1540, of P/W KESSLER, Ulrich, Gen. der Flieger*, RG 165 – 390 35/10/05 box 495

30   Geoffrey Brooks and Wolfgang Hirschfeld, *Hirschfeld: The Story of a U-boat NCO 1940-1946*, p. 203; see also Wolfgang Hirschfeld, *Feindfahrten*, p. 358

31   U.S. National Archives II, Top Secret ULTRA intercepts, 12 April 1945 and 13 April 1945, RG 38 – 370 01/04/07 box 113

Albert Speer (left) was Bormann's (middle) nemesis within Hitler's elite, challenging but losing to Bormann often – including over Hitler's support of who would lead the Nazi atomic bomb project..

Chapter Ten

A Pig Digging For A Potato

I studied Bormann's technique with Hitler and realized he controlled the Fuehrer!"

Chief of Nazi foreign intelligence Walter Schellenberg

Bormann was "the secret master of a despot."

Hitler courtier, Hans Frank.

Everything had to be done through this sinister guttersnipe (Bormann)."

Hitler's General Chief of Staff, Heinz Guderian

Bormann stayed with Hitler night and day and gradually brought him under his will so that he ruled Hitler's whole existence."

Hitler's heir-apparent Hermann Goering

B eneath the city of Berlin, under the Reichstag building, burrowed in a hole like a frightened rabbit seeking desperately to avoid being torn apart by hungry, angry wolves, quivered the once invincible Adolf Hitler. By the last days of April 1945, the Russians had pressed their advance to the outskirts of Berlin, almost completely surrounding it, and with the winking approval of the Americans and English had begun pummeling the symbolic center of Nazism under a steady, 24-hour-a-day barrage of artillery fire. The war-wearied, ghost-faced resident survivors huddled forlorn and resigned in the subway tunnels as the constant thunder of shells rattled whatever structure was left overhead that separated the destitute and despairing from destruction.

The warren-like underground bunker that constituted the Fuehrer Headquarters seemed little more than a living tomb. To make matters worse, in the claustrophobic confines Hitler's moods swung desperately between raging paranoiac psychopath and drugged derelict. Numb from the imposing reality of abysmal failure, exhausted by unremitting pressure, partially paralyzed from a minor stroke suffered while in the bunker, though still officially in command, the Fuehrer was far from the commanding figure he once had been. Holed up almost continuously in his private quarters inside the bunker, the inner sanctum of the warren – actually a cell within this self-induced prison – he alternately snarled at and viciously attacked what remained of his loyal commanders and staff, and drifted in and out of exhausted and drug-induced stupors.

The former great ones – ministers, generals and admirals, territorial governors – shuttled in and out, putting on the faces of devout supporters sacrificing their all to sustain their leader. Actually, behind his back they were planning to flee the terminal tomb at the first opportunity and slide silently from the heinous history they had helped write into some foreign backwater where they would be forgotten.

Those who remained were the ill-fated, lower-level staffers who shuffled quietly up and down the dimly lit concrete corridors in support of their beleaguered Fuehrer and a few of his closest high-ranking courtiers. Most of them were there under orders, but they were loyal to the last. The atmosphere, emotionally as well as physically, was hardly breathable. News was never good. In the final stand for Nazism the old men, young boys, and walking wounded who defended the city were experiencing few successes but massive desertions. The city was being given up inch by inch at great cost – half a million people would die in the battle. Reports of Russian atrocities, rape and torture of the captured were legion. The inevitable – the unspeakable – was morosely moving toward these doomed dependents of Hitler and there was little one could do to halt the inescapable. Everyone despaired.

Everyone except Martin Bormann.

Martin Bormann's legacy is simply that he is a legend – in the truest, most nebulous sense of the word. Within a few short years, far fewer than your average myth, he became an apparition in history. He is no more, in many people's minds, than the fearsome fantasy of conspiracy theorists and the lugubrious legend of Nazi hunters. To talk seriously about Martin Bormann with any earnest historian of World War Two is to learn that Bormann is *persona non grata*. Not that they contend he never lived, but since they cannot clearly define the limits of his life and power, despite volumes of hard evidence that he was, indeed, among us here on earth, and that he personally managed most of the Third Reich for Hitler, there is almost nothing identifying the palpable essence of the man, or the extent of his reach and power, or the limits of his life.

Often the evidence that does exist holds Bormann in such rare circumstances and implausible activities as not to be credible at all, such as his financing, masterminding the development of, and building Berchtesgaden, Hitler's matchless mountain retreat; or his clandestine control of such Reich royalty as Heinrich Himmler and Rudolf Hess; or his monetary manipulation of the Reich's entire economy; or his oversight of one of Germany's most innovative – and secret – research and development laboratories. But these are what is known about Bormann's life.

The differences between what was real but fantastic in Martin Bormann's life, compared to what is believable but fantasy, are minimal and often interchanged. This is why present-day historians turn pale when Bormann is spoken of or written about in more than passing terms. What is less sure than the above, but for which there exists a considerable body of evidence not easily refuted, is the apparent extension of his life beyond his reported death – and the ensuing inculcation of his influence into areas many do not want to consider, much less admit. They can find few cold facts that cut through the quagmire of questions about his existence and the number and depth of the roles he played.

But perhaps now the question is, was this murkiness the object for which Bormann and others were striving? Was there value to be gained in making unknowable or unbelievable his actions during and after the war?

On the other hand, if one truly wants to understand the Third Reich, its causes and culmination, can the seminal influence of a man known to have been one of the most powerful potentates of Nazism be ignored simply because we have not been able to quantify the extent of that influence?

Martin Bormann's mystique over the years has been so widely nurtured and grown to such immense proportions that one must remind ones' self the actual man named Martin Bormann did exist – not just as a mysterious motivating force during the war and a phantom fugitive after – but that he was, indeed, responsible for much of the Third Reich. His alleged escape at the end of the war, and the mystery that has since arisen around him and his ultimate fate, have grown to cast a shadow far bigger than the even larger-than-life man. His is a shadow that absorbs any light that present-day researchers try to shine into it – so, feeling foolish for trying, they choose not to.

Perhaps that was the intent. Martin Bormann's aura feels similar to, though greater and deeper than, those of other fugitives who floated through that Nazi netherworld before their elusive existence was drawn back into human form as they were drug, kicking and scratching, back to the real world – becoming once again flesh and blood. Adolf Eichmann, Klaus Barbi, Franz Stangl, even the "Angel of Death," Dr. Joseph Mengele, now almost inexorably proven and universally believed to have survived his Holocaust role to live in Paraguay for many years after the war, all appeared from those mysterious mists to again take their human forms – dragged back by daring and persistent seekers of truth and justice. Will the efforts of such researchers as Paul Manning, Hugh Thomas and even Ladislas Farago – once castigated for having accepted fraudulent evidence, much of which recently has been proved to have merit[1] and thus greatly rehabilitated post-humously Farago's titan effort – be amalgamated at a future inflection point to bring Martin Bormann back to life, as well?

During the last days of Nazism, this Machiavellian minister to the Fuehrer – hardly known outside the close cortège of Hitler's inner circle at the time – with characteristic energy, focus and determination, in contrast to and quite unconcerned about those around him, was constantly sending and receiving radio transmissions from the bunker communications center. True to Bormann's pragmatic proclivities, in addition to, or as part of, working out his escape, he is known to have been undermining or negotiating with others of Hitler's henchmen for control of the Reich – apparently confident there would be fragments of the Reich worth controlling after the war, despite the bleak outlook for Germany.

Most students of these events have considered Bormann's machinations as madness, given the Reich was in its death throes. But upon closer scrutiny of his actions and review of the evidence and outcomes, it appears Martin Bormann was working a master plan for a resurrection of Germany after the war, with Hitler's consent – and within which U-234 played an important part.

Understanding the low-profile but very powerful Martin Bormann in Hitler's court is a vital key to understanding Hitler and his power over the masses, not to mention the Nazi Party and the Third Reich – and, more important, to understanding events at the end of the war. Bormann's post-war activities – for the evidence is very strong that he did survive the war, with American help – and the impact they had on the Nuclear Age must be considered in the light of his prior behavior, so we can begin to pierce the shadows of his mystery.

British historian Trevor Roper-Smith called Martin Bormann, "Hitler's Mephistopheles," his "alter ego," his "evil genius." Bormann was known in Hitler's inner circle as "The Brown Eminence" behind the Fuehrer's throne.[2] The very fact this one-time farm supervisor should, with Hitler's approval, climb to manage the barbarous Nazis' affairs of state speaks volumes of the exceptional political and financial acumen this sinister Shylock possessed. Hitler eventually came to rely on and appreciate his most trusted lieutenant's talents so much that Bormann – despite almost no military experience – was not only made an honorary major general of the SS but he was awarded SS number 555 – Hitler's own original SS number.[3]

Bormann in return fawned on his Fuehrer embarrassingly yet unapologetically; writing nearly every word of Hitler's on small white cards he carried at all times. He seldom took vacations or trips of any kind that would separate him from the Fuehrer for more than just a few days, for fear of losing court status. "Bormann stayed with Hitler night and day," Hermann Goering later recounted,[4] "and gradually brought him under his will so that he ruled Hitler's whole existence."

Even though serving as his master's slavish lap dog – in fact, because of it – Bormann came to wield complete authority over the Reich. He accomplished this accumulation of power in a variety of ways, virtually all of them stemming from his position with Hitler. He had access to and kept copious files of evidence and materials aimed at exposing for some misdeed or another, if needed, almost every person of authority in the government, military or the party – including Hitler himself. He also discreetly distributed low-interest or no-interest loans from party coffers,[5] some that did not require repayment, to those whom he felt it would be advantageous to have indebted to him, such as SS leader Heinrich Himmler, who accepted from Bormann millions of reichsmarks per year.[6]

The powerful group of 41 Gauleiters, the "governors" – actually virtual dictators – of the Reich's various "states" or "provinces," reported directly to Bormann as head of the Nazi Party. He cultivated and maintained a strong relationship with this group collectively and many of its most powerful members individually, throughout his tenure until the end of the war.

Bormann's position as Reichsleiter of the Nazi Party also made him, in theory at least, the second most powerful man in the Reich. At party rallies as early as 1934, Hitler had declared the party gave orders to the government, not the other way round.[7] Later interrogations that were part of the Nuremberg Trials verified this relationship.[8] The party, therefore, controlled the government, and Bormann controlled the party.

The Reichsleiter underpinned his power-base by duplicating within the party almost every function required and operated by the viable government. In essence, Bormann created and held the strings to a very powerful "shadow bureaucracy,"[9] complete with its own police force – the Gestapo – and its own army – the Wehrmacht SS – both under the direction of one of Bormann's chief accomplices, Heinrich Himmler – and the 1 million-man-strong Volksturm.[10] Bormann was ruthless in his quest for power to the point that his one-time boss, Nazi Party Treasurer Schwartz, compared him to Joseph Stalin lurking behind Lenin, saying, "Bormann was the most pernicious egotist around.... He would kill, like Stalin."[11] Author William Stevenson echoed that sentiment in his book, *The Bormann Brotherhood*, also comparing Bormann to Stalin,[12] as have many authors and historians since.[13]

In truth, Hitler and Bormann were complementary pieces to the same perverted puzzle. Their personalities and psyches fully understood and intermeshed with one another across the complete spectrum of power-over-the-masses leadership they practiced – and recognized in each other the exceptional counterbalances of their strengths and weaknesses. Where Hitler's highly effective political acumen and charisma failed, Bormann would use his web of intrigue and bureaucratic power to achieve the desired end, explains Bormann biographer Jochen von Lang.[14] Whether Bormann on his climb to the top astutely identified Hitler's deficiencies and determined consciously to fill them himself, or whether the marriage was simply a fortuitous match of fate personality-wise, will probably never be known.

Eventually this symbiotic compact – whether spoken or unspoken nobody knows, either – gave Bormann the confidence he needed to take the bold step of cordoning off the Fuehrer from all others, to be accessed only through him who would become "the dictator of the ante-room"[15] – Bormann himself. In 1943, Bormann successfully convinced Hitler – based on their co-dependent relationship and the fact Hitler, who had appointed himself Supreme Commander of the Army and was spending all of his time and energy personally running the German war effort – to sign a decree appointing a Committee of Three,[16] composed of Bormann and two others, to oversee the everyday operations of the Reich *and* to screen the Fuehrer from unwanted distractions. All communications, reports and requests intended for Hitler had to pass through the Committee of Three first. In typical Bormann fashion, he then subjugated the other two committee members and controlled all information coming to and going from the Fuehrer.[17] Combined with his position as head of the Nazi Party, which was already operating in proxy for the federal government, which in turn was now nothing more than a shell, Martin Bormann had solidified his hold as the second most – some said the *most* – powerful man in The Third Reich.

"I studied Bormann's technique with Hitler and realized he controlled the Fuehrer!" recorded the chief of the Nazi foreign intelligence service Walter Schellenberg.[18] Bormann was "the secret master of a despot," according to Hitler courtier Hans Frank.[19] "Everything had to be done through this sinister guttersnipe," complained Hitler's own General Chief of Staff Heinz Guderian.[20]

Following years of careful conniving and sinister strategies, Bormann had realized his dream – he was, many who were there at the time and some later historians agree, in substance if not in title, the leader of the Third Reich.[21]

While Martin Bormann's name, position and the profound power he wielded in Nazi Germany are almost unknown to the average person – and such was the case even when Bormann was enjoying unequaled fraternity with Hitler as his Nazi Party chief, administrative right-hand man and personal paladin – those close to the Fuehrer at the time, to a man, understood that the key to Hitler during the mid- to late-war years, and possibly earlier, was clenched firmly in Bormann's fist.

To understand how Martin Bormann possessed the power at the end of the war to negotiate away Nazi Germany's developing atomic arsenal in order to sustain himself and the Nazi cause after the war, one must understand this symbiotic relationship between him and Hitler. The defining elements of their lives, sometimes detailed in mirror-like reflections and then sometimes balanced by what seem like polar opposites, while at other times punctuated with bizarre and unequaled uniqueness, are as striking as the surprisingly complementary nature of their beings. That two men who were so well fitted for forwarding each others' individually rare ambitions along with their mutual objectives found one another hardly seems probable. Yet the peculiarities they shared and the differences that filled the holes where each was lacking resulted in two remarkably compatible counterparts – although not psychologically healthy ones.

Adolf Hitler was born the son of a low-ranking Austrian bureaucrat, a customs official who was a drunken sadist, already 52 years old when

Adolf was born – the result of a third marriage – who beat his son and wife, squandered the family money on alcohol, and taught through his actions that "right" is always in the hands of the most powerful. Adolf Hitler learned this lesson – and how to hate – from his father, for whom he grew great loathing and animosity.

Martin Bormann was the son of a civil servant, too, a German postal worker.[22] But while Hitler hated his father and had only one sister, younger than he, Bormann adored his father and paid homage to him, often to the point of heaping upon his memory blatant and unearned exaggeration of his achievements. Holding his father in such reverence was undoubtedly the result of Martin not really having known his father, who died when Martin was less than three years old.[23] The elder Bormann had actually lived a simple, ordinary life, had been married once previous to his marriage to Martin's mother and had sired three children (one died in infancy) from that early union. Upon his death his widow, to support her two natural children and the two step-children she had inherited from her husband's previous marriage, quickly remarried her own dead sister's widowed husband.

Bormann's new stepfather brought five children of his own, Martin's cousins, into the now hodgepodge family. Martin immediately disliked this intruder and his gaggle, whom he considered was trying to take his father's place. The feeling was later exacerbated when, during the hardships caused to all Germans during World War One, rather than serving in the armed forces, his stepfather, the town banker, gloated over the money he was making from war lending. Martin's enmity for the man and his unseemly behavior, however, did not keep Bormann during the next world war from indiscriminately emulating similar war profiteering conduct, but on a much grander scale. The two men remained distant throughout their lives.

The lack of respected father figures, the eclectic and tangled family trees and the distorted relationships these conditions fostered must have been the source of much unusual and perverse psychological programming for young Adolf Hitler and Martin Bormann. Thus in the Petrie dish of dysfunctional families and flawed fatherhood were the psychotic psyches of these two men born.

Both Hitler and Bormann, in a society that valued highly the Germanic ideals of education and intellectual achievement, dropped out of high school, neither one achieving consistently good performances in their matriculations but both showing flashes of real genius in the disciplines they personally enjoyed. Hitler, molded by the heavy hand of his abusive father, extended the unmitigated malice resulting from this excessive behavior to all authority figures he faced, which caused him trouble in the classroom. He was ejected from a catholic school for defying a no-smoking rule, overbearingly insisted on being the leader among his classmates despite any hint of trying to earn such a position or the respect that goes with it, and openly "sabotaged completely," in his own words, any school endeavor not to his liking.[24] He later vilified or otherwise repudiated as stupid or crazy educators in general, making exception only for Dr. Leopold Poetsch, a fervent German nationalist, among all the teachers of his childhood.

German pre-World War One schools, most particularly in the Weimar region where Bormann grew up, were teaching a searing brand of nationalism, pan-Germanism and German cultural superiority, too.[25] Like every other youngster in Germany at the time, Martin Bormann was steeped in this doctrine whose spirit swept the German nation right up and into the first worldwide conflict. Bormann absorbed the nationalistic fervor and carried it within him throughout his life, though in his case, as in Hitler's, it would grow in a monstrous, mutated form.

A patriotic appreciation, in whatever form and however important to his later life, was one of the few benefits Bormann would receive from his schooling. While later events proved he was anything but stupid, in the classroom, for whatever reason, he appears to have struggled. Using dates he later provided in government documents and applications, it appears Martin Bormann took eight years to complete seven grades, apparently also sabotaging his own education; and he exited high school without having graduated, as had Hitler before him, after the eleventh grade.[26]

Driven by visions of grandeur and a staunch belief in his own genius, at the age of eighteen years Hitler left his widowed, incurably ill mother in the town of his childhood and moved to Vienna to become an artist. He wandered the streets of the metropolitan city, painted, dreamed and starved. He was rejected for acceptance at the Vienna Academy of Fine Arts when he failed the entrance examination, a rejection he never forgave, and for the next half-decade he wandered the streets, took small jobs, painted and sold his artwork in the streets when he could, and panhandled for food and shelter when he could not.

His "genius" unrecognized and expectations of riding his talent to easy wealth and fame thus unfulfilled, Hitler looked outside himself for the reasons for his failure. The blame, he decided, lay in Germany's weak parliamentary democracy, that, in his estimation, allowed Jews to control and therefore to own its economy and thus disenfranchise the rightful heirs of the fruits of that government, those of Germanic blood. Because so many Jews at the time supported Marxist ideals, he deduced the two parties were colluding on a grand scale to control the world. Communism joined Jewry and democracy as causes for his failings and, in his mind, the failings of the German race. Where his father taught him to hate, and disemboweled dreams magnified this malevolence, Hitler now had a focus upon which to aim his virulence.

Wallowing in his misery, penniless, often homeless and usually sick, his life was in an unpromising, spiraling descent when "The War To End All Wars," World War One, erupted to send much of the civilized world into the depths of hell – and to save Adolf Hitler.

On the crucible of the battlefield he found the vehicle to vent his rage – war. Serving in no less than forty-seven battles in a four-year span, Hitler was wounded twice, for which he spent several months in recuperation and earned the Iron Cross, both First and Second Classes.[27] Although never rising above the rank of corporal during the four war years in which he served, he showed an inkling of the boldness for which he would later become known when he captured an enemy officer and fifteen of his men.

During the war, a new vision began to form in Hitler's fevered head. The images that once he placed on canvas were now being replaced with a skewed vision of how the world should be ordered. Soon his brushes, pencils and painter's palette would be replaced by more formidable media – death and destruction: grenades and guns and tanks, with which he would paint a new and very real picture of what he thought the world should be. "Might is right!"

If there is an opposite to living the daring, Bohemian, but inspired existence of the artist, as had Adolf Hitler, it is living the structured, precise, but ample life of a bureaucrat. So it was with Martin Bormann.

Quitting school during the closing months of the war, Bormann joined the army and spent what few months remained of the already-lost conflict avoiding a useless death by serving as an officer's orderly. Here he learned not only how to evade placing himself in harm's way but, enamored with his proximity to important people – in fact, tutored by them – he began his lifelong avocation, which migrated into a vocation, of licking the boots of those higher than he in order to get ahead. The instinct was one that Hitler, who as a school boy had insisted all others follow him in the game of Follow the Leader, would later enthusiastically acknowledge was the most essential characteristic of his most valued and trusted lieutenant, Martin Bormann.

One should not consider Bormann's position that of weakness. The power that flowed through him from his master and protected him by virtue of his slavish alignment with his master's wishes was unequivocal and untouchable by all others save the source of that power. As long as Bormann remained unquestionably attentive, the powerful host would continue to feed the parasite. And the parasite would continue to feed on the throng that was drawn to his master while at the same time forcing that throng to do their master's bidding. Bormann's parasitic behavior was toward the throng, not his master; the relationship with the master was symbiotic, each benefiting from the behavior of the other.

Bormann's innate and infallible instincts for survival served him well after World War One. With the country in ruins, the economy in chaos and the populace impoverished and starving, Martin Bormann, revealing a latent predilection for always incisively cutting to the kernel of a problem, quickly divined that if lack of food was the problem he faced, going to the source of food was the solution. His instincts drove him not only to get work on a farm but also to achieve a position of control on the farm. Immediately upon being mustered out of the army he found work as an estate manager trainee in Meklemburg, North Germany.[28] He appears to have done well, for he recorded that less than two years later he had worked his way up to general manager of the von Treuenfels estates, which, combined, totaled almost 8,000 acres.

Some historians question Bormann's assertion he became general manager in two years based on the idea he could not have learned the entire farm business in 18 months, and the fact he was still a minor. But such a rise does not stretch the imagination given Bormann's later proven and remarkable skills of administration – and the tell-tale lapdog relationship he quickly cultivated with the lady of the estate, Ehrengard von Treuenfels, the Baroness von Maltzahn. Indeed, the friendship was maintained at least until his escape from bombed-out Berlin a quarter-century later and his disappearance into the back alleys of history. Martin even named a daughter Ehrengard after the Baroness. In any case, Bormann honed and further integrated the skills of administration and vassalage into a potent power base while serving the Treuenfels' at Mecklemburg.

The experience of the victors of the war placing the reckoning of accounts at the vanquished's door caught crosswise in both Hitler's and Bormann's throats. Consumed by hate and inspired by the power of carnage, Hitler took bitter umbrage to the mountainous war reparations despite the country's then non-existent economy and starving population, which the Allies demanded of the German people. In the act of demanding such onerous reparations alone did the Allies incite World War Two. For had the reparations been less burdensome it is doubtful Hitler would have had the fuel he needed to ignite with his private rancor the fires of vengeance in the German people that would propel the Nazi cause.

Bormann shared Hitler's convictions, although he probably had not actually heard of Hitler by then; but for this cause both Hitler and Bormann, during their early political activism, spent time in prison. Hitler for his part in the Munich Beer Hall Putsch of 1923 that would serve as a catalyst to bring the Nazi party to power (he wrote *Mein Kampf* while in prison for the crime), and Bormann for his part in the murder of a man who had betrayed the nationalist cause.

During his time as a land agent, Martin Bormann became involved in political activism. In 1923, the year of the Beer Hall Putsch, Bormann joined the Nazi's predecessor and early competitor, the *Freikorps Rossbach*, where he quickly rose to become one of the leaders of the Mecklemburg chapter's organization. While functioning in this position, Bormann was an accomplice in the murder of another member of the organization, Walther Kadow, a former elementary school teacher of his.[29] Kadow had been suspected of betraying a third party member, the soon-to-be Nazi martyr Albert Leo Schlageter, to the French during the occupation of the Ruhr. Bormann and others recruited a gang to execute Kadow – a mob that included Rudolf Franz Hoess, the future commandant of Auschwitz. Kadow was dragged into a forest and beaten with clubs and heavy branches before having his throat cut and being shot twice.[30]

There is no evidence Bormann had a hand in the actual killing; the mob under his direction performed the deed. But Bormann was tried and condemned to a year in prison for providing the weapons and leadership for the act. Years later, Adolf Hitler would award Bormann the Blutorden (Blood Order) for his part in the murder and the time he paid in prison because of it.[31]

Hitler, too, was implicated for murder when he was a young man, long before he made cold-blooded killing a component of official government policy. Hitler's suspected homicidal action, unlike Bormann's calculated, pragmatic act, was alleged to be the result of jealous and unthinking rage. Depending upon who one believes, he appears to have viciously murdered his niece – with whom he was having an incestuous, turbid relationship – following a violent, jealousy-driven argument. The niece, Angela

"Geli" Raubal, was trying to break off their relationship.[32] She disclosed he had forced her to urinate on him and to perform other heinous obscenities.[33] He also reportedly completed a number of erotic artistic renderings of Geli executed with questionable taste. Bormann, or a cohort under his direction, is said to have later located all of these pictures and quietly bought them back to avoid future controversy.

As Geli tried to extricate herself from the affair – she not only detested her relationship with Hitler but she was interested in another man – Hitler is thought to have confronted her in his apartment in Munich. Possibly she threatened to reveal his perverted predilections but it is not known for certain what led up to the killing or how it was committed. According to William Stevenson in *The Bormann Brotherhood*, there were witnesses to the crime – Gerhard Rossbach and Dr. Otto Strasser – but they were close Hitler cronies who refused to reveal what they knew. All that is known is Geli's dead body was found naked on the floor, her nose allegedly broken, killed by a bullet from Adolf Hitler's pistol.[34]

For Hitler, her death, whether murder or not, was a disaster about to be unleashed that would not only ruin his career but probably his life as well. While he had consolidated his position as leader of the Nazi Party, he was not yet a citizen of Germany much less its uncontested leader. Three more years would pass before he could protect his murderous madness with that shield. By now, September 1931, Bormann had been released from prison, joined the Nazi Party, and in six short years had burrowed his way into the party leadership and was looking for opportunities to demonstrate devotion to his demigod, Adolf Hitler. According to at least one chronicler, in the death of Geli Raubal Bormann recognized an opportunity to prove to his master his allegiance and his shrewd, if immoral, penchants.

Author William Stevenson described how Munich's intelligent, hard-working chief inspector, Heinrich Mueller, who up to that point had been working hard to eliminate the Nazi Party, had assigned one of his detectives to investigate the apparently open-and-shut case. Bormann stepped in. When he stepped back again, so the story goes, the chief inspector dropped the case, Hitler walked free, and Mueller was soon on a train to Moscow to learn from Stalin's goons the black art and septic science of running a secret police department, all at Nazi Party expense.

The net result of Bormann's arbitration? Adolf Hitler escaped that most desperate personal and political predicament to eventually become arguably the most powerful man in Europe. Heinrich Mueller was installed on a career track that would propel him to the pinnacle of the German police state – the police state of all police states – as chief of the vaunted and feared Gestapo. In fact, Mueller would eventually carry to his grave the nickname "Gestapo" Mueller.

And Martin Bormann would grasp Hitler's attention and allegiance in a way that would create a mechanism for perpetual expansion of Bormann's power base through the Master's increasing trust and appreciation. Add to this the power that would later flow to Bormann from his co-opting of Heinrich Mueller and the massive intelligence and control mechanism that would soon be supplied to him through the Gestapo, and Bormann's position had, indeed, increased by several orders of magnitude as a result of this single affair – if events did, indeed, happen as recounted.

According to many Hitler biographers, the story of Hitler's murder of Geli Raubal is anecdotal and has been proven to be false. Their account says Hitler had checked into a hotel two hours from Munich on the day Geli was killed and therefore could not have committed the crime. This, in fact, may be true. His signature is in the guest registration book and on a speeding ticket received while racing back to Munich after he was told about the death. But if Stevenson's version that Bormann and Mueller 'fixed' the outcome of the investigation is correct, this evidence may be part of the cover-up rather than an accurate account of events.

A full day passed between when Hitler left the scene of the murder where Geli was last seen alive, and when Geli's body was found; plenty of time for him to drive the two-hour trip to the hotel and register. And it is well known that Hitler was being chauffeured almost everywhere at this time in his life. Why would he have received and signed the speeding violation rather than his driver, unless to create an alibi?

And what of the reason officially given for Geli's death – suicide? Would a young woman really kill herself while completely nude? The case for Hitler being innocent of Geli Raubal's murder is far from conclusive. Perhaps whatever occurred will never be known. Whether he committed the crime or not, he was certainly under heavy suspicion and the investigation was discontinued in a mysterious manner very favorable to Hitler. Had Bormann worked his magic?

During the six years between Bormann's release from prison in 1925, when he joined the Nazi Party, and his alleged bold intercession on Hitler's behalf in Geli Raubal's death, Martin Bormann had already climbed a considerable distance within the Nazi party hierarchy. Presumably his stature was elevated upon his very entrance into the party as a result of his already-proven commitment to the ideals and operational methods of the Nazi Party as confirmed by time spent in prison for the Kadow murder. Within two years he was the regional press officer for the Nazi Party in Thuringia and the following year was elevated to chief business manager in the same regional party chapter, as well as being made Gauleiter (Nazi Party governor) of Thuringia.[35] He was also promoted to the supreme command of the party's military arm, the S.A. (Sturmabteilung).

By the end of that same year, 1928, Bormann was working for Hitler's personal secretary and right-hand man, Rudolf Hess.[36] Bormann had been referred to Hess by Nazi Party Treasurer Franz Xavier Schwartz,[37] who recognized in Bormann a shrewd and astute financial manager and efficient commissar who could bring the party's business dealings into control, which Hess had been unable to accomplish. Because of Bormann's penchant for working quietly in the background, throughout his career his versatile nature went largely unnoticed despite his latent genius for finance – magnified and unbridled by a complete lack of moral or ethical circumspection. His versatility revitalized party coffers. It made Hitler a rich man. And it made Bormann a rich man.

The following year, Bormann married the daughter of another ardent party member who would soon become the top judge in Nazi Germany, Reichstag Deputy Walther Buch, who enjoyed Hitler's respect (Hitler was a witness to the Bormann wedding, being friend of both bride and groom).

With his new wife Gerda, Bormann began a family that would eventually include ten children and would, if possible, in some respects be even more perverse than the family in which he grew up. He openly and with Gerda's blessing, and, in fact, with her assistance, carried on multiple sexual relationships simultaneously with a bevy of other women, despite universal agreement that Bormann, in the "looks" department, had little to offer the fairer sex. Physical attraction not withstanding, his oily charm and powerful position made him a desired coup to many ladies. Between these liaisons and his official duties he was seldom home, and when he was he ruled his wife and family with an iron fist. Yet he wrote Gerda lovingly almost every day, ensured she was always well taken care of, and, despite his otherwise secretive nature, he entrusted her in writing with his innermost thoughts and feelings on almost every subject. The Bormann's relationship is an enigmatic paradox that makes a fascinating study in and of itself of the man and the manner in which he operated.

Another endeavor that really made Hitler take notice was when Bormann negotiated with Wilhelm Ohnesorge, the Minister of Posts, a royalty to be paid to Hitler whenever the Fuehrer's likeness was used – as it was on stamps.[38] While the income per transaction was small (the cost of a stamp, after all, is minimal), the volume of transactions was huge. The resulting income from this clever contrivance alone made Hitler a wealthy man. More important, Bormann's negotiations with Ohnesorge appears to have opened a long relationship between the two men that culminated in an alliance that contributed to the political fortunes of both; and that was central to Bormann's later escape from Berlin and his post-war survival.

Hitler, who enjoyed his new-found wealth but disliked the details of accumulating it, and who in fact, for political purposes carefully promoted an image of austerity, quickly recognized and appreciated Bormann's astute perceptions; taciturn, confidential nature; and 'fiscal' talents. Bormann would go on to devise and execute a great many other schemes through the years, legal and otherwise, that lined the Fuehrer's pockets as well as his own.

Hitler soon appointed Bormann to be Hess's chief of staff. The appointment came, no doubt, not only as a reward for Bormann's assistance with the Geli Raubal incident and other past accomplishments, such as the Kadow murder, but because Bormann was also piling up a body of work that aided Hitler in a wide variety of other functions. In 1930, for example, recognizing party coffers were in dire straits, Bormann created the Hilfskasse, a compulsory 'accident insurance' fund for party members who were injured while brawling with communists.[39] All party members had to pay into the fund. This capital not only supported the wounded but also generated a substantial surplus that allowed the party to fulfill significant financial obligations and still provide funding for future operations.

Shortly after Hitler took office as chancellor, Bormann also founded the Adolf Hitler Endowment Fund of German Industry.[40] The endowment strong-armed companies that enjoyed success as a result of Hitler's economic policies into making contributions to his government. The funds were then hoarded in Hitler accounts managed by Bormann or dispersed according to Hitler's and Bormann's directions.

By the end of 1934, Hitler had been in power a year, Bormann was serving as his personal secretary and business manager, and considerable advances had been made in Bormann's efforts to weld himself to the man he could now, with the rest of the nation, call his Fuehrer. Bormann had become inseparable from the Fuehrer, following him night and day and writing nearly his every word on little white sheets of paper, to be acted upon immediately or to be treasured up for a future history he was certain would one day be chronicled in a tome that would glorify his Master.[41]

In 1935, leaning on his estate management experience, Martin Bormann initiated construction of and oversaw the management and building of the immense, now nearly mythical, multi-million reichsmark Bavarian complex at Berchtesgaden that Hitler would come to regard as his home and sanctuary from the demands and pressures of public office.

In May 1941, Bormann's position rose again when Rudolf Hess, Bormann's direct superior, in an act that stunned the world, secretly flew his personal Messerschmidt airplane to Scotland. Hess's self-appointed purpose – which he hoped would bring him back into the good graces of Hitler, with whom he felt a rift forming – ostensibly was to sue for peace and a united German/British front against Bolshevism. Hess was immediately rewarded with imprisonment in the United Kingdom. As a result, Bormann was given on a silver platter exactly what he was prepared to work – and conspire – hard for: the chancellorship of the Nazi Party.

Some have suggested Bormann may have been responsible for inspiring Hess's deranged attempt[42] – may have, in fact, suggested it to his superior with foreknowledge of the probable results – in order to remove Hess from blocking Bormann's path to greater power. Whether true or not, Bormann did ascend to the position of Nazi Party Chancellor by Hitler's command, which was added to his responsibilities of personal secretary and manager to the Fuehrer that he had already held before Hess's defection.

Hitler also discovered in 1941, through one of the greatest spy coups ever, that Roosevelt and Churchill had established a secret transatlantic telephone connection.[43] Charles Howard Ellis, possibly one of the Nazis most valued undercover agents as second-in-command to the remarkable Sir William Stephenson (who ran the combined intelligence efforts of Britain and the United States, reporting directly to Winston Churchill) had received information about the hotline and passed it to Heinrich 'Gestapo' Mueller, his Nazi controller. "Gestapo" Mueller was the same Heinrich Mueller who was chief inspector for the city of Munich with whom Martin Bormann had allegedly negotiated a resolution of the Geli Raubal murder case. Mueller was now, perhaps as a result of those negotiations and the path Bormann had put him on, the head of Germany's feared secret police, the Gestapo.

On hearing of the Roosevelt/Churchill hotline, Hitler quickly passed an order to Bormann to break into it and have the 'confidential' conversations decrypted, at whatever cost necessary. Bormann turned again to Wilhelm Ohnesorge, the postal minister.

The Ministry of Posts, of all unlikely places, maintained a research and development institute that worked on an eclectic assortment of scientific problems. The work was well funded from the regular postal service. When, several months later Ohnesorge's program successfully decrypted its first transatlantic conversation, Hitler was delighted, and, from then until the end of the war, he gleefully read the transcriptions of these conversations only hours after the words had been breathed from the mouths of his two great adversaries.

The research institute of the Ministry of Posts was not working on cryptology only. Great amounts of reichsmarks were being invested in atomic bomb development,[44] as well. Ohnesorge – who, as a doctor of physics and mathematics was on the Reich Research Presidential Council,[45] the organization that oversaw nuclear development for Hitler – was a great proponent of atomic bombs. At least twice Ohnesorge personally reported before Hitler the progress and merits of the German atomic bomb programs. That no one has asked why Ohnesorge made these reports rather than the traditional history's nuclear leaders, Heisenberg, Hahn, Weizsacker or many others who were all working on the official German bomb project, seems compelling and telling regarding the real truth of the Nazi bomb's development. Undoubtedly Bormann, as proxy head of the government in his Party Chancellor role, as Hitler's secretary and personal manager, and later as his secret overseer, as well as through his relationship with Ohnesorge, was central to these events. True to his shrewd nature, Bormann must have divined the worth of the nuclear efforts.

Hitler's admiration and dependence upon Bormann grew to immense proportions – noticed, but with little concern until too late, by the court elite. None of them appeared to see in the crude, bulbous, smarmy Martin Bormann the cunning and dangerous threat he represented to them.

The men Bormann considered his competition for Hitler's attention and as the Fuehrer's possible eventual successor, Goering, Goebbels, Himmler, Speer, and at one time even Hess, were men, like Hitler, who championed the grand design of Nazism in overblown speeches, sweeping dramatic demonstrations of their power, and open adulation of their Fuehrer, for which they enjoyed in return the adulation of the crowds over which they lorded. They echoed Hitler but, with the possible exception of Speer, added little to him and therefore they added little to their own potential as well.

Bormann was an altogether different animal. Instead of assuming the voice of Hitler, which after all was Hitler's greatest strength and needed little assistance, Bormann was Hitler's hands and feet, his eyes and ears. He did the details and dirty work Hitler detested, with an eye dedicated to the same purposes the Fuehrer espoused. Bormann did the Fuehrer's bidding, anticipating his wants and requirements without being told, and then fulfilling them with force and power without being directed, or restrained. Bormann served Hitler so loyally and completely that years later, when Bormann started to plant in Hitler's mind his own ideas and then act upon them as Hitler's orders, Hitler did not perceive the transition. As a result, Bormann to a large degree eventually became Hitler's heart and mind as well as his eyes, ears, hands and feet; controlling him and the empire he governed without the master ever suspecting control had slipped from his hands.

Bormann had positioned himself specifically for this task. Not only had he catered to Hitler slavishly to create an unbreakable bond of appreciation, trust, and dependence – it is important to note here that Bormann's allegiance to the Fuehrer was always genuine and total – but Bormann continually cultivated and expanded his resources to forever widen his web of control on behalf of himself and the Fuehrer.

According to biographer William Stevenson, Bormann's great talent was a genius for "what really mattered in a bureaucracy."[46] Stevenson goes on to explain how Bormann dredged police, military and political organizations to form alliances, either by force or by finesse, that he would later manipulate to fill his purposes. Add to this his great propensity for navigating in and, in fact, forming, molding and operating bureaucracies, and one sees a master who controlled all the strings that ran the party and the government. His mind "thrived upon this kind of nutrition,"[47] Stevenson wrote. "Where the Fuehrer's genius and aura failed to work, (Bormann) would step in and exert power,"[48] wrote Jochen von Lang in his biography of Bormann, *The Secretary*.

Bormann used the bureaucracies around him to consolidate his position and control the forces – pro and con – against and within which he had to operate. These bureaucracies were his source of all control through the currencies they commanded. Hard currencies such as the millions of reichsmarks stashed in his, Hitler's and the party's various accounts, funds and business operations; and soft currencies, like the personal intelligence collected on various leaders inside and outside the party and the country.

The constitutional government of Germany controlled the country's legal administration; in the early years of Hitler's chancellorship the party, on paper, held little power. Bormann, as primarily a functionary of the party, therefore, could only administer in party matters, not government policy. To circumvent this inconvenience Bormann created and constantly grew his "shadow bureaucracy"[49] over the ensuing years that duplicated each crucial government function and then allowed him to control the strings he desired to pull. The state police was shadowed by the Gestapo, with Bormann's alleged protégé Mueller at its head. The province chiefs and mayors were shadowed by Nazi Party Gauleiters (district governors), and their administrative regional structures, who vied with the province chiefs and mayors for control of their jurisdictions. Bormann would usually side with the Gauleiters, or convince Hitler to do so, thus empowering them over their counterparts and expanding the influence of Bormann, leaving Gauleiters and other party officials in his debt.[50] These party officials would eventually virtually run the country when Hitler later placed the Nazi Party in control.

Bormann also placed large numbers of key officials under his bondage through bald-faced bribery, providing discrete distributions of loans from party coffers to whoever he deemed would be a valuable leader to own.[51] "Almost all the top party functionaries received gifts from this fund,"[52] wrote Speer, who added that such gift giving, though innocuous, had the very real effect of conferring more power upon Bormann than almost any other person in the land.

Himmler approached Bormann for one such loan of 80,000 reichsmarks so he could buy a house near Berchtesgaden for his mistress and their illegitimate child.[53] Bormann not only produced the loan but he encouraged Gerda to befriend Himmler's mistress. The women would share cozy conversation and children's clothing in the years ahead, until Bormann severed Himmler's relationship with the Fuehrer in the waning days of the war. But the Bormanns' and Himmlers' "pseudo-friendship,"[54] and Bormann's ongoing contributions to Himmler's personal cache thereafter – totaling millions[55] – was a valuable protection for Bormann later when the real extent of his power became apparent among Hitler's coterie and envious courtiers tried to destroy him.

"Again and again I have come to terms with Bormann although it is my duty really to get him out,"[56] complained Himmler. Knowing Bormann had "the goods" on him, there was little Himmler could do to dethrone the Fuehrer's Iago. In fact, it is doubtful Himmler really wanted to topple Bormann, since 30 to 40 percent of his personal income would be lost if Bormann fell.[57]

Bormann, using Mueller and his Gestapo as well as other vehicles, had access to a comprehensive collection of files, reports and dossiers that provided a solid engine of power by blackmail[58] to drive Bormann's schemes.

The files included virtually every ranking member of the Nazi Party, including possibly Hitler himself, if Hitler's murder of Geli Raubal is true.

As Hitler pushed his foreign policy toward war with the rest of the world during the mid- to late-thirties, Bormann increasingly and on his own volition dominated domestic affairs. By the time the war actually broke out in 1939, the party was firmly in control of the government.

The official mantel of German domestic power now placed upon Bormann, as Nazi Party Chancellor, combined with the very real puissance he practiced through bureaucracy, blackmail and bribery, placed Bormann at the pinnacle of power. Only Goering, Goebbels and Himmler could hope to unseat him; and Himmler, as has been described, was in a poor position to do so.

Bormann did not stop. He continued to increase and fortify his position throughout the next year. In the winter of 1942, the others distracted by the war and Hitler increasingly relying on Bormann to manage administrative affairs while he pontificated military strategy, Bormann slapped his fellow courtiers with a most direct, jolting and revealing blow that for the first time unveiled him openly as a contender for the throne. In an alliance with General Keitel, Hitler's military second-in-charge, and Hans Heinrich Lammers, Chief of the Reich Chancery, in other words the government's chief legal minister, Bormann created the Committee of Three through which all business directed for Hitler must pass, effectively cordoning off Hitler from all others.[59] Hitler, appreciative as always that distracting details were being lifted from his busy schedule, supported the arrangement.

Barely half a year later, in July 1943, Bormann again redefined his role as Party Chancellor and paladin to the Fuehrer, again with Hitler's consent, by proclaiming himself the sole mediator between the government, the party and the Fuehrer,[60] thus eliminating even Keitel and Lammers from the picture. Bormann was now the sole link between Hitler and his chiefs. Speer noted with disgust how important issues and programs could not reach Hitler without first going through Bormann's hands and first having his blessing before even being considered by the Fuehrer.[61]

With Hitler insulated from opposing views on critical affairs, Bormann could now set and execute agendas, needing Hitler only to rubberstamp his plans. Speer asserted that, as an important military minister, he was not among those excluded from Hitler's presence, but in reality even Hitler's most favored associates were dealt the indignities of having to crawl to Bormann for access to the Fuehrer. Often an audience was denied and Bormann responded alone. For example, it was Bormann, not Hitler, who answered in the negative Speer's request of Hitler that he be awarded ultimate jurisdiction over the important V-1 and V-2 rocket projects and other research and development programs based at Peenemunde.[62]

Simon Wiesenthal wrote that many orders bearing Hitler's signature showed obvious evidence of being the product of Bormann's mind. And Goering stated flatly that many documents issued from Hitler bore the unmistakable stamp of Martin Bormann's heavy hand.[63]

Bormann now controlled Hitler and guided from him the decisions that were running the Reich. According to biographer Paul Manning, "Martin Bormann was now the leader in fact of Germany."[64] William Stevenson agreed that Bormann covertly governed the Third Reich, adding that historians have consistently misunderstood both "Bormann's role and his character." Bormann was not interested in the fame and glory the rest of Hitler's courtiers desired, according to Stevenson, he craved the real power.[65] Jochen von Lang, another of Bormann's biographers, asserted, "Bormann now considered himself the actual heir of the Third Reich,"[66] if not the one so stated in Hitler's will. Bormann at last was looking down from the top of the heap, and carefully watching his quibbling cohorts.

"Those who were Bormann's rivals and even enemies always underestimated his abilities,"[67] lamented one of those enemies, Walter Schellenberg. "They spoke about (Bormann), calling him a bootlicker and often a pig," described Hitler Youth Leader Baldur von Schirach, continuing, "If cartoonists had drawn his picture, his shape, bulk, short legs, mug – it actually would have turned out to be a pig."[68] Schellenberg, too, likened him to a wild pig digging for potatoes.[69] Most of Hitler's retinue simply called him "Hitler's evil spirit,"[70] or "The Brown Eminence"[71] – behind his back, of course.

By now Bormann was a general in the SS commanding a 1 million man army; he controlled vast sums of money that he used freely for his own legal and illegal purposes; he had at his fingertips enough information to pull down any party or government leader in the Reich; and he held in his hands the strings that controlled Adolf Hitler. At the end of the war, nobody in Nazi Germany held more power than Martin Bormann.

Endnotes: Chapter Ten – A Pig Digging For A Potato

1 Hugh Thomas, *Dopplegangers*, pp. 260-268

2 Trevor Roper-Smith, *The Letters of Martin Bormann*, p. IX

3 Jochen von Lang, *The Secretary*, p. 285

4 William Stevenson, *The Bormann Brotherhood*, p. 39

5 William Stevenson, *The Bormann Brotherhood*, p. 28

6   U.S. National Archives II, War Crimes Records, *Interrogation Summary #1739, of General Karl Wolff, Nuremberg, p. 2, 8 April 1947*, RG 238 – M1019 Roll 80; also *Interrogation Summary #2797, of General Karl Wolff, Nuremberg, p. 2, 25 June 1947*, RG 238 – M1019 Roll 80

7   William Stevenson, *The Bormann Brotherhood*, p. 28; Jochen von Lang, *The Secretary*, p. 109

8   U.S. National Archives II, War Crimes Records, *Interrogation Summary of General Karl Wolff interrogation, Nuremberg, 5 September 1945*, pp. 1, 2, RG 238 1270 Roll 22

9   Jochen von Lang, *The Secretary*, pp. 87, 107

10   Jochen von Lang, *The Secretary*, p. 271

11   Jochen von Lang, *The Secretary*, p. 86

12   William Stevenson, *The Bormann Brotherhood*, p. 285

13   Paul Manning, *Martin Bormann: Nazi In Exile*, pp. 38, 46

14   Jochen von Lang, *The Secretary*, p. 297

15   Jochen von Lang, *The Secretary*, p. 176

16   Albert Speer, *Inside The Third Reich*, pp. 300, 301

17   Jochen von Lang, *The Secretary*, p. 23

18   William Stevenson, *The Bormann Brotherhood*, p. 23

19   Jochen von Lang, *The Secretary*, p. 108

20   William Stevenson, *The Bormann Brotherhood*, p. 65

21   Paul Manning, *Martin Bormann: Nazi In Exile*, pp. 29, 46; Jochen von Lang, *The Secretary*, pp. 108, 109, 328; H.R. Trevor-Roper, *The Letters of Martin Bormann*, p. ix; William Stevenson, *The Bormann Brotherhood*, p. 18

22   Dr. Louis L. Snyder, *Encyclopedia of the Third Reich*, p. 36; Jochen von Lang, *The Secretary*, p. 16

23   Jochen von Lang, *The Secretary*, pp. 16-18

24   Dr. Louis L. Snyder, *Encyclopedia of the Third Reich*, p. 151

25   Jochen von Lang, *The Secretary*, p. 19

26   Jochen von Lang, *The Secretary*, p. 20

27   Dr. Louis L. Snyder, *Encyclopedia of the Third Reich*, p. 152

28   H.R. Trevor-Roper, *The Letters of Martin Bormann*, p. ix, x; Jochen von Lang, *The Secretary*, p.22

29   H.R. Trevor-Roper, *The Letters of Martin Bormann*, p. x; Dr. Louis L. Snyder, *Encyclopedia of the Third Reich*, p. 36

30   Alan Levy, *The Wiesenthal Files*, pp. 212, 213

31   Alan Levy, *The Wiesenthal Files*, p. 319

32   William Stevenson, *The Bormann Brotherhood*, p. 30

33   Ron Rosenbaum, *Explaining Hitler*, pp. 129, 134

34   William Stevenson, *The Bormann Brotherhood*, p. 31

35   Dr. Louis L. Snyder, *Encyclopedia of the Third Reich*, p. 36

36   Alan Levy, *The Wiesenthal Files*, p. 319; Dr. Louis L. Snyder, *Encyclopedia of the Third Reich*, p. 36

37   Ladislas Farago, *Aftermath*, p.218, 219

38   Albert Speer, *Inside The Third Reich*, pp. 103, 104; Jochen von Lang, *The Secretary*, p. 51

39   Dr. Louis L. Snyder, *Encyclopedia of the Third Reich*, p. 36; Albert Speer, *Inside The Third Reich*, p.103

40   Albert Speer, *Inside The Third Reich*, p.104

41   William Stevenson, *The Bormann Brotherhood*, p. 39; Albert Speer, *Inside The Third Reich*, pp. 104, 114

42   William Stevenson, *The Bormann Brotherhood*, pp. 50-55

43   Paul Manning, *Martin Bormann: Nazi In Exile*, pp. 76, 77; David Irving, *The German Atomic Bomb*, p. 150

44   David Irving, *The German Atomic Bomb*, pp. 77, 78; Albert Speer, *Inside The Third Reich*, p. 271

45   David Irving, *The German Atomic Bomb*, p. 256

46   William Stevenson, *The Bormann Brotherhood*, p. 49

47   William Stevenson, *The Bormann Brotherhood*, p. 45

48   Jochen von Lang, *The Secretary*, p. 297

49   Jochen von Lang, *The Secretary*, p. 107

50   Jochen von Lang, *The Secretary*, p. 109

51   William Stevenson, *The Bormann Brotherhood*, p. 28

52   Albert Speer, *Inside The Third Reich*, p. 104

53   Jochen von Lang, *The Secretary*, p. 273

54   Jochen von Lang, *The Secretary*, p. 256

55   U.S. National Archives II, War Crimes Records, *Interrogation Summary #1739, of General Karl Wolff, Nuremberg, p. 2, 8 April 1947*, RG 238 – M1019 Roll 80; also Interrogation Summary #2797, *of General Karl Wolff interrogation, Nuremberg, p. 2, 25 June 1947*, RG 238 – M1019 Roll 80

56   William Stevenson, *The Bormann Brotherhood*, p. 50

57   U.S. National Archives II, War Crimes Records, *Interrogation Summary #1739, of General Karl Wolff, Nuremberg, p. 2, 8 April 1947*, RG 238 – M1019 Roll 80; also *Interrogation Summary #2797, of General Karl Wolff, Nuremberg, p. 2, 25 June 1947*, RG 238 – M1019 Roll 80

58   William Stevenson, *The Bormann Brotherhood*, p. 45

59   Albert Speer, *Inside The Third Reich*, p. 301

60   Jochen von Lang, *The Secretary*, p. 238

61   Albert Speer, *Inside The Third Reich*, p. 301

62   Paul Manning, *Martin Bormann: Nazi In Exile*, p. 64

63   Simon Wiesenthal, *The Murderers Among Us*, p. 319

64   Paul Manning, *Martin Bormann: Nazi In Exile*, p. 29

65   William Stevenson, *The Bormann Brotherhood*, p. 18

66   Jochen von Lang, *The Secretary*, p. 38, 39

67   Paul Manning, *Martin Bormann: Nazi In Exile*, p. 39

68   Jochen von Lang, *The Secretary*, p. 253

69   Jochen von Lang, *The Secretary*, p. 285

70   Paul Manning, *Martin Bormann: Nazi In Exile*, p. 39

71   H.R. Trevor-Roper, *The Letters of Martin Bormann*, p. ix

Chapter Eleven

Operation Fireland

Bury your treasure, for you will need it to begin a Fourth Reich.[1]
Adolf Hitler to Martin Bormann in 1943

*When the story of Martin Bormann is written it will reveal him to be
the man largely responsible for West Germany's postwar recovery... "*[2]
Paul Manning, *New York Times*, March 3, 1973

The turning point against Germany during World War Two was not the loss of the Battle of Britain or the mounting of D-Day on Normandy's shores. While the air battle over London was an important German defeat that allowed Britain to fight on – alone at the time – other than as a moral victory, taking the islands of the United Kingdom would have had little strategic value to Germany before the United States joined the conflict. And by the time Allied soldiers stormed the beaches of northern France, the tide of war had already turned against the Nazi horde. D-Day, while imperative and impressive, was actually the beginning of massive mop-up operations.

During the autumn and winter of 1942, Germany suffered the most pivotal defeat of the war at the Battle of Stalingrad. From that day on, the outcome of the war was almost fixed. And almost everybody knew it. Until the moment when Hitler looked up from the strategic objective he was pursuing in the Soviet Union, the oilfields and refineries of Ukraine to fuel his war machine, Germany was winning the war. But the Fuehrer could not resist the moral victory that taking "Stalin's City," now so close, would be. Planning a quick campaign that would take mere weeks, he swung his Sixth Army from its course southward toward the oilfields and refineries, and instead turned them to the northeast and attacked. The bold move was at first successful and Stalingrad was captured. But in the frozen winter of 1942-43, a four million-man Russian army surrounded the 330,000-man force of Field Marshal Friederich von Paulus.

The Soviets laid siege. They starved the Germans. They ran them out of ammunition. They ran them over on the rock-hard frozen snow under the treads of their heavy tanks, the Wehrmacht infantry unable to dig foxholes into the steely ice to avoid being crushed. By the time Paulus surrendered, SS forces had barely been able to break through and rescue only 5,000 survivors. The rest were force-marched to Siberia and most never heard from again. After the moral loss at Stalingrad and the loss of oil to feed the hungry Nazi war machine, ultimate surrender for Germany was just a matter of time, barring an unforeseen miracle.

Martin Bormann, true to his proven, pragmatic ways, was uniquely prepared to deal with the former eventuality, and possibly capable of providing the latter. Through his old friend at the Reichspost, Wilhelm Ohnesorge, he was supporting a program that could furnish the miracle needed – Manfred von Ardenne's uranium enrichment program. The program just required enough time to be successful.

On the other hand, if time should run out, the last thing Martin Bormann would allow his Fatherland to endure was another rapacious war reparations assessment like that forced upon it after World War One. The Allies could kill the people, plunder the land, rape the women, and level the cities, but in his shrewdly insightful way, Bormann knew that they could not own Germany itself if they did not own Germany's wealth. In the spring of 1943, Bormann began to look for ways to conserve the Reich's riches if the war was lost.

He started with *Aktion Feuerland*, 'Operation Fireland.' As German forces had overrun country after country, storm troopers followed behind advance waves and plundered each nation's valuables[3] while the Gestapo gathered its Jews into ghettos and concentration camps and relieved them of every gram of valuable property they owned; including the gold and platinum in their teeth. The treasure consisted of hundreds of millions of reichsmarks; boxes and boxes of gold and platinum, pearls and diamonds; crates full of the priceless art of Europe; and billionaire bundles of stocks and other securities.[4] The loot was amassed in a series of bank safes and underground vaults throughout the Reich – until Martin Bormann was made aware of its existence by one of his many internal intelligence conduits. In late 1943, he took control of much, though not all, of this booty and informed Hitler of its existence and a plan he had formulated for its conservation.

"Bury your treasure, for you will need it to begin a Fourth Reich," Hitler had responded. With this blessing, Bormann took control of as many as six U-boats,[5] some of them unmarked, from Gross Admiral Karl Doenitz, and garnered the support of Generalisimo Francisco Franco to headquarter the U-boats in the Spanish port cities of Cadiz and Vigo. The U-boats for the next two years, supplied by cargo planes from Germany that transported the treasures to the coastal towns on the Atlantic, began a non-stop circuit transporting the treasure to the far southern reaches of Argentina – the region known as Tierra del Fuego, or Land of Fire. At their destinations they were unloaded by Bormann's mysterious minions and deposited into a variety of international bank accounts controlled by a cryptic cabal of Bormann partners. This was *Operation Fireland*.

But Bormann was not satisfied just to rob the SS of the treasure trove it had stolen from murdered Jews, plundered citizens and overrun countries. Earlier in 1943, he had recognized for himself the value of masterpieces hung in museums and those owned by Catholic and other churches and held in cathedrals, monasteries and convents throughout the Reich. He initiated a program to collect all that could be gathered and even ordered high-ranking members of the party who had already assimilated such artwork into their own collections to turn them over to the Party Chancellery.[6] From this time to the end of the war, one-third of Italy's great art treasures, and much of the rest of Europe's masterworks collections, were lost to the Nazis; a fair share of it going into Bormann's South American hideaway.

Bormann appears to have *laundered* some questionable treasures of his own through *Operation Fireland*, as well. For example, in 1942 Bormann started heading a Nazi project designed to weaken the British war economy while providing currency to pay for German armaments production. The British currency-counterfeiting program overseen by Bormann was printing 400,000 notes a month, which eventually totaled $600 million.[7] Bormann deposited the money into foreign banks through his mysterious partners. Later, he exchanged the funds into a more stable currency, often dollars, and then, instead of using the funds for the munitions for which they were intended, he would often hold them in one of his "ghost" accounts for his own future use. Of the $600 million of counterfeit currency processed, approximately $300 million has never been accounted for, presumably lost to Bormann's enigmatic interchange.

Bormann also generated huge sums of money through a vehicle that he had already utilized at least twice before to the benefit of the Fuehrer and the party – the creation of a fund designed to finance a specific task and to which all able Germans were compelled to contribute. In this case, the

"Winterfund" was established ostensibly for the welfare of the soldiers and civilians impoverished by the war.[8] Besides mandatory donations, the fund was also supported by wealthy industrialists who were wined and dined at concerts they were expected to attend, all the while being coerced into contributing huge amounts of money, sometimes as much as 100,000 reichsmarks in a single donation.[9] Eventually the fund accumulated over 3 billion reichsmarks but little of it was used for the support of the needy. Presumably, at least part, if not a great percentage, of these funds may have been included in *Operation Fireland*.

Estimates of the value of *Operation Fireland* range from the unbelievably low $17 million, considering the sheer volume of non-stop transport voyages of the six U-boats over two years, and subsequent value of the treasures, into the more probable hundreds of millions and possibly even billions of dollars. But *Operation Fireland* was small change compared to the blockbuster business venture Bormann would soon unveil.

As the Thousand Year Reich began to crumble barely a decade after its inception, memories of Germany's World War One failure were still fresh in Martin Bormann's mind. A devastated citizenry impoverished by the war had been saddled with yet even more hefty burdens than what the country had already lost in the conflict. From the scant assets that had survived, the Germans were forced to pay the costs of the losses of the victors, as well; to replace their burnt out cities and towns, the sunken ships and shot down airplanes, their industries and lost revenues. Because the conquered had so few resources left that there were insufficient assets with which to make recompense, their futures were mortgaged – a whole generation was indignantly indentured to its mortal enemy of yesterday. While the fighting had ended, the war smoldered on in the angry hearts of the vanquished, to erupt again two decades later in World War Two. Now the pattern was repeating itself. But the bitter gall of the last defeat was not going to be repeated in this one. Not while Martin Bormann had a hand in the outcome.

Reichminister Hermann Goering was responsible for the Reich's economic Four-year Plan and, as a result, the economic heads of all the occupied countries (and surreptitiously, many of the neutral nations, also) reported to him. These countries included France, Belgium, Holland, Czechoslovakia, Denmark, Norway, Yugoslavia, Austria, Poland, Spain, Sweden, Switzerland, Turkey, Portugal, Finland, Bulgaria, and Romania – virtually all of Europe except Russia – and also included many Latin American countries.

What is little known, however, according to *Martin Bormann: Nazi In Exile* author Paul Manning, is that Martin Bormann was the Party Minister of Economics[10] and therefore he oversaw all economic issues for the entire Reich, even outranking Hitler's then-chosen heir, Goering, in financial matters. In this role, on the heels of the Stalingrad defeat, Bormann had already begun to plan for the economic protection and resurgence of Germany following the war. *Wall Street Journal* reporter Greg Steinmetz writes of how top Nazis prepared for German post-war emergence by calling together a meeting of the leaders of many of Germany's top companies in August 1944. The meeting, held in a hotel in Strasbourg, France, was convened expressly "to discuss financing plans for the Fourth Reich,"[11] according to Steinmetz.

Steinmetz's article also included information about *Operation Fireland*. By the time the Steinmetz article ran in April 1997, however, it was very old news. Decades before, Bormann biographers Paul Manning, William Stevenson and Ladislas Farago had already written in detail about Nazi exporting of plundered treasure and the secret economic summit in Strasbourg. What *was* new was the fact no rebuff of Steinmetz or the *Journal* appears to have followed for revealing the information. In the past, accounts printed about *Operation Fireland* and the Strasbourg Conference had been squashed, quickly debunked or ridiculed into historical oblivion. For example, when this author initially proposed *Critical Mass* to a publisher using only *Operation Fireland* documentation cited by Farago in his book *Aftermath*, I was told Farago had forged and planted the documents within the top secret files of foreign governments in order to support his "fictitious" claims. Apparently there had been quite an international row in the publishing world over this deception, which occurred when I was too young to have taken notice. At any rate, Farago and his book had been publicly and acrimoniously denounced and Farago died unvindicated a few years later.

The publisher's initial assertions convinced me of the correctness of the dismissal of Farago's claims, thus stopping me from pursuing this book further – at least for a time. I later came across Paul Manning's treatise of the despoiled Nazi loot and the Strasbourg meeting in his book, *Martin Bormann: Nazi In Exile* and again in William Stevenson's book, *The Bormann Brotherhood*. The same events that Farago had revealed in his book were proven in these accounts, as well as some very important new information, but in many cases using different documentation.

I contacted a member of the intelligence community with whom I had connections and whom I was told had researched the subject matter of these Nazi business dealings. Without mentioning Manning or Stevenson by name, he asserted that what they had written about Nazi involvement in post-war international business preparations was true and that United States government intelligence agencies – he mentioned the CIA and its predecessor the OSS by name – had conducted a full inquiry into the issue. He asserted these agencies had identified all of the relevant business dealings, had broken up the German cartels and stripped the Nazi owners of their financial properties and placed those instruments in the hands of the United States Alien Property Custodian program. He "shared" this information with me in the spirit of proving that, while certain German businessmen and high-ranking Nazis – he mentioned Bormann specifically – tried to survive the war supported by Nazi funds invested by clandestine means, the United States had found and uprooted the deception. Therefore, he insisted, there was no story and no need for me to research further.

But if what Paul Manning and William Stevenson had written about Nazi international business activity was true, as he said it was, then the same assertions that Ladislas Farago had earlier written about it are likewise true, as is other very essential information about who they all agreed initiated the Strasbourg Conference. The effort to vilify Farago, therefore, was probably a smokescreen. With the knowledge my original premise was intact and there was now an effort being put forth to fog the truth, I put forth, more carefully, once again on this book.

The fact the Nazi scheme had supposedly been put down was of no account to me; the confirmation that the Strasbourg plan was made and initially carried out is the cogent point for this volume. In later research I discovered, however, that the story about the financial properties being expropriated once and for all by the United States government, while true in form, was not true in reality. It was yet another effort to create a fog behind which the truth could be hidden.

In any case, the fact Steinmetz was allowed to run his article unchecked was an important event that begins to blow the haze away from the central truth of these events. Perhaps the reason the article ran unassailed was the irreproachable reputation for integrity of the *Wall Street Journal* and its sheer stature in the world of journalism. Perhaps the article was allowed to run because it was triggered by a United States Senate investigation initiated in response to Nazi victims who are now United States citizens trying to retrieve personal property originally looted from them by the Nazis. Probably both reasons are true to some degree.

But likely the most important reason the Steinmetz article was allowed to run uncontested was that it still hid the issues at heart – issues some elements of the United States government want protected. What are these issues? The first is that Martin Bormann was the central player in the Strasbourg Conference. The second is that Bormann escaped Germany at the end of the war and lived for many years rebuilding and controlling the economy of West Germany and much of Europe and Latin America, and that he did this all with the protection, support and collusion of the United States government.

While Steinmetz's article does not say so, Manning's and Stevenson's stories both have a central point in common regarding the Strasbourg Conference; and Farago's work, illustrated by other events, although not detailed in the specifics of the conference itself, supports the point: Martin Bormann initiated the conference, controlled it and oversaw its results for many years following the war. Bormann's yet unborn Fourth Reich, by war's end, had already ratholed $800 million plus 95 *tons* of gold.[12] And that was just by war's end.

The Strasbourg Conference where Bormann introduced a new economic initiative was convened under the highest secrecy and security in August 1944. The purpose was to discuss post-war preparations between the Nazi government and the major German industrialists,[13] strengthening a partnership that had grown increasingly stronger since the end of World War One and particularly strong since the Nazi Party had gained control of the government. Bormann assigned Lieutenant General Sheid to conduct the conference in his behalf,[14] along while Dr. Bosse, of the Ministry of Armaments.

"German industry must realize that the war cannot now be won," Bormann told Sheid, continuing, "and (Germany) must take steps to prepare for a postwar commercial campaign which will in time insure the economic resurgence of Germany."[15]

What Bormann then proposed was devious, conspiratorial and illegal within Nazi Germany. To avoid security breaches, therefore, he ensured in every possible way that the strictest secrecy was maintained. The meeting was held in a hotel conference room insulated from visual or audio surveillance by having rented all of the rooms above, below and on all sides of the chamber. All attendees and their personal possessions were thoroughly inspected physically and electronically by SS technicians.[16]

High-ranking industrialists from a spectrum of German firms listened intently to the amazing proposal: All companies that agreed with Bormann's plan to conserve their businesses for post-war operations and to share post-war revenues with selected underground Nazi operations would, until such time as the Third Reich failed, be protected by Bormann from the "Treason Against the Nation" law.[17] This law required death for all those who subverted currency regulations, traded in foreign currency or concealed ownership of foreign currency. The law also precluded firms from being involved in almost any type of partnership, joint venture or licensing agreement with any country outside of the Reich or the boundaries of its allies.

In reality, many of Germany's largest companies were already engaged in relationships with businesses neutral to or hostile to the Reich, including Germany's largest conglomerate, I.G. Farben. But the government had been turning a blind eye in order to keep the huge amount of capital rolling in that these companies generated. The waiver of the Treason Against the Nation law proposed at Strasbourg was therefore not only an incentive to those German companies that desired to survive the war but were not yet participating in such activities, but it was a veiled threat to those that were already circumventing the law. To them Bormann was saying, in essence, if you do not share the wealth you are already gaining, we will have your heads by enforcing the law. The Strasbourg announcement, for these companies, amounted to a form of blackmail; which they were glad to pay not only to save themselves but to save their companies from the post-war commercial blood bath that was sure to come.

According to Dr. Bosse, participating companies' funds were to be invested in foreign financial institutions while the Party maintained access to them, "in order that a strong German empire can be created after defeat."[18] Bosse went on to explain: "Industrialists with government assistance will export as much of their capital as possible, capital meaning money, bonds, patents, scientists and administrators."[19] While hard currency was valuable, the currency with real potential was the 'soft capital' the industrialist firms held: the trade agreements, patents and braintrusts that generated colossal revenues in perpetuity. The potential income of such intellectual and proprietary properties as international licenses sold to use the patents on stainless steel, synthetic fuels and rubbers and other commercial advances, and control of the braintrusts who created them, was huge, generating millions, possible tens or hundreds of millions of dollars per year. Many international companies, such as Bayer, Winthrop Chemical, AGFA-ANSCO, Hoescht and DuPont to a large degree owed their existence and continuing prosperity to exclusive use of I.G. Farben patents and licenses alone.[20]

In addition to exporting technologies, the German firms were directed to borrow against these and other assets to obtain more hard capital and thus be able to more quickly export additional hard currency[21] into what was now being called Bormann's Flight Capital program. Technical and business bureaus were to be established for each industry and in each foreign office of each company, with a covert Nazi liaison officer in each office to oversee and, where possible, personally manage the operations.[22] From among these liaisons German economic specialists successfully penetrated eleven nations' economies, in addition to Germany's, and eventually controlled them in the post-war period.[23]

Bosse reported to Bormann after the meeting that the terms of the Strasbourg conference had been agreed to by all involved and therefore the new Flight Capital Program had been successfully initiated.[24] Bormann in turn established 750 camouflaged corporations under the names of companies or individuals for which he held power of attorney, and therefore over which he had total control,[25] as vehicles for managing the income of the Flight Capital Program. These businesses were scattered across countries throughout Europe, the Mid-east and Latin America. Holdings were even kept in bank accounts in the United States of America,[26] some of which eventually were in Martin Bormann's own name, including accounts with Manufacturers Hanover Trust, The Chase Manhattan Bank, and First National City Bank, according to author Paul Manning.

Although not listed as a company represented at the Strasbourg conference, Germany's largest industrial cartel, the chemical concern I.G. Farben, was active in the Flight Capital Program as well. In fact, it had not been necessary for Farben representatives to attend the program's introduction at all because its leader, chairman, and president, Hermann Schmitz, had been integral to creating the Flight Capital Program. I.G. Farben had supported the Nazi cause from the beginning of its climb to power, having donated generously through Farben's intelligence, propaganda and political economic operations, known as I.G. NW7.[27] In his *Wall Street Journal* article, Steinmetz unwittingly hints at this involvement – and particularly at the Flight Capital Program – in a portion of the article that reviews reports that Germany's Bosch A.G. company during the war allied with the wealthy Wallenberg family of Sweden to camouflage German funds. Robert Bosch, the founder of Bosch A.G., was the uncle of Carl Bosch,[28] the founder of I.G. Farben. Close relationships were maintained between the companies.

Before taking Carl Bosch's place at the head of I.G. Farben, Schmitz had been Bosch's top lieutenant and handpicked successor.[29] He had overseen all of I.G. Farben's international business, and, between the wars, was responsible for concealing Farben's huge global income from German tax administrators through the use of foreign "blinds" he had created. These camouflage devices operated remarkably like the alleged arrangement between Bosch A.G. and the Wallenberg's.[30]

Before the war, Schmitz took over the helm of I.G. Farben and had become a close "confidant and advisor to Martin Bormann,"[31] writes Paul Manning in his book *Martin Bormann: Nazi In Exile*. Manning added that Bormann was a student in a sort of personal, and confidential, tutelage under Schmitz.[32] Bormann, in fact, surreptitiously gave the title of "Secret Councilor to the Nazi Party and Martin Bormann," to Hermann Schmitz,[33] in return for the latter's intellectual contributions and mentoring. Under Schmitz's direction and with the complicity of Bormann, I.G. Farben looted the chemical properties of the nations Germany had conquered: Austria, Czechoslovakia, Poland, Norway and France.[34] By the end of the war, Farben had interests in over 700 companies, not including operations within its own corporate structure that stretched across 93 countries.[35]

In all, Schmitz, in league with Bormann, who cleared the path of government constraints, expanded Farben's foreign investment to at least 7 billion reichsmarks during the war.[36] As the two men wove their web they made many pacts; among them one that ensured all Farben leaders overseas were Nazi Party members accountable to Martin Bormann – a precursor to the Flight Capital Program. Working together, the two men expanded this relationship to other German firms in the form of the Flight Capital Program.

The objective of the Flight Capital Program was not to make money in and of itself. The objective – Bormann's master plan – was to save and protect Germany's industries and economy from being looted at the hands of the conquerors as had happened at the end of the First World War. After the war, the Flight Capital Program would control and direct not only the German economy, but also other economies linked to the underground Fatherland, in an effort to produce a quick German rebirth and eventual European economic domination by Germany. Bormann and Schmitz met on multiple occasions while developing the Flight Capital Program.[37] So thoroughly did Bormann capture all of the funds transferred out of the Reich that when Hermann Schmitz died in 1960, at the age of 79, he was nearly a pauper.[38] "To this day no one has been able to explain what happened to his fortune. Few who knew him can believe it doesn't exist," wrote Joseph Borkin, author of *The Crime and Punishment of I.G. Farben*.

Strategies for covertly redeploying the economy included the implementation of a "foreign trade offensive," according to Peter Hayes' book *Industry and Technology: I.G. Farben in the Nazi Era*.[39] They also included a "'European economic community'" that positioned Germany as the hub and "flag bearer" of a confederated Europe that would "predominate by 'elastic political methods'... not with brutal force." These elements are certainly recognizable in the history of post-war Europe as it actually unfolded, and, in fact, Germany continues with a high profile in the European economic model of today – which is named the "European Economic Community." The evidence reflects that the Flight Capital Program and Bormann's partnership with I.G. Farben not only paid off as planned, but it set the foundation for the Europe we know today, and by extension all of the world.

But in April 1945, with Berlin succumbing to the Russian siege, a hysterical Hitler visibly crumbling in front of him, and the Reich reeling in its death throes, Bormann, true to his brutish, realistic, pragmatic nature and leaning heavily on his incomparable bureaucratic proclivities, was focused on escaping. Bormann was willing, able and self-authorized to negotiate any agreement that secured his – and presumably, at one time, the Fuehrer's – escape. Signals from "The Brown Eminence's" radios bounced to and from various German generals authorized to negotiate with Russian and United States military leaders. The Allies, in complete control and determined to achieve nothing but total and unconditional surrender – outwardly at least – would not negotiate. Escape was the only option.

Endnotes: Chapter Eleven – Operation Fireland

1   Paul Manning, *Martin Bormann: Nazi In Exile*, p. 29

2   Paul Manning, *New York Times*, March 3, 1973, p. 31, column 2

3   Ladislas Farago, *Aftermath*, pp. 201-203; Paul Manning, *Martin Bormann: Nazi In Exile*, p. 207, 208; Alan Levy, *The Wiesenthal File*, p. 222

4   Greg Steinmetz, *The Wall Street Journal*, 28 April, 1997; Paul Manning, *Martin Bormann: Nazi In Exile*, p. 207; Ladislas Farago, *Aftermath*, pp. 201-203; Alan Levy, *The Wiesenthal File*, p. 222

5   Greg Steinmetz, *The Wall Street Journal*, 28 April, 1997; Harry Cooper, *Sharkhunters KTB 104*, p. 8; Paul Manning, *Martin Bormann: Nazi In Exile*, p. 207; Ladislas Farago, *Aftermath*, p. 202

6   Jochen von Lang, *The Secretary*, pp. 172, 173, 183

7   William Stevenson, *The Bormann Brotherhood*, pp. 150-152

8   Ladislas Farago, *Aftermath*, pp. 220. 221

9   Paul Manning, *Martin Bormann: Nazi In Exile*, p. 44

10   Paul Manning, *Martin Bormann: Nazi In Exile*, p. 114

11   Greg Steinmetz, *The Wall Street Journal*, 28 April, 1997

12   William Stevenson, *The Bormann Brotherhood*, p. 66

13   Paul Manning, *Martin Bormann: Nazi In Exile*, p. 23; William Stevenson, *The Bormann Brotherhood*, p. 67

14   Paul Manning, *Martin Bormann: Nazi In Exile*, p. 23

15   Paul Manning, *Martin Bormann: Nazi In Exile*, p. 24

16   Paul Manning, *Martin Bormann: Nazi In Exile*, p. 24

17   Paul Manning, *Martin Bormann: Nazi In Exile*, p. 25

18   Paul Manning, *Martin Bormann: Nazi In Exile*, p. 26

19   Paul Manning, *Martin Bormann: Nazi In Exile*, p. 27

20   Paul Manning, *Martin Bormann: Nazi In Exile*, p. 117

21   Paul Manning, *Martin Bormann: Nazi In Exile*, p. 25

22   Paul Manning, *Martin Bormann: Nazi In Exile*, p. 26, 27

23   Paul Manning, *Martin Bormann: Nazi In Exile*, p. 114

24   Paul Manning, *Martin Bormann: Nazi In Exile*, p. 27

25   Paul Manning, *Martin Bormann: Nazi In Exile*, p. 11; William Stevenson, *The Bormann Brotherhood*, p. 68

26   Paul Manning, *Martin Bormann: Nazi In Exile*, pp. 139, 205

27   Raymond G. Stokes, *Divide and Prosper*, p. 24

28   Joseph Borkin, *The Crime and Punishment of I.G. Farben*, p. 56

29   Joseph Borkin, *The Crime and Punishment of I.G. Farben*, p. 165

30   Joseph Borkin, *The Crime and Punishment of I.G. Farben*, p. 180

31   Paul Manning, *Martin Bormann: Nazi In Exile*, caption, second photo section

32   Paul Manning, *Martin Bormann: Nazi In Exile*, p. 114

33   Paul Manning, *Martin Bormann: Nazi In Exile*, p. 114

34   Joseph Borkin, *The Crime and Punishment of I.G. Farben*, p. 2

35   Paul Manning, *Martin Bormann: Nazi In Exile*, p. 153

36   Paul Manning, *Martin Bormann: Nazi In Exile*, p. 28

37   Paul Manning, *Martin Bormann: Nazi In Exile*, pp. 157-162; Peter Hayes, *Industry and Technology: I.G. Farben in the Nazi Era*, p. 368

38   Joseph Borkin, *The Crime and Punishment of I.G. Farben*, p. 166

39   Peter Hayes, *Industry and Ideology: I.G. Farben in the Nazi Era*, p. 368

Chapter Twelve

The Pig Finds a Potato

Irrefutable proof exists that a small plane left the Tiergarten at dawn on April 30, flying in the direction of Hamburg. Three men and a woman are known to have been on board. It has also been established that a large submarine left Hamburg before the arrival of the British forces. Mysterious persons were on board the submarine... [1]
– From a Soviet intelligence commission of inquiry report, as quoted by James Mc-Govern, CIA agent in charge of researching the post-war survival of Martin Bormann

Stalin told Harry Hopkins in Moscow that he believed Bormann escaped. Now he went further and said it was Bormann who got away in the fleeing U-boat. More than that Stalin refused to disclose. [2]
– William Stevenson, author *The Bormann Brotherhood*

For over seventy years a debate has raged about whether Martin Bormann escaped from Berlin in the spring of 1945, whether he was killed in a fiery explosion on Weidendammer Bridge in that city, or whether he mysteriously died a few hours later at the Lehrter Station Bridge a few miles away. Over that nearly-three-quarters-of-a-century, so many accounts of his last days in Berlin have been generated, fabricated, amended, modified, denied, rebutted, investigated, expunged, reborn, re-shaped and abridged that nothing is certain but a black mist of confusion and suspicion that hangs over the whole affair like a thick pall. Indeed, the truth may never be known. Not just because the evidence supporting any outcome is inconclusive, but because there seems to be few participants who were or are objective on the matter, and therefore the testimony and evidence they provide must, of prudence, be viewed with varying degrees of skepticism. What is known, despite the bleak picture that is always painted of these events, is that 90 percent of those who were in the bunker at the end of the war survived. [3]

The only "eyewitnesses" to Bormann's death did not actually verify either that they were certain they saw him die, or that they were sure they saw him in death. All eyewitnesses were avowed Nazis headquartered in the bunker and therefore may have had vested and agreed upon interests in the world thinking Bormann was dead. And therefore, the argument goes, they may have provided misinformation in evidence of his death. Additional "proofs" of Bormann's demise beyond the eyewitness accounts did not surface until decades later. The veracity of their provenance has been effectively argued pro and con since.

Those who argue for his death, most notably the German government and, in a more innocuous manner, certain United States agencies, almost invariably have important interests of their own to protect. Many of those who say he survived seem to have their reasons for maintaining his ongoing existence, as well, sometimes based on only the flimsiest evidence to support their claims, but often with substantially more confirmation. Frankly, this book falls in the latter category.

The evidence, in fact, is significant in support of both theories and, despite claims of certainty by both camps, a detailed study of all the data available tends to muddy the already shadowy history beyond ever finding certain resolution. Indeed, there is strong evidence the waters were muddied intentionally by those who merely had to make his fate questionable, in order to win their objective of invalidating any reference to Martin Bormann in post-war history. But by filtering all of the available information through two criteria, one may possibly gain, if not a crystal clear understanding of the outcome of events, at least the most probable outcome of Bormann's last days in Berlin.

One of these criteria is to look at disassociated stories surrounding these events and see what parallels might verify each other to validate details. The more numerous and specific the evidentiary pieces paralleled, the more probable they are true – assuming they are not totally invalidated by known facts.

The other criteria is that of judiciously weighing the evidence against who presented and/or supports it – and why – in an effort to identify and properly interpret political and other influences that may have colored the information presented. By combining these two methods of analyzing the information, a more coherent and believable – in fact, this author believes, probable, though disturbing – case for Bormann's survival is formed, rather than the accepted scenario of his death. Let it be noted here, however, that this author does not believe this evidence is conclusive. I believe only that the evidence is significant and superior for Bormann's survival over evidence of his demise in Berlin. And that evidence shows U-234 was related to his escape and, therefore, additional research should be completed.

The official version of Bormann's last days ends with his death at the Lehrter Station Bridge. Or possibly he died not far away at Weidendammer Bridge a few kilometers north of the Reichs Chancellery building, under which Adolf Hitler's bunker was hidden. The "eyewitness" accounts disagree.

According to reports later provided by occupants of the bunker, in the late hours of 1 May 1945, the small gaggle of survivors still burrowed in the Fuehrer Bunker after Hitler's suicide separated into a few small groups and, at intervals, sneaked out of the ground and into the frightful night. Artillery and tank shells were falling indiscriminately around them. A few hundred meters away, the sounds of gunfire could be heard as firefights occurred in the darkness, splashing the acrid, smoky air with bursts of red and streaks of light. Each group was responsible to find its own way to safety.

In one of these pathetic patrols reportedly stalked the potbellied, short-legged, bull-necked profile of Martin Bormann, commander of the Nazi party and Hitler's closest confidant. According to the provided scenario, the small group slowly picked its way through the bombshells, bodies and debris littering the streets to a local subway station, where, once again, it slipped under cover of earth. Walking the rails in the dark subway tunnels, the silent group made its way north, where it again surfaced to find means to cross the Spree River.

At Weidendammer Bridge the group ran into heavy fighting between German tanks and Russian forces. One story asserts that Bormann tried to cross the bridge under cover of a German tank navigating the narrow span. The tank was shelled by a bazooka and exploded in a violent burst of flame, killing Bormann[4] according to "eyewitness" Erich Kempka, Hitler's chauffer and a member of the Fuehrer Bunker escape party. Kempka admitted during his Nuremberg testimony at Bormann's *in absentia* trial, that he did not approach the body to confirm Bormann had been killed but was certain from the extent of the violent blast and the manner in which Bormann's body was seen "flying away,"[5] that the Reichsleiter was dead. At least four others of Hitler's trusted insiders reported seeing virtually the same event, but again, none had inspected the body or could declare with certainty he was dead, though all were convinced of it.[6]

Not to worry, a sixth eyewitness later claimed to have observed the events at Weidendammer Bridge also, and to be able to verify Bormann was killed by the tank blast. Except this witness, the Spaniard Juan Roca-Pinar, who, as an avowed Nazi was fighting near the bridge as part of a small SS unit defending the bunker, later reported that Bormann was not at the side of the tank *but riding inside* the tank when it was hit by the bazooka shell.[7] Roca-Pinar reported that he was ordered to board the tank and save Bormann, but when he opened the hatch to rescue survivors, he found Bormann dead from the blast, though easily identifiable. He reported that he pulled Bormann's corpse from the tank before being forced to abandon it in the street under pressure of enemy fire.

Harry Mengerhausen, a member of Hitler's bodyguard, told a story that agreed with Roca-Pinar – Bormann had been inside a tank. But he declared firmly that Bormann was not killed in the blast because he was not in the tank hit, but in an entirely different tank.[8]

The conflicting stories, while containing significant discrepancies, at least agreed, with the exception of Mengerhausen, that Bormann died during a tank explosion on Weidendammer Bridge. But other accounts soon spun these similar scenarios on their heads. Artur Axmann, the one-armed leader of the Hitler Youth, claimed to have run into Bormann *after* the Weidendammer Bridge catastrophe and asserted that Bormann was alive, well and completely unharmed.[9] In fact, the two men, in company of others, tried for some time to escape together before later separating to search for their own passages to freedom. Axmann headed west, but, finding the way blocked, subsequently retraced his steps and claims to have again come across Bormann and Dr. Stumpfegger, one of Hitler's physicians, on a railroad trestle at the Lehrter Fairgrounds train station. Bormann and Stumpfegger were lying side by side on the bridge and appeared to be dead. Axmann leaned close to Bormann's body to check for breathing and could discern none. He later would not swear with certainty, however, that the Reichsleiter had expired.

Indeed their "deaths," if they were dead, were strange. According to Axmann, neither corpse showed any indication of being wounded or injured or showed any signs of violence – quite out of line with the reports from Weidendammer Bridge, even if Bormann had survived the tank blast, and further mystifying given their deaths having taken place during a heavy battle.

They lay calmly next to each other in peaceful repose, their arms resting casually at their sides, as if they had quietly laid down on their own, or somebody had laid them there. Axmann wondered if they had been poisoned or poisoned themselves, but could think of no reason why they should do so, except perhaps that they had lost hope of escape and preferred not to be captured. He left the bodies where he found them and eventually escaped to the Tyrol to command a small band of Hitler Youth determined to keep fighting after the war. American forces captured him there.

And so the semi-official version of Bormann's demise is dubiously documented in a melee of misaligned explanations and seemingly unexplainable inconsistencies. The picture would get further obscured. A rash of post-war Bormann sightings across Europe began to be reported. He was in Sweden, Italy, Spain, Denmark, Germany, Switzerland, Norway, even as far away as Argentina. Many sightings were explained away as misidentifications. Others went unexplained. Stalin was sure he was alive and accused the United States of hiding him.[10]

The evidence for his death was so uncertain that a year after his reported demise, the Nuremberg court convened by the Allies to bring war criminals to justice, tried and convicted Bormann *in absentia*, thinking from the evidence it was probable Bormann had survived the war. With so many sightings and so many unanswered questions, people – and government agencies – began the quest to answer the controversy over Bormann's fate. Or in some instances, it seems the goal was to ensure the question was unanswerable.

Articles and books flooded the media arguing that Bormann died – and arguing that Bormann lived. Searches began for evidence that proved either case. The sightings continued, some by those who knew Bormann and were very convincing, but almost no hard evidence was found, though much was claimed. New theories and additions to the existing stories began to appear, and then even to be reversed; such as that asserted by Simon Wiesenthal.

After firmly assuring the world for many years that Bormann had survived – and strongly hinting that he knew where the fugitive resided[11] – Wiesenthal abruptly reversed himself and asserted that Bormann had committed suicide that night in Berlin when he realized escape was not possible.[12] Historian Hugh Trevor-Roper, considered by many to be the leading expert on Bormann's fate, reported he was dead, then alive, then dead, then alive again. What caused these sweeping reversals is hard to keep track of, but they illustrate the high state of confusion and uncertainty surrounding Bormann's fate.

Journalist Paul Manning, for his part, reported that Bormann was alive, thanks to the help of Gestapo Chief Heinrich Mueller, who had searched for, and found in the Sachsenhausen concentration camp, a man who could serve as a "double" for Bormann. While the claim of a double for Bormann initially seems far-fetched, one must understand it was Mueller who found, arranged and prepared the well-known double for Hitler, presumably under Bormann's orders since Bormann held ultimate responsibility for Hitler's security. The Fuehrer's bodyguard, and even his pilots, were in reporting structures that ended at Bormann. Mueller's assistance with Bormann's double may be attributed to the fact that he not only may have owed his position to Bormann, but Hitler had ordered Bormann to serve as a go-between for Mueller and his direct superior Heinrich Himmler,[13] whom Mueller hated. According to Manning, in the months prior to Mueller's and Bormann's anticipated escapes – both men felt Germany's surrender was a probability that ought to be prepared for – Mueller ordered that the double be coached to behave like Bormann and that his dental work be redone to match that of the Reichsleiter's.[14]

Manning explained that he was provided the initial information about Bormann's double from a highly placed British intelligence source. He received confirmation of the incredible story from one of General Reinhard Gehlen's top aides.[15] Manning subsequently treated the account as accurate and never questioned the story. Indeed, the suggestion that a double actually "stood in" for Martin Bormann during the last known day of his life resolves many anomalies about Bormann's widely noted "un-Bormann-like" behavior during the events of 1 May 1945, as well as other historical inconsistencies.

General Gehlen was Hitler's chief intelligence officer for Eastern Europe before the German surrender, at which time Gehlen became the Central European expert for the OSS and then the CIA, and eventually head of the secret service in the Federal Republic of Germany.[16] He was, nonetheless, still financed by American money and thus provided America with East-bloc intelligence.[17] According to Manning, the Bormann post-war story was at one point even further convoluted when Gehlen was forced by the CIA to write in his memoirs that Bormann was a Soviet spy who had died in Russia in 1969. It was one of the agency's many efforts to obfuscate the facts around Bormann's fate, to make any clear exploration impossible. Gehlen later retracted the claim.

More than two decades passed before the first physical evidence suggesting Bormann's fate surfaced. At that time, a report was uncovered that was written shortly after Bormann disappeared. The report declared the Russians had found Bormann's and Dr. Stumpfegger's bodies where Axmann had said they were, on the Lehrter Station Bridge, and the Russians had the corpses buried a few meters away in the Lehrter fairgrounds just days after the city's surrender.[18] They identified the body from a journal of Bormann's that was found in the pocket of the dead man's overcoat.[19]

On this point accounts parted, however, for according to author and forensics expert Hugh Thomas, as noted in his book *Dopplegangers*, the Soviets had determined neither the body, nor the overcoat on the body, were Bormann's.[20] The Russians quickly surmised that the diary had been placed in the overcoat pocket to lead investigators off of Bormann's trail. If so, the report of the dead man's body being Bormann's must, from the outset, be regarded with skepticism. Bolstering this position is a report that Frau Stumpfegger, the giant doctor's wife, received a letter on 14 August 1945 from the postmaster at Lehrter Station, who had been responsible under Soviet orders for burying the many bodies in the area. The letter stated her husband's remains had been found and buried, and conveyed the sad news and appropriate condolences.[21] Thomas then writes that no such letter was ever sent to Bormann's family, automatically suggesting that Bormann's body had not, at that time, been found – at least not lying next to Stumpfegger's corpse. To put an exclamation point on this, the 14 February 1964 edition of the German magazine, *Der Spiegel* reported on their investigation of events, concluding, "In actual fact…, the corpse of Stumpfegger was found. But of Bormann, who was said to be lying alongside, there was no trace."[22]

In the mid-1960s, the German State of Hesse asked that the area where the bodies were reportedly buried be dug up in search of clues, but when excessive digging came up empty, the quest was abandoned.[23] Then in December 1972, just as two separate series of articles by Ladislas Farago and Paul Manning began being published that convincingly argued Bormann had escaped Berlin,[24] a construction crew "accidentally," and under suspicious conditions, unearthed two skulls and some bones twenty yards from the location previously dug up by the official Bormann search party.[25] The discovery was suspicious because there was a report of a former Hitler Youth member offering a $100,000 deutschmark bounty to any workmen who found the relics.[26]

The skulls were examined to see if one was Bormann's but there was a problem: no records of Bormann's dentistry or any other identification marks existed that could be compared against the skeletal remains. The only record available was a sketch drawn from memory by Bormann's by-then-deceased dentist, Dr. Hugo Blaschke, who drew the sketches during interrogations for the Nuremberg trials.[27] The accuracy of the chart was attested to by Fritz Echtmann, a dental technician who had never actually seen Bormann's teeth, but who had built a dental bridge for a patient he "assumed" was Bormann, based on data given by Dr. Blaschke. Using this data, the pathologists in the case compared the sketch with the unearthed skulls and proclaimed a match with one.

The riddle of the fate of Martin Bormann had been solved: Martin Bormann had died on Lehrter Bridge in Berlin on 2 May 1945, as Artur Axmann had asserted. All of the stories regarding his survival, therefore, were false. The version was made semi-official with a press conference, although it was not certified or recognized at that time by a court of law, therefore disallowing any recovery of Bormann's assets by his family,[28] who, in any case, refused to believe their father was dead despite the vast fortune his legal expiration would mean to them.[29] A great many journalists thereafter reported the search was over and the whole world could breath easier knowing Hitler's closest confidant was dead and gone. One of the great unanswered mysteries of World War Two now was solved.

Except the skull may not have been Bormann's. Or if it was Bormann's, it was not the man's who had been buried next to Dr. Stumpfegger in May 1945. In 1953, almost 20 years before the skull was found and eight years after it was supposed to have been buried, CIA agent James McGovern was working an investigation in Berlin with the assignment of verifying for his agency what had happened to Bormann. He later wrote that in discussions on the matter with the KGB, the CIA had learned a body supposedly Bormann's had been identified by means of the diary found in the pocket of the corpse's overcoat.

Within days of its burial at Lehrter Station, Moscow had ordered the body be disinterred.[30] The corpse was dug up and removed – presumably to conduct forensic testing to see if it was, indeed, Martin Bormann's remains. The remnants were subsequently reburied elsewhere in East Germany, without any report of whose remains they were, thus explaining why the Bormann family received no letter of his demise – apparently it was not him – and why *Der Spiegel* concluded Bormann was not found next to the dead degenerate doctor.

If the Soviet report that Bormann's body was buried somewhere outside of Berlin is true, whether it was Bormann's remains or his double's, it seems highly probable Bormann's remains would not have been at Lehrter Station when the skull was dug up by workmen. Therefore, one would assume, the skull found there could not have been Bormann's and the identification of the skull as his, was, at best, a serious mistake of inefficiency and sloppiness, and at worst, a fraud.

Indeed, Ladislas Farago documents the skull actually went through four iterations,[31] each succeeding cranium becoming more and more aligned with the dental sketch of Dr. Blaschke as succeeding complaints came in about obvious inconsistencies. In fact, writes Farago, Professor Reider F. Sognnaes, a specialist in oral biology and anatomy who had positively identified Hitler's burnt corpse from its dental records, while initially vouching for the identification of the skull as Bormann's being correct, was later so uncertain of the positive identification that he wrote a letter of concern to West German Chancellor Willy Brandt. Sognnaes later stated, according to Farago "that he did not believe that the skull found … was the skull of Bormann."[32] Manning confirms this evidence regarding the skull, writing that one of General Gehlen's aides – one of three independent sources refuting the claims about the skull – confided to him that "the skull is a fraud."[33]

Forensics expert and author Hugh Thomas agreed the skull was a fraud, but disagreed that it was not Bormann's. After carefully studying the forensic dental evidence available and the evidence surrounding the provenance of the skull, Dr. Thomas concluded the skull may be Martin Bormann's – provided for the construction diggers to unearth after his real death.[34] According to Dr. Thomas's forensic analysis and accompanying research, Bormann's death probably occurred in Paraguay in 1959.

Dr. Thomas's clarifications are compelling in that only one of two conclusions can explain the forensic evidence – both forms of fraud. The dental work of the skull, while clearly matching in many aspects that of Bormann's, contained small but often unexplainable anomalies, and additional, in some cases major, dentition not performed on Bormann prior to his disappearance in May 1945. One explanation for the discrepancies would be that an existing person with dental characteristics of his own was fraudulently fitted with additional dental elements in an effort to fit Bormann's dental profile, per Paul Manning's description of Bormann's double. It would be nearly impossible to find a man of Bormann's age to serve as his double who had not already received dental work of his own.

The second explanation would be Bormann actually did survive his Berlin adventure and subsequently had necessary additional dental work done to keep his teeth healthy in the intervening years between his escape and final demise. In either case, it is easy to see why the eminent Dr. Sognnaes, not imagining the latter explanation, could have at first confirmed the identification, and upon further consideration had reservations and reversed his findings.

In the latest development regarding the skull, DNA tests were begun in May 1997 and a positive identification of Bormann was announced in May 1998, though no specific results have been made public.[35] Such a finding would be the final word on the matter if the provenance of the skull was impeccable and the disposition of those who controlled the relic was beyond question neutral and they were protective of it. But, as has been shown, the confusion regarding the skull's whereabouts for almost two decades, the inconsistencies between the dubious dental records and the new-found skull itself, the 50-odd years that had transpired since Bormann's disappearance at age 45 – meaning Bormann almost certainly had died by 1997 and easily could have died in 1959, as reported, and his handlers may have supplied his skull in 1972 or submitted samples from his actual skull for DNA testing thereafter – and the fact he was not even in the grave to begin with, if the Soviet report is true, all combine to cast more than reasonable doubt, not so much upon the authenticity of the DNA tests, which, as demonstrated, could have gone either way, but upon whether they have any relevance in the matter at all.

One last possibility is worth mentioning, however, regarding DNA testing of the skull. If the Soviet report is wrong and the skull tested for DNA was actually that of the person buried in Lehrter Station with Dr. Stumpfegger, it may well have been that of Bormann's double. Mueller had successfully found a double for Hitler in one of the Fuehrer's distant cousins. Might he have done the same for Bormann when developing Bormann's double? James O'Donnell, author of *The Bunker*, noticed on a personal visit to Bormann's hometown that a large percentage of the people there looked like Bormann, and were possible relations.[36]

If the body was that of a Bormann relative, DNA tests quite possibly would have shown similar markers with the DNA provided by the source relative, without the skull actually being that of Martin Bormann. Without full disclosure of the DNA tests, and without ensuring all appropriate testing was completed to isolate Martin Bormann as the individual in question, as opposed to the remains simply being those of someone related to Bormann, the DNA testing proves, once again, less than meaningful.

So what really happened to Martin Bormann?

Among the many reports detailing Bormann's escape, although it was never given weight in the West, was an accusation Joseph Stalin made stating Soviet intelligence had reported Bormann was flown out of Berlin in a small airplane on the dawn of 30 April.[37]

> Irrefutable proof exists that a small plane left the Tiergarten at dawn on April 30, flying in the direction of Hamburg. Three men and a woman are known to have been on board. It has also been established that a large submarine left Hamburg before the arrival of the British forces. Mysterious persons were on board the submarine....[38]

According to author William Stevenson, Stalin then "told Harry Hopkins in Moscow that he believed Bormann escaped. Now he went further and said it was Bormann who got away in the fleeing U-boat. More than that Stalin refused to disclose."[39]

Stalin continuously reiterated this belief, claiming Bormann was being harbored by the United States government in his escape and continued freedom. The Allies, led by the United States, refused to give this story credence and ignored Stalin's demands for an explanation, and, in fact, began claiming in defense that the Soviets held Bormann – a claim they supported with Gehlen's extorted account of Bormann living in Russia. But Stalin insisted until his death that his was the correct account of Martin Bormann's fate.

Why would Stalin make such a claim? What did he stand to gain if it was a lie? What did he stand to loose if it was true? And if it was true, why would the United States discount it out of hand? These seem to be the obvious questions concerning the matter, and will be answered later. But more important for proving the case, though much less glaring, are the small questions – the questions about the innocuous details that make up the fabric of Stalin's very specific, though unlikely, story.

If Stalin was not telling the truth, why did he include such unique, improbable and seemingly contestable details as the fact the airplane carried four people when the only two airplanes capable of using the ad hoc runway – the Fieseler Storch and the Arado – were designed, respectively, to carry three and two people. Depending on which witness tells the story, both the Fieseler Storch and the Arado are said to have been the escape airplane.

Why did Stalin include a woman in the escape party when it was almost inconceivable a woman would be so important as to justify overloading the plane on this desperate and dangerous mission? And why would Stalin assert the escape was continued from Hamburg on a "large"

U-boat? The chances seemed slim that such an escape as Stalin described was ever made.

A series of totally independent facts and accounts, however, corroborate Stalin's very specific story. First, a makeshift runway is now well known to have been operating on the Tiergarten to service the Fuehrer Bunker during the last days of the war,[40] although at the time of Stalin's comment that knowledge was not widespread. Albert Speer, Hitler's Munitions Minister, described flying into the stop-gap landing strip on the occasion of Hitler's fifty-sixth birthday – celebrated a week before Bormann's mysterious escape – when the Russians were still at the outskirts of Berlin.[41] According to Speer, as an airplane prepared to land or take off, a detachment of SS soldiers would light a series of lanterns placed along both sides of the wide avenue that stretched from the Brandenburg Gate to the Reichs Chancellery. The airplane would use the strip and then the lanterns quickly would be extinguished again.

Second, the great German aviatrix, Hanna Reitsch, a contemporary of Amelia Earhart's and close friend of Adolf Hitler, had flown into Berlin only a few days previous to the mysterious escape flight.[42] Reitsch had in the past received personally from Hitler, who adored her, the Iron Cross (the only woman to do so) both first and second class,[43] for bravely test piloting the flying capabilities of a V-1 rocket that had been modified with a cockpit. She was also the first person to successfully fly a helicopter inside a building.

Now, she had accompanied her lover Luftwaffe General Robert Ritter von Greim to Berlin so Hitler could make him the overall commander of the Luftwaffe in place of the recently dethroned Goering. During the flight into Berlin, Greim, who was piloting the Fieseler Storch, was injured by enemy anti-aircraft shrapnel. According to her own account, Reitsch reached over the General and landed the Fieseler. After landing, Reitsch and Greim were harbored in the bunker for a few days while Greim lay in bed recuperating before attempting the exit flight.

Reitsch reported in a 5 December 1945 press interview[44] that she, with a heavily bandaged General von Greim, flew out of Berlin from the Tiergarten at dawn on 30 April, exactly the same time Stalin reported Bormann's mysterious escape flight took off. Later she recorded in her memoirs an odd justification for this event. Instead of flying straight to Austria, their ultimate destination, Reitsch wrote how they flew 400 dangerous miles, partly over enemy territory, with the badly injured and very important General von Greim, to Ploen, Admiral Doenitz's headquarters.[45] She gave the reason for this detour, depending upon which account one reads, as fulfilling an order from the Fuehrer to deliver an important package, or else the desire to wish the Admiral a fond farewell.

Such a detour for such a superfluous reason as bidding farewell seems remarkably improbable given the desperate state of affairs on the military front and the injuries to General von Greim. Would not a radio message have done? What if all the remaining German leaders decided to travel to each other in order to wish one another farewell? There seems to be no indication that Doenitz and Greim had any special relationship beyond two professionals doing their jobs. So this reason for the detour seems highly suspect.

On the other hand, the suggestion that Reitsch and Greim were fulfilling Hitler's orders by delivering an "important package" to Doenitz seems very salient – especially if the package was Martin Bormann with Hitler's last will and testament.

To be sure, other reasons were later given for the strange flight deviation, but, despite their seeming veracity, when subjected to even minimal scrutiny they seem almost as hollow as Reitsch's "farewell" flight. The chief of these later assertions is that Greim was flown to Ploen after Hitler had concluded Himmler was a traitor who had begun separate surrender negotiations with the West. Supposedly Greim was sent to arrest Himmler.[46] But the Fuehrer Bunker was in radio contact with Doenitz many times a day and could have had Doenitz make the arrest at any time. The wounded Greim, with his one-woman retinue, was in far less able condition to arrest Himmler than the healthy Doenitz with his considerable military cortege. Doenitz was a strict and efficient military professional with a strong reputation for carrying out his command.

If Doenitz was not willing to fulfill the order, however, to send the injured Greim to enforce the order over Doenitz's head and in his own headquarters, surrounded by the Admiral's full retinue and in the face of Himmler's substantial SS bodyguard, seems laughable. And if they had, in fact, flown to Doenitz for this purpose, why would not Reitsch have stated so in her memoirs, written many years later? The order for Himmler's arrest was never a secret – not even at the time it was issued, much less decades later when she wrote her book. And in the end, when Greim met Himmler, he only told the Reichsfuehrer SS that Hitler had denounced him,[47] further suggesting Greim was not really sent to Ploen to arrest the SS chief. In short, on the surface, there seems to be no viable official reason why Reitsch and Greim had flown to Ploen.

There is a reason for the huge flight deviation, however, if they were not alone. The traditional history documents well Bormann's intense efforts to make his way to Admiral Doenitz during this time.[48] Bormann had told his family they would be escaping on a U-boat to Japan;[49] and some of Bormann's closest associates, including Gauleiter Erich Koch and others, expected to escape by U-boat as well, with Bormann's help.[50] So strong was Bormann's effort to reach Doenitz that by 3:30 the morning of 30 April, Bormann had convinced Hitler to issue an order to his pilot, Hans Baur, to fly Bormann to Doenitz.[51]

With Bormann and Baur both aware General von Greim and Hanna Reitsch were preparing to fly out of Berlin within hours after Hitler gave the order, it seems highly probable they would take advantage of this opportunity. In fact, it seems especially so considering Reitsch did, in fact, fly to Doenitz's headquarters, although there seems to be no viable reason for her to have gone there – as noted above – and there were many reasons for her not to go to Doenitz. Was Bormann the "important package" intended for the Grand Admiral?

Possible validation of this phantom flight is provided in another flight supposedly made from the Tiergarten, which was reported to have occurred late on the night of 29 April 1945. The provenance of this account is suspect, but if it is true, it certainly adds to the argument Bormann and Gestapo Chief Heinrich Mueller may have escaped together by airplane.

In 1996, author Gregory Douglas published the first of three volumes titled *The 1949 Interrogation of Gestapo Chief Heinrich Mueller*. The books are claimed to have been written from Mueller's own records as provided to Douglas by the Mueller family. Douglas has done a considerable job of proving the information in the documents is true, even if the documents themselves may be suspect. Details of such a limited nature that few people would know them are included in the book, and they have been reviewed by Robert Wolf, who worked for many years as an archivist specializing in World War Two for the United States National Archives and Records Administration.

In a telephone interview with this author, Mr. Wolf, though obviously finding his comments personally disheartening, if not distasteful, admitted that all of the details he could find independent, objective reference to that were claimed in Douglas's account proved to be true. The records are purported to be a post-war interrogation of Heinrich Mueller by the OSS, forerunner of the CIA, when the agency was considering hiring Mueller and the substantial spy apparatus that the former Gestapo Chief operated throughout the Soviet Union and elsewhere.

During the alleged interrogation, Mueller described his escape from Berlin on the night of 29 April, just before or around the same time Reitsch claims to have escaped, in a Fieseler Storch airplane,[52] the same type of aircraft Reitsch claims to have flown. In his account, Mueller is flown out alone, with a male pilot the only other person in the airplane. Instead of flying to Ploen, Mueller contended the Fieseler Storch was flown south to the Austrian/Swiss border, the approximate location Hanna Reitsch described as her and General von Greim's final destination following their detour to Doenitz's headquarters.

There are obvious inconsistencies in this tale compared to Reitsch's – besides the fact that it may not even be true. But the discrepancies may be easily explained if the base assertion is true. First, if Bormann and Mueller did escape secretly in company of Reitsch and Greim, it would seem that as joint conspirators and protectors of the post-war Nazi cause they all would have agreed to eliminate each other's escapes when reciting their own cover stories. For this reason, Reitsch would not have identified Bormann or Mueller as having been on the flight she piloted to Doenitz. Under the same agreement, Mueller would not have identified any of his flying mates either; therefore he reported he flew alone. As regards the connotation of time differences of the flights, it is logical to assume that he considered the pre-dawn hours of 30 April as part of the "late night" of 29 April, a very common conclusion many people would make.

Second, Mueller would not have wanted to reveal the flight to Doenitz was tied to U-234 – considering the inferences of collusion that could later be made from that connection. At the time of his OSS interview, no one knew as well as Mueller how compartmentalized governments, and intelligence agencies in particular, are when it comes to maintaining state secrets. He could not assume the OSS officer interrogating him was aware of a possible Bormann/U-234 deal with the United States; and he would have known that that information was available within the agency on a "need to know" basis only. Since there was no need for his interrogators to know of the U-234 deal – in fact, there were good personal and United States national security reasons for them not to know of it – Mueller simply excluded any reference to it from the interrogation. Those in the know within the OSS/CIA would have expected him to do so, and he knew it.

Third, Mueller was not with Reitsch and Greim when they landed in the Tyrol, but claims, instead, to have landed in a vacant area and to have crossed the frontier from there on foot. Again, this portion of the cover story served his, and Reitsch's and Greim's, purposes of keeping their stories separate.

One last set of observations may be made concerning Mueller's reputed flight. While it was possible to fly in and out of the Tiergarten, it was not an easy thing to do and it was quite dangerous – as illustrated by General von Greim's injuries a few days earlier. Such flights were even more risky during the last two days of April when the Reichs Chancellery was almost entirely surrounded by Soviet forces. The more flights that flew in and out of the makeshift airstrip, the more likely they would draw attention of Russian air or artillery reconnaissance. The Russians would then be more able to locate the airstrip itself, and, as a result, identify the general area of Hitler's bunker.

Given these considerations, it seems unlikely several flights per night were permitted. To suggest two, and possibly three, flights may have lifted off from the airstrip on the night of 29-30 April – Mueller's flight, Reitsch's flight, and the one Stalin claimed – while Hitler was still inside the bunker, seems risky, and therefore improbable, if not out of the question. The chances of multiple flights being allowed to depart seem even less likely when considering Reitsch's and Mueller's flights were supposedly both destined for approximately the same location on the Austrio-Swiss border; and likewise, Reitsch's and Bormann's flights both had a supposed destination of Ploen.

Surely Hitler's headquarters staff would have consolidated flights when possible, rather than let several take off and increase the risk of exposure. The trouble involved in lighting, dousing and relighting the airstrip multiple times and in keeping available a number of the rare airplanes capable of using the short landing strip also would have discouraged such activity.

Even without the questionable Mueller account, using only Reitsch's memoirs, Stalin's story of the mysterious flight appears to have a strong basis in fact. Bormann and Mueller are known to have spent considerable time, energy and resources together prior to and in the bunker making escape plans.[53] If Bormann and Mueller were in the Fieseler Storch with Hanna Reitsch and General von Greim, Stalin's description of four people flying out of Berlin together, one of them Bormann and one a woman, would have been accurate to the tee.

In addition, the description of the small party escaping in a large U-boat identifies itself particularly well with U-234, which, it will be remembered, had received at least one – and possibly two – radio transmissions from Hitler's headquarters; and which led General Kessler to anticipate an important passenger coming from Berlin. And as will be reviewed in future chapters, Captain Fehler appears to have taken U-234 on a convoluted voyage, with each successive twist and turn apparently intended, at first, to keep U-234 within a few days proximity to Germany, and later, to hide the U-boat's movements and activities as it tried to surrender only to the United States.

The description of Bormann's getaway boat as a *large* U-boat links the escape to U-234 even closer, not just because U-234 was by comparison extremely large, but even more so because it appears to have been the only boat of its mammoth size left in the Atlantic theater.

U-234 was originally built as a mine-laying Type XB U-boat. These double-hulled, triple-sized U-boats were designed to seed strategically chosen bodies of water with high-explosive mines. The Allies became so adept at detecting and eradicating these mines before any harm was caused, however, that the Type XB quickly became obsolete.[54] There was but a handful of Type XBs ever built: U-116 through U-119, U-219, U-220, U-233 and U-234.[55] When the Type XB proved not to have the impact for which it was designed, the boats were refitted as supply vessels for the "Wolfpack" boats sinking Allied convoys on the battlefront in the Atlantic.

Compared to the Wolfpack boats, however, Type XB U-boats were huge, more than 2100 tons displacement, while the ubiquitous Type VII U-boats that constituted 75 percent[56] of Germany's submersible fleet, were just over 700 tons – one-third the size of a Type XB. The other popular U-boat, the Type IX, was larger than the Type VII at anywhere from 900 to 1400 tons. But the Type XB was 50 percent larger than even these more common front boats that, combined with the smaller Type VII, constituted almost the entire remaining U-boat fleet. Russian agents reporting on U-boats were accustomed to both the Type VII and the Type IX and probably would not have differentiated them by size as out of the ordinary, as opposed to the probability of identifying U-234 as a *large* U-boat.

Type XBs were comparatively very large and almost unknown. As noted, there had been only eight of them made. U-116 through U-220, with the exception of U-219, were all sunk in the year between the first of October 1942 and the end of October 1943.[57] U-219 had fortuitously avoided this fate by being stationed in the Pacific immediately upon commissioning.[58] In the South Pacific it was far away from Europe and Bormann and the fierce Atlantic fighting when the war in Europe ended. When Germany surrendered, U-219, still in the Pacific, was turned over to the Japanese Imperial Navy to continue the war under the flag of the Rising Sun.[59] U-233 had been sunk before commissioning, leaving U-234 as the only remaining *"large"* Type XB U-boat available in Europe that fit the description of Bormann's alleged escape.

The Type XIV U-boat was the only other U-boat larger than the popular Type IX and comparable in size to the Type XB. Like the XB, few of these boats were made – only ten – which were all built and operational by the end of 1942.[60] Unlike the Type XB, none survived to the end of the war. They were designed and used as a refueling boat for the Wolfpack vessels, and, as a result, like the XB, had a very high mortality rate. The sinking of a single Type XIV shortened the combat patrols of approximately twelve fighting U-boats, so Allied anti-submarine hunters concentrated on what the German U-boaters affectionately called their "Milk Cows."

The process of refueling was dangerous, requiring the Type XIV fuel supply boat and its recipient lie still in the water for hours on end during the fuel transfer process. During this time, both boats were vulnerable to attack, which happened often, at which the panicked crews would quickly detach the umbilicals and both boats would execute emergency dives. The smaller fighting boat, with its more compact size, greater maneuverability, and with its more disciplined, battle-seasoned crew, would invariably be the first to maneuver out of harm's way, leaving the clumsy behemoth Type XIV at the mercy of the enemy. It was an easy target.

Of the 39,000 German sailors who fought on U-boats during the war, 28,962 were killed and an additional 4,000 captured. A total of over five out of every six U-boaters, therefore, was lost in the war. Remarkably, despite these numbers, Germany's U-boat service was the only one of its military services that had more volunteers than it could use throughout the entire duration of the war.[61] Type XIV U-boats had an abnormally high mortality rate compared to even these chilling statistics, making it apparent survival of a Type XIV U-boat for even a few months was miraculous. As noted, none of the Type XIV U-boats survived to the end of the war, all ten had suffered the fate of the majority of Type XBs by the end of 1943.[62] The only other *large* U-boat built was the Walther U-boat, which was designed and under construction, but not operational, before the end of the war.

U-234, therefore, was the only "large" U-boat left in the Reich's fleet that would most closely fit Joseph Stalin's escape boat description. And, as already mentioned, it is known that U-234 had received at least one radio transmission directly from the Fuehrer Bunker, and quite possibly more; and that Bormann, apparently, had some connection or even control over the U-boat. Apparently, the "wild pig routing for a potato" had dug up the morsel that would save his life.

Endnotes: Chapter Twelve – The Pig Finds A Potato

1   James McGovern, *Martin Bormann: 100,000 Marks Reward*, p. 127

2   William Stevenson, *The Bormann Brotherhood*, p. 164

3   James P. O'Donnell, *The Bunker*, p. 20

4   James McGovern, *Martin Bormann: 100,000 Marks Reward*, pp. 135, 136; Louis Snyder, *Encyclopedia of the Third Reich*, pp. 36, 37; Ladislas Farago, *Aftermath*, p. 143; Simon Wiesenthal, *The Murderers Among Us*, p. 321

5   James McGovern, *Martin Bormann: 100,000 Marks Reward*, p. 136

6   James McGovern, *Martin Bormann: 100,000 Marks Reward*, p. 147; Ladislas Farago, *Aftermath*, p. 143

7   James McGovern, *Martin Bormann: 100,000 Marks Reward*, p. 147

8   Ladislas Farago, *Aftermath*, p. 140

9   James McGovern, *Martin Bormann: 100,000 Marks Reward*, p. 121; Ladislas Farago, *Aftermath*, p. 144

10   William Stevenson, *The Bormann Brotherhood*, pp. 177, 178

11   Ladislas Farago, *Aftermath*, p. 65 and p. 65 note

12   Alan Levy, *The Wiesenthal Files*, p. 226

13   Paul Manning, *Martin Bormann: Nazi In Exile*, p. 90

14   Paul Manning, *Martin Bormann: Nazi In Exile*, pp. 17, 179-183

15   Paul Manning, *Martin Bormann: Nazi In Exile*, p. 180

16   William Stevenson, *The Bormann Brotherhood*, p. 19

17   William Stevenson, *The Bormann Brotherhood*, p. 113

18   James McGovern, *Martin Bormann: 100,000 Marks Reward*, p. 168; Hugh Thomas, *Dopplegangers*, pp. 233

19   James McGovern, *Martin Bormann: 100,000 Marks Reward*, p. 168

20   Hugh Thomas, *Dopplegangers*, pp. 213

21   Hugh Thomas, *Dopplegangers*, pp. 232

22   *Der Spiegel*, 14 February 1964, *as* recorded by Hugh Thomas, *Dopplegangers*, pp. 232

23   Ladislas Farago, *Aftermath*, pp. 361, 362; Paul Manning, *Martin Bormann: Nazi In Exile*, p. 16; Hugh Thomas, *Dopplegangers*, pp. 234, 235

24   Ladislas Farago, *Aftermath*, p. 31; Paul Manning, *New York Times*, March 3, 1973, p. 31, column 2; Hugh Thomas, *Dopplegangers*, pp. 235

25   Paul Manning, *Martin Bormann: Nazi In Exile*, p. 16; Hugh Thomas, *Dopplegangers*, pp. 235

26   Hugh Thomas, *Dopplegangers*, pp. 235, 236

27   Ladislas Farago, *Aftermath*, pp. 26, 27; William Stevenson, *The Bormann Brotherhood*, pp. 13, 14; Hugh Thomas, *Dopplegangers*, pp. 236

28   Ladislas Farago, *Aftermath*, p. 26

29   Hugh Thomas, *Dopplegangers*, pp. 238

30   James McGovern, *Martin Bormann: 100,000 Marks Reward*, p. 168

31   Ladislas Farago, *Aftermath*, p. 28

32   Ladislas Farago, *Aftermath*, pp. 28-30

33   Paul Manning, *Martin Bormann: Nazi In Exile*, p. 180

34   Hugh Thomas, *Dopplegangers*, pp. 251, 252

35   BBC, Sunday, 3 May, 1998

36   James P. O'Donnell, *The Bunker*, p. 296

37   James McGovern, *Martin Bormann: 100,000 Marks Reward*, p. 127; William Stevenson, *The Bormann Brotherhood*, pp. 94, 163

38   James McGovern, *Martin Bormann: 100,000 Marks Reward*, p. 127

39   William Stevenson, *The Bormann Brotherhood*, p. 164

40   Albert Speer, *The Rise and Fall of the Third Reich*, p. 575; James McGovern, *Martin Bormann: 100,000 Marks Reward*, p. 100

41   Albert Speer, *The Rise and Fall of the Third Reich*, p. 575

42   Hanna Reich, *Fliegen, Mein Leben* pp. 92, 93; Louis Snyder, *Encyclopedia of the Third Reich*, pp. 126, 127, 129; Hans Dollinger, *The Decline and Fall of Nazi Germany and Imperial Japan*, pp. 227; Jochen von Lang, *The Secretary*, pp. 326, 327; William Stevenson, *The Bormann Brotherhood*, p. 81; Paul Manning, *Martin Bormann: Nazi In Exile*, p. 171

43   Louis Snyder, *Encyclopedia of the Third Reich*, p. 291

44   Ladislas Farago, *Aftermath*, p. 41

45   Hanna Reitsch, *Fliegen, Mein Leben*, p. 92

46   Hans Dollinger, *The Decline and Fall of Nazi Germany and Imperial Japan*, p. 228; Jochen von Lang, *The Secretary*, p. 326; Louis Snyder, *Encyclopedia of the Third Reich*, p. 127; William Stevenson, *The Bormann Brotherhood*, p. 81

47   Louis Snyder, *Encyclopedia of the Third Reich*, p. 127

48   Ladislas Farago, *Aftermath*, pp. 155, 158, 340; Hans Dollinger, *The Decline and Fall of Nazi Germany and Imperial Japan*, pp. 228, 237-240; Jochen von Lang, *The Secretary*, p. 367

49   Jochen von Lang, *The Secretary*, p. 281

50   William Stevenson, *The Bormann Brotherhood*, pp. 85, 107

51   Hans Dollinger, *The Decline and Fall of Nazi Germany and Imperial Japan*, p. 240; James P. O'Donnell, *The Bunker*, pp. 297, 298

52   Gregory Douglas, *Gestapo Chief: The 1948 Interrogation of Heinrich Mueller*, p. 219

53   William Stevenson, *The Bormann Brotherhood*, pp. 161, 290, 175, 290; Ladislas Farago, *Aftermath*, p. 158; Jochen von Lang, *The Secretary*, p. 290

54   *Sharkhunters KTB 105*, p.11

55   *Deutsche U-boote 1906-1966*

56   *Sharkhunters KTB 116*, p. 30

57   *Sharkhunters KTB 109*, pp. 7,9,13,19

58   *Sharkhunters KTB 101*, p. 16

59   *Sharkhunters KTB 117*, p.13

60   *Sharkhunters KTB 118*, p. 6

Chapter Thirteen

Escape

That damn Martin made it safely out of Germany.[1]
– Walter Buch, Top Nazi judge and Martin Bormann's father-in-law, upon his deathbed

Of course [Bormann escaped]. He is a natural survivor.[2]
– Colonel General Alfred Jodl, At the signing of the European capitulation when
asked if Martin Bormann made it safely out of Berlin

S o what does the composite story of Martin Bormann's escape look like, taking into account all of the acknowledged tales of Bormann's last days in Berlin and the additional evidence since uncovered?

Even though Doenitz's order to U-234, referred to in Chapter Nine, countermanding the directive from Berlin to ensure the Admiral retained command of the submarine, and then ordering the U-boat to leave as soon as able, was received by Hirschfeld and Fehler on the 14th, U-234 did not actually set sail until two days later, on the 16th – the same day the artillery barrage of Berlin began. Had Bormann, from Hitler's headquarters, set the final attack on Berlin as the automatic signal for Fehler to stealthily set to sea, from where he would await further orders?

On the morning of 22 April, Bormann radiogrammed Helmut von Hummel, his top aide who was now working in Obersalzberg, "agree to proposed overseas transfer south." Convoluted efforts have been made to suggest this message was actually code for Hummel, who was in possession of much of Bormann's records and personal effects, to take the cache and "get out of town" over the south Tyrol and dispose of the goods in a safe place in southern Germany, Austria or northern Italy.

The Soviet Bormann expert Lev Besymenski, however, interpreted this message to refer to a prearranged escape to South America, which seems far more probable given the "overseas" aspect of the message.[3] If that was the case, it is one more piece of evidence Bormann was planning to escape by U-boat, the most probable method for a high-ranking Nazi leader to make an escape to South America.

In Berlin, the Russians were daily tightening their noose around the beleaguered city and the core of Hitler's remaining leaders huddled in the bunker under the Reich Chancellery. During the final three days of April, virtually all Bormann experts agree, Bormann struggled mightily to escape the strangle-hold of Berlin and make his way to the commander of all submarines, Admiral Doenitz. At the same time, he held conference with Heinrich Mueller as they tried to execute their escape plan and finalize the details of fleeing Berlin.[4]

On the night of 28-29 April, when Hitler ordered Hanna Reitsch to fly out of Berlin with new Luftwaffe commander General Robert Ritter von Greim, the opportunity Bormann and Mueller were looking for had arrived. Even the traditional history concedes that at just this time Bormann succeeded in getting Hitler to order that he be flown to Doenitz.[5] Hitler had married Eva Braun and then composed his last will and testament and a political manifesto urging the continuation of Nazism and his legacy. These acts demonstrated he expected Nazism to carry on despite its dismal condition and his absence, for he had already declared he would commit suicide rather than leave a defeated Berlin while still alive.[6] Bormann had waited anxiously for this moment. But up until then, the Fuehrer had not given the Reichsleiter final permission to forever leave his service. Bormann, loyal to the end, would not dream of deserting Hitler if he knew his master might need him yet.

At 3:30 a.m. 30 April, the Fuehrer had concluded his baneful business on earth and all but ended his life. He would kill himself 12 hours later, but not before he had given the order to his pilot, Hans Baur, in no uncertain terms, to make sure Bormann got to Doenitz to deliver The Fuehrer's last will and testament and political manifesto, which Bormann would hand carry and personally deliver to the Admiral.[7] "Bormann has been given several orders which he must take to Doenitz in person.... It is most important that Bormann gets to Doenitz," Hitler told Baur. Bormann's uncanny influence over Hitler had worked one last time.

According to author James P. O'Donnell, Bormann was simply substituted for Hitler in an escape plan Hitler's pilot, Hans Baur, had prepared.[8] O'Donnell reported the original plan, which according to Baur was never completed, was for Baur to fly Hitler out of Berlin, not for Hanna Reitsch to fly the substituted Bormann. Reitsch's and Greim's impending departure, however, and Hitler's refusal to leave, provided a fortuitous opportu-

nity to fulfill Hitler's order of getting Bormann to Doenitz by implementing Baur's plan, with Bormann and Mueller as the escapees instead of the
Fuehrer, and Reitsch as the pilot instead of Baur.

Baur was extremely loyal to Hitler and he was a staunch Nazi[9] to his
dying day, and he reported directly to Bormann.[10] Given Hitler's order to
get Bormann to Doenitz, Bormann's mission to preserve Nazism and the
Fuehrer's legacy, and Baur's loyalty to the cause and subservience to Bormann, Baur would have done everything in his power to fulfill the order,
despite the traditional history's later ignoring the whole affair.

At dawn of the same day as Hitler's order to fly Bormann to Doenitz,
30 April 1945, Martin Bormann and Heinrich Mueller most likely departed with Hanna Reitsch and General von Greim toward Admiral Doenitz's
headquarters in Ploen. Bormann's double remained to play unwittingly
his farewell role in a final fraud performance.

Given the convergence of so many disparate details – Bormann's and
Mueller's escape preparations on the night of 28-29 April, which coincided with the timing of Hitler's order that Bormann be flown to Doenitz during that exact timeframe; the report that Mueller had flown out
to freedom at that time, as well; Reitsch's admission that she flew a small
plane to Doenitz the morning of 30 April to deliver an "important package;" and, again, Stalin's insistence that Bormann escaped in a small plane
at exactly the same time – the fact of all this disparate evidence converg-
ing despite such unlikelihood, seems far too compelling to be overlooked
in favor of the historically entrenched but seriously conflicting versions of
what occurred in the bunker, and of Bormann's fate.

The escape described above would have given Bormann and Mueller
a day or two head start from the others in the bunker and ensuing searches for them, provided they leave behind a viable alibi that would resolve
their fates for the outside world and eliminate post-war investigations.

Bormann's and Mueller's detailed plans at first worked. Indeed, five
staunch Fuehrer bunker Nazis all testified they saw Bormann killed on
Weidendammer Bridge, an assertion now considered by even hardened tra-
ditionalist 'Bormann died' evangelists to be a conspiratorial lie. And other
would-be observers provided slightly different versions of the same story.

The death-by-tank-explosion alibi, designed to end later searches,
would insist Bormann and Mueller escaped Hitler's headquarters with
the others in the bunker the night of the breakout. Upon exiting the
bomb shelter, the alibi went, Mueller and Bormann were separated and
Bormann made his way to a location; possibly Weidendammer Bridge

had already been selected. Possibly it was left to the vagaries of the fluid condition of the battle for that to be decided.

The story would describe how, once at Weidendammer Bridge, Martin Bormann was killed by a blast to a tank he was using to cross the bridge. Both Paul Manning and James O'Donnell site accounts of a tank having been pre-arranged specifically to be at the bridge at the fateful moment to complete the illusion.[11] Manning believed the story, O'Donnell did not.

To further validate the death, Bormann's double would be taken to the bridge and killed, to later be found with Bormann's diary placed in the unfortunate corpse's pocket.[12] The journal would make identification of the body as Bormann's easy, and conclude the illusion – much as Hitler's double had been killed in the bunker and stuffed in a water cistern to throw investigators off of finding, exhuming and violating the Fuehrer's remains. The "Bormann" body would validate the "eyewitness" reports of the Reich Minister's demise: Bormann's death would be assured and he would fade into the shadows of history.

Once the machinations had been put in place to conclude the alibi and the story was disseminated to Hitler's remaining top aids for post-surrender circulation, Bormann and Mueller flew with Reitsch and Greim to Hamburg, where U-234 would pick up Bormann just as Stalin insisted he had been gathered by a "large U-boat." Reitsch, Greim and Mueller then flew on to Ploen to report to Doenitz, and then finally made their way to the Tyrol.

If the above is true, the most probable scenario is that upon disembarking Hamburg, the U-boat took Bormann, under protection of the watchful eyes of the Allies – remember the planes that did not attack – through the English Channel to a prearranged rendezvous point in the Bay of Biscay. There he boarded another vessel and was ferried to the north coast of Spain, where OSS reports say he stayed with one "Leon DeGrelle."[13] In Spain, Bormann quietly completed his European business affairs behind the scenes and under the protection of Spain and, by secret extension, the United States' watchful eyes, as he consolidated the underground economy that would soon reinvent Europe.

The plan was a good one, detailed and well thought out considering all of the possibilities. But the unpredictability of battle, the serendipitous nature of fate, and the persistence of people who refused to let justice go undone, undid it. First, the integrity of the cover story was not kept after the key bearers of the alibi were captured. Erich Kempka, Hitler's chauffer; Hans Baur, Hitler's pilot; Heinz Linge, Hitler's valet; Johann Rattenhuber, chief of Hitler's detective bodyguard; and Otto Gunsche, Hitler's SS adjutant, were the survivors of Berlin who were closest to Hitler and Bormann during the final days in the bunker. They all asserted they saw Bormann die in the tank explosion on Weidendammer Bridge. As the keepers of the cover story, this was what they were expected to say.

But others swore to different events, both on the bridge and off. As noted in the previous chapter, Roca-Pinar and Harry Mengerhausen testified to very significant variances in the Weidendammer Bridge episode. These versions possibly were the result of junior soldiers trying to support the prescribed alibi but either not having all of the details or taking some liberties in order to carve themselves a small niche in history.

The later identification by Axmann of the dead Bormann on the Lehrter Station Bridge, not the Weidendammer Bridge, further undid Bormann's and Mueller's caper. Axmann may not have been lying, despite the bizarre details, when he told his odd story of calm corpses lying uninjured in the midst of the great battle. The peculiar circumstances of these deaths amidst the heavy action actually seem too strange to be fictitious, especially if one were trying to create a convincing alibi. Axmann probably had, in fact, checked the breathing of the poisoned body not of Bormann, but of Bormann's double lying peacefully next to that of Dr. Stumpfegger. Stumpfegger's presence may be a central clue in what happened, for it was Dr. Stumpfegger who had poisoned the Goebbels children and Hitler's dog, Blondi. Here we have a man killing innocent victims left and right. Presumably, Stumpfegger was in on the phony Bormann escape scenario and it was his task to poison Bormann's double – as he had poisoned the Goebbels children – to conclude the desired illusion.[14]

With the Russians already in control of Weidendammer Bridge, where the cover story was supposed to take place, Stumpfegger may have decided to "do in" the counterfeit Bormann on Lehrter Station Bridge instead. The Doctor probably calculated the Lehrter trestle was as close as he was going to get to fulfilling the details of the cover story and so committed the execution there. Once the deed was done, seeing he was on his own and devoid of hope of escaping the tightening Soviet ring himself, Stumpfegger concluded his grotesque killing spree by taking his own despicable life as well; following Hitler, Goebbels, General Burgdorf and others, in suicide.

Or possibly he knew too much and was done in by others. The skull later claimed to be his was missing a considerable portion of the top of it – it having been surgically removed – suggesting, according to forensic expert Dr. Hugh Thomas, that Stumpfegger was shot and the skull altered to remove the evidence.[15] This point does not jive, however, with how Axmann reported he found Bormann – or actually Bormann's double – and Dr. Stumpfegger lying dead, peacefully reclined side by side on Lehrter Station Bridge but otherwise unmarked.

Bormann was supposed to have been escaping Berlin primarily to deliver Hitler's will and political manifesto, which he was personally carrying to Admiral Doenitz.[16] The body found by the Russians was initially identified as Bormann's when the Reichsleiter's personal journal was found in its overcoat pocket. Hitler's last will and testament and his political manifesto were never mentioned as having been found on the body, however, despite the obviously detailed scrutiny the Russians visited upon the remains – proving the Russians did not, in fact, have Bormann's actual body, as they quickly surmised they did not.

The second series of scenario-crippling conclusions to the traditional history came when additional facts began to arise. For instance, although the disappearance of Heinrich Mueller was lost on many in the confusion surrounding the escape attempt, a grave reportedly containing his remains was later identified in the Kreuzberg garrison cemetery in Berlin.[17] Supposedly, he had been killed in street fighting during the escape. After that, each year for 18 years, flowers were placed lovingly at his headstone – presumably by family members. Later reports were received, however, suggesting that possibly the Gestapo Chief's remains were not in the coffin under the headstone bearing his name and at which flowers were regularly being placed. By order of the West Berlin District Attorney's office, the remains were exhumed and forensics tests performed. The findings showed that bits and pieces of three men shared the grizzly grave, but none of them was Heinrich Mueller.[18]

The depth and breadth of the escape plan was beginning to become clear. Had Bormann and Mueller made plans so complete, so airtight, that they included detailed, carefully prepared camouflaging tactics to conceal the escapes, and carried out macabre charades for decades after to ensure their safety? The answer, viewed against the conflicting testimonies and cryptic anomalies linked to the supposed demise of Martin Bormann, caused those who suspected Bormann might not have died in Berlin to look even closer at the evidence.

Especially interested were those investigators, such as Paul Manning,[19] Ladislas Farago,[20] and William Stevenson,[21] who believed Bormann and Mueller together carefully worked out their escapes. Manning quoted an unnamed Bormann expert as saying "Bormann planned this flight with extreme care and part of the grand design was a scheme to lead future forensic and dental specialists astray."[22] The journalist later sited Mueller's skill and considerable professionalism at such endeavors,[23] which was evidenced by Hitler's convincing double and the phony grave Mueller had had created for himself. Even Jochen von Lang, who ultimately insisted Bormann died in Berlin, intimated Bormann and Mueller worked on their escape plans together.[24] If Mueller and Bormann went to such pains as Mueller's fake grave to hide the escape, the investigators started asking, what had they done to prepare for it? As the investigators found and started pulling on loose threads, the carefully constructed tapestry began to unravel.

Many will assert it was impossible for Bormann to have escaped Berlin because the testimony of witnesses who were with him and the long litany of radio transmissions he authored to Doenitz from the Fuehrer Bunker prove he was intact in Hitler's headquarters until just hours before the escape attempt. A careful, chronological review of the messages and of his actions, however, reveals some interesting irregularities that demonstrate Bormann was not participating in events that were under his responsibility but that he almost assuredly would have been controlling had he been in the bunker. By deduction, therefore, they subtly suggest he was not in the bunker at all.

Up until the night of 29 April, the historical record seems fairly unassailable except for one large, perplexing fact. The record shows Bormann was paying particular attention to keeping Admiral Doenitz informed of events in Berlin. Bormann's constant contact with Doenitz is now accepted thoughtlessly, seventy years later, and is unquestioned. But in its contemporary political context, such activity on Bormann's behalf is bewildering. Interface with Doenitz on the military situation in Berlin was undoubtedly needed, but it would have been a military matter and should have been carried out through Hitler's military chain of command – which was still intact – not through a civilian office, which was Bormann's domain. Instead of Bormann in constant contact with Doenitz, it would have been expected, just as easy, and more efficient and appropriate for the Supreme Command to keep the ranking naval officer in the loop.

Why Bormann, of all people, was in contact with Doenitz is an interesting question. Hitler had not yet announced his unexpected appointment of Doenitz as his successor, so it was too early for Bormann to initiate government business with the Admiral. Despite all Bormann's machinations in the past, through which he at times had influenced military matters,

Hitler had never allowed Bormann to participate directly in military affairs. And Bormann seldom showed more than passing interest in doing so. Despite these conditions, Bormann, for some reason, now was in regular contact with the U-boat supreme commander, constantly updating him on the state of the battle.

On 29 April, the Reichsleiter wired Doenitz, "Situation very serious.... Those ordered to rescue the Fuehrer are keeping silent.... Disloyalty seems to gain the upper hand everywhere.... The Reichs Chancellery a rubble heap.... We are staying on."[25] Other than the incongruous chain of command, such an update, though abnormal, would not have – and has not – been considered remarkable. But considered in its full political context and in light of later developments, such communications appear to have been part of a more specific agenda, rather than a simple update on the state of affairs.

Earlier that night, the last gasps of Hitler's Thousand Year Reich had begun in earnest. The Fuehrer married, concluded all his worldly affairs, and began his last day on earth awaiting the moment to ignominiously end his life. Shortly before midnight on the 28th or in the early morning hours of the 29th, he asked his old friend Hanna Reitsch to fly General von Greim out of Berlin.[26]

While it is fairly certain Hitler gave the order for the flight on the night of the 28th, historical accounts vary as to when the flight actually occurred. Some, such as General Koller in his account of events, claim the flight took place on the night of the 28th.[27] His conclusion was possibly based on the assumption Hitler's order was fulfilled on the night it was given. Such a supposition is probably in error, however, since it was already late night when the order was given and it would still take planning and logistical effort to arrange the flight. Whether that could be accomplished in the hours of darkness that remained in the night of the 28th/29th is questionable. If time was not available, it would have been necessary to wait for the following night, in order to use the cover of darkness for the plane's escape.

Other witnesses claimed the flight occurred the following night, on the 29th, which seems logical given the argument above. Hanna Reitsch herself stated in a news report that she and Greim flew out at dawn on the morning of the 30th.[28] As has been noted already, it is a common conclusion to include all the dark hours of a single night as the night of the previous day. Thus it is reasonable to conclude that a flight whose passengers left for the airstrip, and that was prepared to leave in the pre-dawn hours, and then took off at the crack of dawn on the 30th, would have been considered to have departed on the night of the 29th.

At about 1 a.m. the morning of the 29th, Hitler married Eva Braun in a short civil ceremony witnessed by Bormann and Goebbels and attended by a few others.[29] He then sequestered himself with a secretary and dictated his last will and testament and political manifesto, which he completed and signed about 4 a.m. A few moments later, at 4:17 a.m.,[30] Bormann sent his message to Doenitz informing the Admiral of the dire state of the military situation in Berlin and of the Reichs Chancellery being "a rubble heap," but that they were determined to "stay on." He mentioned nothing of Hitler's marriage or preparations for his death, although Hitler had already made his absolute decision to die in Berlin, as attested by granting Eva Braun her last wish of marriage to him and preparing his will.

Despite this decision, later the same day Bormann sent another message to Doenitz challenging him to prove his loyalty by immediately relieving the Fuehrer.[31] What this meant is uncertain. Doenitz had already sent two divisions[32] and a contingent of sea cadet trainees[33] – most of whom were slaughtered – to defend Berlin. Knowing Hitler vehemently had refused days earlier to escape to Bavaria, and that he had now determined and started the preparations to die in Berlin,[34] it seems improbable Bormann was encouraging Doenitz to invest more men in yet another massive military rescue attempt of the Fuehrer, undefined as that may be.

Hitler had determined his only acceptable rescue would be if German troops could win the Battle of Berlin and save him while he heroically led the battle from his hopelessly surrounded position. Otherwise, the Fuehrer did not want to continue his life. But winning the battle was unthinkable with the forces at Doenitz's disposal and both Bormann and Doenitz knew it. The message to the Admiral, therefore, implying that Doenitz could somehow rescue the Fuehrer, seems to be the beginning of a series of deceptions Bormann would play against Doenitz for some mysterious end.

The next 24 hours in the bunker must have felt hopelessly macabre for the subterranean survivors, with the final hours interminably passing and the incessant rumbling of heavy guns and artillery constantly jarring the earth overhead. Hitler's generals sent communiqués far and wide, continually trying to save the desperate, gradually growing hopeless, situation. But Soviet forces were too strong and held a stranglehold on the city.

At 3:15 a.m. 30 April, approximately one half-hour before Reitsch's mysterious flight out of Berlin with her unknown passengers, and within hours following Hitler's command to fly Bormann to Doenitz, Martin Bormann sent the Admiral another message.[35] He described briefly how the Wehrmacht's rescuers were "stubbing their toes," inferring that a rescue by them was doubtful, and then added a postscript of sorts: "Addition from Berlin. Attempts will probably be made to jam radio transmissions. Do not let it upset you. Future communication will be forwarded to Ploen." The message appears to be instructions to expect communications from Bormann to come from different transmission centers than from the bunker itself, or possibly by a different manner of communication altogether.

At dawn a half-hour to an hour later,[36] depending on which account one chooses to believe,[37] Hanna Reitsch and General Robert Ritter von Greim flew out of the Tiergarten in a small aircraft, probably accompanied by Martin Bormann and Heinrich Mueller. Despite the plane's short take-off capacity – it was designed to be airborne in only 60 meters[38] – and the fact the section of the Tiergarten being used as a runway was 750 meters long, the airplane barely cleared the statuary atop the Brandenburg Gate at the end of the take-off strip.[39] The reason given for the dangerous near miss was the aircraft had taken off with the wind. Perhaps so. But perhaps the aircraft, which was designed to carry only two people, was carrying four people and whatever important baggage they could squeeze in. Such a scenario would explain the over-long takeoff and would certainly add credence to Stalin's determined assertion that three men and a woman took off in a small airplane at the same time and place as this flight.

Twelve hours later, around 3:30 in the afternoon of the same day, Adolf and Eva Hitler killed themselves. Two hours after that, Bormann informed Doenitz that the Admiral had been chosen the Fuehrer's successor, but, mystifyingly, he did not tell the Admiral that Hitler was dead.[40] Doenitz asked Bormann for verification from witnesses, apparently suspecting Bormann might be playing him for a dupe, as Bormann had played Goering.[41] Bormann made no effort to provide the requested witnesses – probably because he was no longer in the bunker to receive and fulfill the request; nor had he been for almost 12 hours. This may be the reason, in fact, that Bormann failed to tell Doenitz that Hitler was dead – he may not yet have known the suicide had occurred. But even if he did know the Fuehrer was dead, Bormann would have feared that witnesses would tell Doenitz, which would have ruined his plan.

Fourteen hours after that, at 7:40 a.m. on 1 May, Bormann again contacted Doenitz, this time to tell him Hitler's testament was in force, but once again he did not share that the Fuehrer was dead.[42] The Reichsleiter then recommended to Doenitz that he not publish this information; probably a cover to keep others from finding out Doenitz was in the dark and then informing him of the Fuehrer's death, which would have undone Bormann's plans.

Historians for over seventy years have tried to explain in the context of the traditional history Bormann's strange, outwardly unnecessary and seemingly meaningless efforts to not tell Doenitz of the Fuehrer's suicide. In the context of the traditional history, regardless of the tortuous and convoluted explanations, Bormann's messages seem to make little sense, though many writers have strained to read meaning into them. But against the background of the earlier reported radio signals to U-234 from the Fuehrer Bunker compared to Doenitz's struggle to maintain command of the U-boat, Bormann's strange convolutions begin to be clear.

The Reichsleiter, as only he could, appears to be playing a game of cat-and-mouse with the Admiral. There is evidence Doenitz was concerned Bormann was manipulating him, such as Doenitz's request for witnesses to his being Hitler's successor. Indeed, later Doenitz issued an arrest order for Bormann should he make it to Ploen.[43] Apparently anticipating this, Bormann felt it necessary to manipulate the U-boat service commander, convincing him to commit to help Hitler's escape – even though he, Bormann, would be the one escaping. Presumably, Doenitz's thinking he was helping Hitler escape kept him from barring Bormann's break out.

Once U-234 was on its way back to Germany, Bormann appears to have kept Doenitz "on the string" by hanging the bait of being post-war leader of Germany in front of the Admiral, which Doenitz was guaranteed when Hitler named him his successor. But with everything lost, the real reason Bormann convinced Hitler to appoint Doenitz was so Bormann could manipulate the Admiral's cooperation to affect his escape. Bormann needed that U-boat.

With Doenitz feeling he was on the verge of leading the nation, Bormann knew the Admiral would be careful not to step outside of protocol, especially if he thought Hitler was still alive and had heard how the Fuehrer dealt with Goering and Himmler upon learning of their separate bids for control. But once Doenitz knew Hitler was dead, the Admiral's command would be law and Bormann would be one of the first of Hitler's paladins he would seek to bring down, and Bormann knew it. Until Doenitz became aware of Hitler's death, however, Bormann would have the upper hand. So Bormann kept the death a secret. He flew out of Berlin, not to Ploen, as he told Hitler he would, but to Hamburg, where he waited for "a large U-boat" to land, while Mueller, Greim and Reitsch went on to Doenitz.

At 5 p.m. on 30 April, Bormann probably was hidden safely away not in the besieged bunker in Berlin but in Hamburg, awaiting the arrival of U-234, when he sent the message informing the Admiral that he had been chosen Hitler's successor. Doenitz would not have been surprised if the message was identified as coming from Hamburg, or any other point for that matter, since Bormann, in anticipation of his escape requirements, had warned the Admiral to expect communications to come from almost anywhere because of possible signal jamming.

Bormann sent his last message to Doenitz on 1 May, informing the Admiral the testament was officially in force, while Bormann was safely ensconced in Hamburg but Doenitz thought he was in Berlin. Bormann was still careful not to let the Admiral know of the Fuehrer's demise. As already mentioned, possibly Bormann did not know of Hitler's death himself since he had not been in the bunker for hours before the suicide. In any case, Bormann appears to have tried to make Doenitz think he was in the Admiral's control: "Testament in force. Will join you as soon as possible. Advise delay publication until then."

With that sketchy information, Doenitz would be careful not to overstep his bounds and would wait patiently for an explanation when Bormann arrived – and then he would arrest him. But Bormann never showed. The suggestion he was coming to Doenitz was a ruse, not just to neutralize Doenitz while Bormann waited for the U-boat at Hamburg, but it would work as well to camouflage his escape when investigators later pursued his whereabouts. While Doenitz later was told Bormann had been killed in the street fighting, actually Bormann set out to sea on the U-boat, which by then Doenitz thought was well on its way to Japan.

Champions of the traditional history will assert there are serious flaws in these conclusions. They will ask, how could Bormann be in Hamburg waiting for the U-boat while he is known to have been participating in Hitler's death and burial and the unsuccessful surrender negotiations with the Soviets during the early morning hours of 1 May? Or they will state Bormann's alleged signing, with Goebbels, of the message later informing Doenitz that Hitler was dead, sent some time between 2:15 and 3:15 p.m. 1 May, long after Bormann would have been in Hamburg waiting for the U-boat, according to the thesis of this book.

The serious flaws in these accounts are actually in the traditional history. For despite assertions Bormann oversaw the Soviet surrender negotiations, General Krebs, who was sent to the Soviets to parlay, stated he could not agree to the Soviet demand for unconditional surrender because he did not have Goebbels' authorization to do so.[44] He never mentioned Bormann in this context, even though Bormann signed the authorization to initiate negotiations.[45] Bormann probably signed the negotiation release, and any other anticipated documents, before leaving the bunker. As the signatory of the release, Bormann almost certainly would have been expected to provide approval of the negotiations, too, had he still been present.

And although the traditional history insists Goebbels forced Bormann to sign the document notifying Admiral Doenitz of Hitler's death that afternoon,[46] a photograph of the actual document as shown in Dollinger's *The Decline and Fall of Nazi Germany and Imperial Japan* shows Goebbels alone signed the communiqué to Doenitz – Bormann's signature is not on it.[47] This is an important and very telling discrepancy, since up until then all Fuehrer bunker communications with Doenitz had gone through Bormann. The telephone exchange of the bunker was also under Bormann's direct command up until 30 April, after which Goebbels took control of the system[48] – an event not likely to have been allowed by Bormann had he still been there. Apparently, judging from the evidence, the Reichsleiter seems to have left the premises.

Bormann still had a presence in the bunker, though – in the form of his Gestapo-supplied double, who would soon be sacrificed on Lehrter Bridge. Undoubtedly those who did not know any better continued to account for the Reichsleiter in this inconsequential counterfeit. But those who knew Bormann was gone gave the double no consideration. That is why Krebs and Goebbels failed to take him into account in their dealings with the Soviets and Doenitz.[49]

Many witnesses who were still present in the bunker later reported Bormann's attitude changed profoundly, from overbearing to feckless, after Hitler's death.[50] Researchers have explained this as Bormann's survival reaction to the loss of his protector, Hitler. But such a behavior swing seems out of character with the persona of the man, as illustrated by his radiogram sent after Hitler's death in which Bormann, while informing Doenitz he was Hitler's successor, was still forceful and confident in his position. This confidence does not square with the description of the man in the bunker at the time. The aberrant behavior of the 'Bormann' observed in the bunker, however, could be expected of a common man thrown into such bizarre circumstances as playing the role of a very important international leader during the catastrophic collapse of the empire that leader served.

Admittedly, the scenario above assumes much in certain areas of the account. No direct proof has surfaced that Bormann and Doenitz ever actually communicated specifically about U-234 or that any of the transmissions from Bormann to Doenitz originated from any other location than the Fuehrer Bunker. Nor is there any infallible documentary evidence that U-234 was part of an escape plan or that Bormann was ever aboard her.

But the preponderance of circumstantial evidence – especially when viewed through the two filters of comparing disparate stories to find specific similarities and patterns, and of weighing evidence, or lack of it, against the possible vested interests of its sources – certainly tends to validate this scenario above the traditional history. Had the traditional history not been ingrained in the psyches of today, but instead was introduced objectively, side by side, compared against all the details above, as if in a court of law, the author believes the above scenario must be chosen, of the two, as the more likely course of events. And the explanations for the conflicts and incongruities of this scenario are much less strained than those leaps of faith presently accepted in the traditional history.

Is it really more credible that Bormann was blown up inside a tank, or next to a tank that exploded so violently that his body went flying, and then he walked away without a mark before deciding to just give up and lay down and peacefully poisoned himself, even though 90 percent of his cohorts in the bunker survived, many of them much less survival committed than he? Is this the result of the elaborate and detailed plans Bormann and Mueller are known to have made for their escape?

Or is it likely the Soviets would have found Bormann's body, hauled it away for forensic investigation, but not found several large documents sewn into his clothing? Would the Soviets then have gone to the trouble of hauling the body back to its original makeshift grave and reburied it, rather than inter it in a location convenient to the forensic location?

Stalin's report of the flight from Berlin and Bormann's boarding a U-boat in Hamburg, the Hanna Reitsch flight and Bormann's determination to get to Doenitz, and Hitler's order that Bormann be taken to Doenitz, all combine to present the most credible, compelling account for Bormann's escape. It is hard to believe Hanna Reitsch departed to fly to Doenitz at the same time Bormann was trying to get to the Admiral, by order of Hitler, and yet Bormann was not on that airplane.

There appears to have been little reason for Stalin to lie about such an episode, for what could he have hoped to gain from it? If he had made it up, the Western Allies would have paid little attention to him, which is what they did, so such a concoction would be of little value. If it were true, however, especially considering the implications to the Soviet Union of the cargo U-234 carried, if Stalin knew about it, then Stalin would be upset and insist the mystery be resolved. Even if he did not know about the U-boat's contents, but only that the United States helped with Bormann's escape and in later harboring the fugitive, he could be expected to never let the subject die, which he did not during his lifetime.[51]

Certainly to protect its advantage gained by a Bormann deal for U-234, the United States would deny and minimize any such accusation – which it did and has done ever since. In fact, the United States threw the same complaint of harboring Bormann into the Soviet's face – with the help of General Gehlen – in order to belittle the accusations against it, and confuse and defuse the public mind regarding Stalin's claim.

If Stalin was telling the truth about the flight from Berlin, as the details he included tend to demonstrate he was when compared against other, separate accounts, then why not about the large U-boat, as well? British Field Marshal Bernard Montgomery was reported in early September 1945 to have said British Intelligence received a report of Bormann in Hamburg the night of 1 May,[52] apparently verifying Stalin's assertion Bormann had been flown to Hamburg.

That Bormann flew to Hamburg and escaped in a submarine is further supported by an episode Ladislas Farago described when he asked British Intelligence about a report that Bormann escaped in a U-boat. He was told by one of Britain's highest ranking intelligence officers that they had investigated the report immediately after the war, but that the inquiry was more interested in the U-boat he escaped in than in the missing Reichsleiter himself.[53]

Two points are of interest in this response. The first is there was no denial Bormann had escaped by U-boat. On the contrary, the connotation drawn from the response is that the report was true and there seemed to be additional specific knowledge about the escape and the escape vehicle, which would tend to validate the U-boat escape story. The contact noted that the investigation was later dropped, which is quite possibly a telling event, as well. The investigation would have been dropped once it was discovered the U-boat wound up in American hands, and probably not until then or until the safe disposition of the wayward U-boat and Bormann had been determined.

The second point is almost all German U-boats had surrendered by this time, and, with the war over, held little more value than as surplus submarines for the Allied navies. Most were sunk as target practice shortly after the war. On the other hand, the Allies knew by then Bormann controlled all of Hitler's vast wealth as well as the Nazi Party's massive funds and properties and several colossal government accounts. In addition, he had untold knowledge about the workings of the Third Reich, its intelligence services and what they knew about the Soviet Union, which would be valuable in the upcoming Cold War; and international business dealings that were worth tens of billions of dollars.

These were the spoils of war, and under the guise of reparations the Allies were intent on claiming them, if they could identify them. For that, it would be most helpful to have Bormann. What Bormann controlled, therefore, was far more valuable than a single submarine. Certainly Hitler's missing lieutenant would take top billing over any single U-boat and its cargo, which British intelligence seemed so interested in – with the possible exception of the world-molding critical cargo of U-234. That cargo held a political premium unequalled at that time, or since.

The mysterious activities of U-234 – which will be reviewed in the next chapter – overwhelmingly support the idea Bormann was picked up by the U-boat in Hamburg. Indeed, William Stevenson noted a direct link between Bormann and U-234 when he described how Bormann "had at his fingertips all the details required for ... moving special cargoes like the dismantled rockets shipped by U-boat to Japan"[54] as well as the scientists who developed Germany's atomic bomb.[55]

Endnotes: Chapter Thirteen – Escape

1   Paul Manning, *Martin Bormann: Nazi In Exile*, p. 45

2   Paul Manning, *Martin Bormann: Nazi In Exile*, p. 185

3   Jochen von Lang, *The Secretary*, p. 332

4   William Stevenson, *The Bormann Brotherhood*, pp. 175, 290; Ladislas Farago, *Aftermath*, p. 158; Jochen von Lang, *The Secretary*, p. 290

5   James P. O'Donnell, *The Bunker*, pp. 297, 298; Jochen von Lang, *The Secretary*, p. 329; Hans Dollinger, *The Decline and Fall of Nazi Germany and Imperial Japan*, p. 240

6   James P. O'Donnell, *The Bunker*, p. 254

7   Hans Dollinger, *The Decline and Fall of Nazi Germany and Imperial Japan*, p. 240; Jochen von Lang, *The Secretary*, p. 329

8   James P. O'Donnell, *The Bunker*, p. 309

9   James P. O'Donnell, *The Bunker*, p. 380

10   James P. O'Donnell, *The Bunker*, p. 298

11   Paul Manning, *Martin Bormann: Nazi In Exile*, p. 180; James P. O'Donnell, *The Bunker*, p. 301

12   James McGovern, *Martin Bormann: 100,000 Marks Reward*, p. 168

13 U.S. National Archives II, *State Department Telegram From Madrid to Secretary of State*, 29 August, 1947, RG 59 862.20252/8 - 2947

14 Jochen von Lang, *The Secretary*, p. 332

15 Hugh Thomas, *Dopplegangers*, p. 256

16 Jochen von Lang, *The Secretary*, p. 332

17 James McGovern, *Martin Bormann: 100,000 Marks Reward*, pp. 152, 153; Ladislas Farago, *Aftermath*, p. 177; William Stevenson, *The Bormann Brotherhood*, pp. 290, 296, 300

18 William Stevenson, *The Bormann Brotherhood*, pp. 290, 296, 300; James McGovern, *Martin Bormann: 100,000 Marks Reward*, pp. 152, 153; Ladislas Farago, *Aftermath*, p. 177

19 Paul Manning, *Martin Bormann: Nazi In Exile*, pp. 175, 179-183

20 Ladislas Farago, *Aftermath*, p. 158

21 William Stevenson, *The Bormann Brotherhood*, p. 161

22 Paul Manning, *Martin Bormann: Nazi In Exile*, p. 17

23 Paul Manning, *Martin Bormann: Nazi In Exile*, pp. 179-183

24 Jochen von Lang, *The Secretary*, p. 290

25 Hans Dollinger, *The Decline and Fall of Nazi Germany and Imperial Japan*, p. 237

26 Hans Dollinger, *The Decline and Fall of Nazi Germany and Imperial Japan*, p. 228

27 General Koller, *The Last Month*, as quoted by Hans Dollinger, *The Decline and Fall of Nazi Germany and Imperial Japan*, p. 228

28 as quoted from the news report by Ladislas Farago in *Aftermath*, p. 41

29 Hans Dollinger, *The Decline and Fall of Nazi Germany and Imperial Japan*, pp. 228, 239; James McGovern, *Martin Bormann: 100,000 Marks Reward*, p. 104

30 Hans Dollinger, *The Decline and Fall of Nazi Germany and Imperial Japan*, p. 237

31 Hans Dollinger, *The Decline and Fall of Nazi Germany and Imperial Japan*, pp. 228, 238

32 James P. O'Donnell, *The Bunker*, p. 147

33 James P. O'Donnell, *The Bunker*, p. 343

34 James P. O'Donnell, *The Bunker*, p. 169

35 Hans Dollinger, *The Decline and Fall of Nazi Germany and Imperial Japan*, p. 239

36 James P. O'Donnell, *The Bunker*, pp. 216, 302 note

37 Ladislas Farago, *Aftermath*, p. 41

38 David Donald, *The Complete Encyclopedia of World Aircraft*, p. 415

39 Peter Padfield, *Himmler*, p. 600; James P. O'Donnell, *The Bunker*, p. 154

40 Hans Dollinger, *The Decline and Fall of Nazi Germany and Imperial Japan*, p. 240; Ladislas Farago, *Aftermath*, p. 139; Jochen von Lang, *The Secretary*, p. 330

41 Jochen von Lang, *The Secretary*, p. 331

42 Hans Dollinger, *The Decline and Fall of Nazi Germany and Imperial Japan*, p. 241; Ladislas Farago, *Aftermath*, p. 139; Jochen von Lang, *The Secretary*, p. 331

43 Jochen von Lang, *The Secretary*, p. 332

44 Hans Dollinger, *The Decline and Fall of Nazi Germany and Imperial Japan*, p. 228; James P. O'Donnell, *The Bunker*, p. 364

45 Jochen von Lang, *The Secretary*, pp. 330, 331

46 Jochen von Lang, *The Secretary*, p. 331

47   Hans Dollinger, *The Decline and Fall of Nazi Germany and Imperial Japan*, p. 241

48   James P. O'Donnell, *The Bunker*, p. 348

49   Jochen von Lang, *The Secretary*, p. 331

50   Jochen von Lang, *The Secretary*, pp. 331, 332

51   William Stevenson, *The Bormann Brotherhood*, pp. 177, 178

52   Ladislas Farago, *Aftermath*, p. 40

53   Ladislas Farago, *Aftermath*, p. 64

54   William Stevenson, *The Bormann Brotherhood*, p. 64

55   William Stevenson, *The Bormann Brotherhood*, p.5

Chapter Fourteen

Riddles

16 April 1945 Top Secret Ultra GP MN
From: 15th U-Flot
To: Comsubs Op, Com Adm U/B's
U234 – Fehler put out of Kristiansand south at 2200B under escort of
'Krieger' ((58.01.3N, 06.40.5E)).
Kptlt Bulla was appointed 1st WO beginning 1/4 ON.[1]
    – Intercepted German U-boat command radio transmission 16 April 1945

18 April 1945 Top Secret Ultra GP W
The following are on their way out of port at present: U-2511--Schnee,
Type 21; U-234 – Fehler, Type 10B.[2]
– Intercepted German U-boat command radio transmission 18 April 1945. These
    intercepts show U-234 apparently left port twice in two days.

There are more mysteries about U-234 than its enigmatic passengers and cargo. The whereabouts of U-234 from 16 April until 12 May 1945, almost a month, is a mystery hidden by a series of riddles, to the point of almost being a conundrum – an unsolvable puzzle. Review of the U-boat's logbook itself reveals a perplexing collection of contradictions when compared against intercepted radio transmissions, other accounts of the voyage, and even other information within the same logbook, suggesting at least part of its record is falsified. In fact, even a cursory glance at what are purported to be various pages of the war log reveals astounding inconsistencies in the physical nature of the book and the handwriting therein, leading to questions and doubt regarding its very provenance. In addition, the few apparently clear facts provided by the war log reveal a bizarre and unexpected travel routine for a fleeing U-boat. And the actions taken by the U-boat commander in the final days prior to its surrender are duplicitous and deceitful – and apparently in coordination with United States Navy activities.

In short, the evidence suggests U-234 may not have left Norway under the conditions reported, probably did not cruise the course across the Atlantic it was claimed to have traveled, and definitely did not surrender when, where and to whom it was ordered to capitulate. Instead, in almost every case its commander, Captain Lieutenant Johann Heinrich Fehler, appears to have been intent on achieving a different, unknown end.

Even before the U-boat cast off from the pier at Kristiansand, its presence was generating considerable interest – both in Germany and across the Atlantic in the United States. A captured German Enigma radio encoder/decoder had allowed the Allies to decode U-boat transmissions describing U-234's secret mission and other aspects of its operations.

As noted in Chapter Nine, U-234 had received important radio transmissions that indicated a struggle was taking place over chain-of-command of the U-boat between Hitler's headquarters and U-boat Grand Admiral Karl Doenitz. There is some interesting additional information that reveals the level of effort later made to hide the truth about what really happened on U-234.

Hirschfeld had written two accounts of his experiences on U-234. The first book, *Feindfahrten*, was written in German and contained several cogent details that are not included in his second account, *Hirschfeld, The Story of a U-boat NCO – 1940-1946*, written in English several years later with Geoffrey Brooks as co-author.

From my contacts I learned that after the first book was published, Fehler and others "in the know" about the mysterious Fuehrer Bunker dispatch, vigorously censured Hirschfeld for having revealed anything about it. Ensuing claims were made that Hirschfeld had, in fact, falsely elaborated his account of events. Such after-the-fact editing seems suspect, however, given the proven veracity of many other elements of Hirschfeld's accounts – which will be pointed out as our narrative continues – and the chain of anomalies and enigmas unrelated to Hirschfeld, but unexplainable except in the context Hirschfeld's account provides.

As noted before, in Hirschfeld's early version Kessler surmised Berlin was sending another passenger to travel in the U-boat and he guessed the unexpected traveler would be Hermann Goering. Fehler was horrified when Kessler called him "The Fat One." But Kessler went on to explain that Goering's presence in the U-boat was unacceptable to him because he and Goering had had a falling out and so he was not prepared to spend several months confined in small U-boat quarters with the Luftwaffe Reich Marshal.

Hirschfeld's account of the order from Admiral Doenitz telling U-234 to leave only on his command is also validated by a signal intercept held in OSS files in which Doenitz does indeed order Fehler to leave on no one's order but his own.

Shortly after receiving the message from Hitler's headquarters about a secret passenger en route to U-234, according to Hirschfeld's German account, the second dispatch came through the flotilla communications center from "Bubbi." Hirschfeld's later English version of events once again excludes any mention of this transmission. Hirschfeld claimed in his first book, however, the communiqué again was sent on a leadership-dedicated frequency just as the first had been.

Although no back-up documentation is available to validate the "Bubbi" message claim, this dispatch, too, I believe may be taken at face value because Hirschfeld so consistently provided accurate details as the enmity between Kessler and Goering, and the OSS documentation validating the struggle for control of the U-boat[3] in virtually every other episode he shared that can be documented.

It seems in every provable detail Hirschfeld has told what he saw and his account should be given credence.

Upon receiving Doenitz's order to follow only his command and to leave port when ready, according to Hirschfeld, Fehler openly acknowledged the clash over chain-of-command of the U-boat by joking about Doenitz's willingness to take on the top brass – a direct confirmation of the struggle over control of U-234 at the highest levels. But this put Fehler in a tight spot. How could he execute conflicting orders from both the supreme commander of the U-boat navy and from the Fuehrer's headquarters itself? Should he fail to do either, the personal consequences promised to be catastrophic.

Certainly, at the very least, an order directly from the Fuehrer's headquarters to Fehler had to have a profound influence personally on U-234's commander. He must have felt great pressure as he was ground between his two powerful leaders. Fehler would need to work magic to squeeze out of this, his very first pickle as the U-boat's captain – and he had not even left friendly shores! Yet he seems to have worked some effective sleight of hand, for radio transmission intercepts record that U-234 apparently fulfilled *both orders*! The massive U-boat is actually documented by intercepted radio transmissions to have left port twice

> 16 April 1945 Top Secret Ultra GP MN
> From: 15th U-Flot
> To: Comsubs Op, Com Adm U/B's
> U234 – Fehler put out of Kristiansand south at 2200B under escort of 'Krieger' ((58.01.3N, 06.40.5E)).
> Kptlt Bulla was appointed 1st WO beginning 1/4 ON.[5]

And two days later:
18 April 1945 Top Secret Ultra GP W
The following are on their way out of port at present:
U-2511--Schnee, Type 21; U-234 – Fehler, Type 10B.[6]

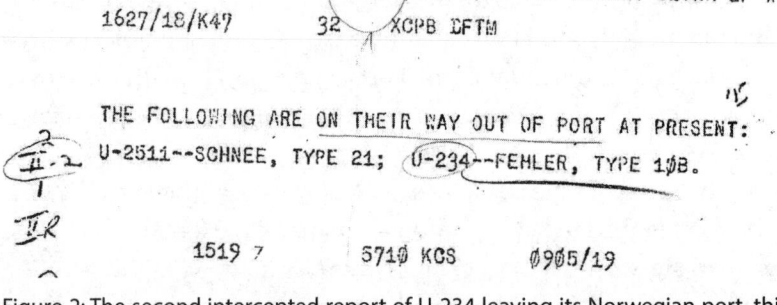

```
                      16 APRIL 1945 TOP SECRET ULTRA GP MN
    0003/16Y        50    PCTM CFCG
    FROM:  15TH U-FLOT
    TO  :  COMSUBS OP, COM ADM U/B'S
    U-234--FEHLER PUT OUT OF KRISTIANSAND SOUTH AT 2200B
    UNDER ESCORT FOR 'KRIEGER' ((58.01.3N, 06.40.5E)).
    KPTLT BULLA WAS APPOINTED 1ST WO BEGINNING 1/4 ON.

    0127 Z            3800 KCS     1215/17
```

Figure 1: The first intercept of two conflicting reports regarding when U-234 left Norway. In this report the U-boat departed 16 April.

```
                      18 APRIL 1945 TOP SECRET ULTRA GP W
    1627/18/K47      32    XCPB DFTM


    THE FOLLOWING ARE ON THEIR WAY OUT OF PORT AT PRESENT:
    U-2511--SCHNEE, TYPE 21;  U-234--FEHLER, TYPE 10B.

        1519 7          5710 KCS        0905/19
```

Figure 2: The second intercepted report of U-234 leaving its Norwegian port, this time on 18 April. By this time, according to U-234's logbook, the U-boat was already 200 miles north of its port of departure and had been sailing for two days.

Here is the second of our circular puzzles. The intercepts record U-234 had "put out of Kristiansand south" on 16 April, according to one transmission. But the other transmission states clearly it was in process of leaving port two days later, on 18 April. How could the U-boat have left port on the 16th and still be leaving Kristiansand two days later on the 18th? Possibly Fehler had changed plans and returned. According to U-234's "official" log, however, on 18 April the U-boat was already approximately 200 miles away, heading north in the opposite direction reported on the 16th, and was then in the latitudes around Bergen, Norway.[7] Apparently the U-boat had not changed plans or been called back, if the log is correct. But then, we shall see that the log, itself, is suspect. For it ends abruptly on 18 April,[8] the very day of the second report of U-234 exiting port. And in its stead there appears another "official" log – a second log, which we shall consider later.

The strange contradiction of the two messages regarding U-234 leaving port twice may be answered once again by radioman Hirschfeld, in another of his cryptic, abstruse passages that shines a wavering light on these mysterious movements. In both of his accounts of the journey he writes that, once U-234 was clear of Kristiansand, U-boat Commander North Captain Hans Rosing sailed to and boarded the U-boat from a "*communications* launch" [italics added].

As a researcher this event perplexed me because, although Hirschfeld inferred it occurred off of Kristiansand, I knew Rosing was headquartered in Bergen. True, he may have been visiting Kristiansand on official or personal business. The U-boat base was certainly within his jurisdiction. But why would he wait for U-234 to leave Kristiansand and then chase it down in a small craft rather than address its crew at the pier, safely ensconced in the U-boat bunker, which was the much more common, efficient, time saving and far safer practice for both the high-ranking officer and the U-boat?

After reviewing Hirschfeld's writings and the intercepted "second exit from port" message, combined with the evidence of the strangely truncated logbook on the one hand, and the position of U-234 near Bergen, as posted in that logbook on the other hand, it seemed to me U-234 secretly had detoured to Bergen for an unknown purpose. If this was the case and the detour was supposed to be kept secret, it would explain the mystery of one logbook having been discontinued abruptly on that very day, rather than record the fact of the Bergen visit. A "mock" or replacement logbook would then have to have been created – possibly right away, possibly at a later date – to hide the clandestine detour and future deceptive events.

Thus we have the second "official" logbook that documented the rest of the journey.

Hirschfeld's description of the meeting with Rosing strongly supports the idea U-234 visited Bergen. According to Herbert Werner, author of the classic U-boat account *Iron Coffins*, and himself a U-boat commander serving in Norway at the time, Rosing was, in fact, in Bergen during 16 through 19 April, 1945.[9] Rosing himself asserted he did not remember his whereabouts at the time,[10] although the event seems so singular one would expect the key details to remain in his mind.

Whether at Kristiansand or Bergen, almost certainly Rosing and Fehler did not risk their one-of-a-kind U-boat, priceless cargo and important passengers and crew sitting openly in the dangerous waters off port – where British submarines and anti-submarine aircraft regularly prowled to interdict U-boat activities – just so Rosing could give three cheers for captain and crew. One can only speculate what the purpose of the detour might have been. The few small clues Hirschfeld provided, and knowing Fehler was caught in the middle of a perilous game of cat-and-mouse between Doenitz and the Fuehrer's Headquarters, surely must be considered as a context for any masquerade. There must have been an important operational reason for this secret side trip. Probably that reason is revealed in Hirschfeld's description of the boat that brought Rosing to U-234's side – he described it as a *communications* launch [italics added].

Apparently certain communications were of such high importance or of such a secret nature they were encrypted at sophistication levels where their decryption was possible only at properly equipped high-level communications centers, and not on the average U-boat. At least, such seems to be the case here. Possibly Rosing was hand delivering one of the special-frequency dispatches from the Fuehrer Bunker that U-234 was not equipped to receive; so this detour was U-234's "at sea" version of Hirschfeld's visits to the Kristiansand communications center on land.

We may speculate such a message most likely was operational orders for U-234, possibly resolving the struggle between Doenitz and Berlin over who would control the U-boat, or perhaps giving instructions on how to deploy until time to pick up its secret guest from Berlin. Or the communications launch itself may have been sent to transfer to U-234 the equipment required to receive the special-frequency messages from Berlin. This is conjecture, but certainly not outside the realm of possibility. Hirschfeld makes it clear in his writings the radio components of the boat were modular and easily changed in and out of the console;[11] and that the boat was equipped with the very latest instrumentation and every possible technical advantage.

Rosing's final words to captain and crew may be telling about what he knew of the mission of U-234. He said, "Comrades, when you return from this mission, we will have our final victory." Given the desperate situation for Germany – it would fall within two weeks – the crew rightfully, though quietly, questioned his sanity. But given the purpose of U-234's mission, if there was hope of victory at all for the Third Reich, it was in the success of this mysterious made-over minelayer – and, tellingly, Rosing knew it.

While our first puzzle – was there a mystery guest from Berlin and if so, who was it, is still a mystery to be considered more fully later – it would seem our second puzzle, U-234's leaving port twice, is solved.

But what of the strangely truncated logbook – which leads to our third puzzle? Why does one logbook end abruptly and its supposed sequel not jive with the rest of the evidence regarding U-234's movements? When I first requested a copy of the captured war log of U-234 from United States archives at the beginning of my research, I was told by an archivist the logbook had been thrown into the sea by U-234's captain. He asserted Fehler got rid of the journal prior to the U-boat's surrender to avoid compromising the document and the boat's movements.

But U-234 carried Nazi Germany's greatest secret weapons, I reasoned, including the V-4 rocket, the Messerschmidt 262 jet fighter, all of the plans and documents required to manufacture them, atomic bomb components and presumably plans to build those weapons, as well. If Fehler did not know the important details about his freight, which seems improbable despite his later claims, he at least knew the basic reason for and deep importance of his cargo, passengers and mission, and yet he surrendered them all intact. I reasoned this was a significant incongruity.

Had Captain Fehler surrendered the valuable Nazi secrets and personnel, supposedly with little more than a second thought, but had refused to surrender his comparatively trivial logbook? The journal, presumably, simply reported the course he cruised prior to surrender. What could be damning about that if the story was as simple as suggested? No, the logbook itself apparently held important secrets Fehler did not want revealed, and thus Fehler had indeed consigned it to the deep and we would never know U-234's whereabouts between 16 April and 12 May 1945. Or, possibly, the book was intact but held damning evidence, and thus was being kept in some separate archive out of circulation from prying eyes.

When during a research session in Washington in 1997, I was told the Library of Congress held a collection of captured German documents, I raced over to the venerable old building in hopes of locating the missing log. I was informed the captured documents did indeed contain a journal from U-234, but that all the documents had been microfilmed and returned to Germany to be archived there. Satisfied with the opportunity to read the microfilm rolls, I began searching for traces of a logbook from U-234.

Microfilm roll 18 held what I was looking for – almost. A logbook identified as that of U-234 began on 24 March 1945, the day before the U-boat's departure from Kiel to Kristiansand. As noted previously, it ended abruptly on, of all days, 18 April 1945 – the same day of U-234's mysterious "second exit" from port. I use the word "abruptly" because, while the U-boat's activities were meticulously detailed throughout the days and weeks leading up to and through 17 April, including leaving Kristiansand on 16 April – corroborating the first intercepted message of it leaving port on that day – the heading "18 April" is written in longhand halfway down the page, but the rest of the page is empty, as if it was then forgotten about. There are no entries in the half-page underneath the date. No course coordinates, no weather reports, no times, no bearings. The remaining half-page is just blank. And there are no entries for the 19th or 20th – the microfilmed portion of the log does not pick up again until 12 May, the day U-234 first transmitted its intent to surrender to Allied forces.[12]

Baffled by the inconsistencies and the gargantuan gap in the record, I approached a Library of Congress archivist, who informed me the original logbook, of which I now had access only to a microfilm copy, had been sent to the Bundesarchiv in Germany. He suggested perhaps I could get the missing portion of the record from there. I faxed the Bundesarchiv, requesting a copy of the missing portion of the log, if they had it. In return I was mailed a photocopy of record RM 98/676, with the words "Uboot U234" written in blue fountain pen ink on the front cover. Nowhere throughout the entire document is U-234 identified as an organic, photocopied part of the journal as the U-boat of record. The copy of the log begins on 19 April, per my request (I now wish I had requested it from 24 March, when U-234 left Kiel. I wonder if the record would have started then or abruptly on 18 April?) and ends on 12 May, the day Fehler radioed his intent to surrender.

There are two problems with this logbook; the positions, speeds, bearings and coordinates given for the last day before surrender show a course materially different than that actually sailed by Fehler, as revealed by Allied radio direction-finding coordinates, and as substantiated by Hirschfeld and even Fehler himself, during later accounts of events. And the Bundesarchiv logbook is neither the same printed layout nor are its entries written in the same handwriting as that of the logbook micro-filmed by the Library of Congress, of which the Bundesarchiv copy was supposed to be part and parcel.

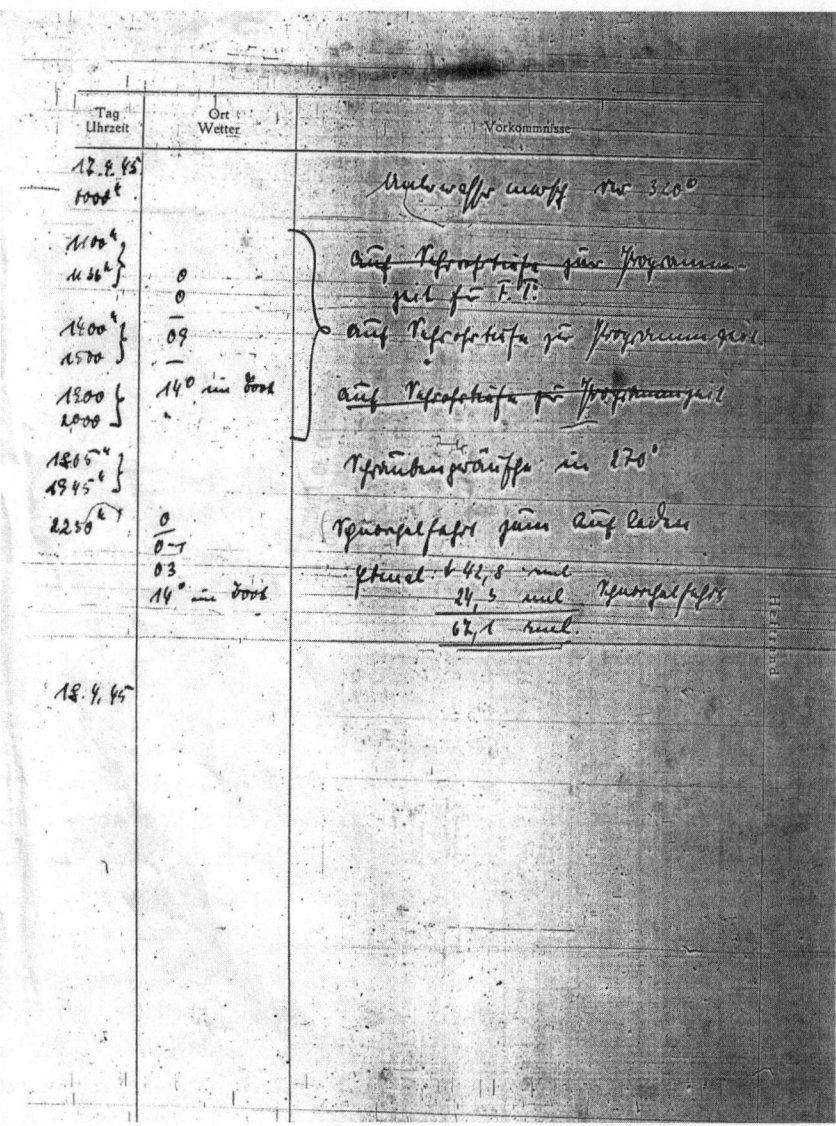

Figure 3: U-234's first logbook ends abruptly on 18 April, the same day the U-boat is supposed to have been leaving port a second time. The replacement logbook shows demonstrably incorrect information about U-234's movements and activities.

In fact, when I had Captain Carl Triebes, USN (Ret), an experienced submariner and navigator, chart U-234's course from the information given in the Bundesarchiv logbook, Captain Triebes stated he did not think the document was a working logbook at all – the handwriting and layout of the log were just too nicely done. He felt someone had copied it over from an actual log to make it more presentable. I had not told him about the project I was working on or what I expected him to find, so this volunteered observation seems a compelling indication that perhaps the logbook was "doctored."

Since both the intercepted transmissions and the Bundesarchiv logbook are primary evidence – authoritative, contemporaneous and organic to the events under study – these conflicts are significant. The inconsistencies in the evidence suggest gross negligence or willful deceit in completing one or both of the records. Radio intercepts are and were dispassionately dated intelligence for the purpose of tracking important events, and there seems to be no reason why anyone would manipulate these particular records.

On the other hand, that there are major inconsistencies between the physical and informative aspects of the Library of Congress and Bundesarchiv logbooks when compared to each other, or to the external, objective, solid data collected elsewhere, casts considerable doubt on the logbooks' veracity, in the opinion of this researcher. The data recorded in the Bundesarchiv logbook in many cases does not fit either the official account or unofficial recollections given of U-234's journey; and on another level, in fact, the entries appear to try to hide the U-boat's actual movements.

One can make a long list of details within the Bundesarchiv logbook that conflict with other data in the log or with other substantive evidence regarding the movements of U-234, or that is incongruous with the U-boat's stated mission and the rest of its activities. Thus all the information together suggests an organized effort to camouflage U-234's movements.

For example, according to its daily noon-time coordinates postings, the fleeing U-boat – specially equipped to sail submerged at eight to ten miles-per-hour, and almost 20 miles-per-hour surfaced,[13] was hardly moving throughout most of the voyage, traveling at between one and two-and-a-half miles per hour – just enough speed to maintain steerage. The average man walks between two and two and-a-half miles-per-hour. The traditional history would have us believe the U-boat was racing for

Japan, but according to the Bundesarchiv logbook the U-boat traveled slower throughout most of the journey than a person walks – until its last few days at sea, when it was no longer going to Japan but preparing to surrender instead.

1.KTB No.48, Page 5.
With an economic underwater speed of less than 2 knots it was certainly anything else but a "dash" to Japan. Depending on which way we could choose, via Cape of Good Hope or round South-america,we calculated the better part of 5 months,and this even if we could with 9 knots over the better parts of the South-pacific and the Southatlantic.

Figure 4: Captain Fehler's explanation of U-234's slow speed, made to 'Sharkhunters International' president Harry Cooper. Despite his reference to "economic underwater speed," the boat was not running submerged at night, nor was it planned to sail to Japan without refueling, although it had the capability.

While Fehler later explained this slow pace by suggesting it was the most economic speed for such a long mission,[14] intercepted transmissions show U-234 planned to refuel in Indonesia,[15] even though it had enough fuel to make Japan sailing under normal conditions.[16] Considered against this information and the nature of the mission, the special capabilities of the U-boat, and reports it later sailed submerged for six days, apparently unnecessarily – using much more fuel than sailing surfaced would have required – the slow speed recorded in the log seems to suggest Fehler was marking time in an effort to remain close to home.

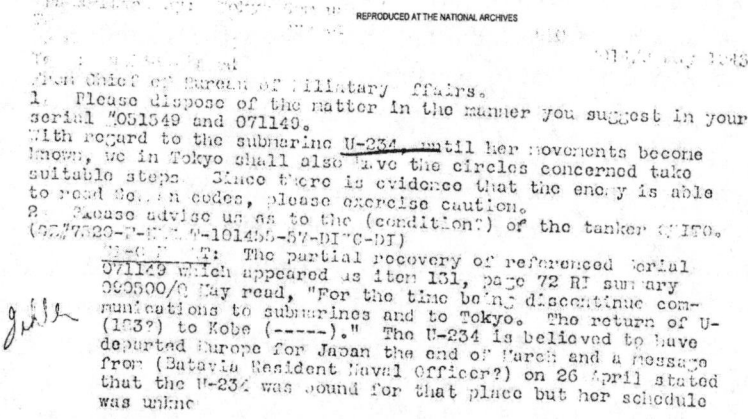

Figure 5: An intercepted Japanese radio transmission discusses the whereabouts of U-234 and its plans to refuel in Batavia (Indonesia).

Another example of deceit? There are other sizeable and apparently un-explainable discrepancies between where the U-boat was at a given time as recorded in the logbook according to celestial or electronic navigation co-ordinates, and where it was plotted to be according to reckoning by distance and compass direction. Records of both techniques were kept in the log. Small disparities between these two forms of navigation are to be expected as they are used to crosscheck one another. But the errors recorded in the case of U-234 occur too often and are too large – off by as much as 200 per-cent or more in speed and almost 90 degrees in direction in a single day's travel. In short, while certain details seem to be accounted for in the log, like a near-miss with a steamer and an electrical fire, which are both recorded in Hirschfeld's account as well, the general plotting in the logbook appears to be patently and inexplicably sloppy and inaccurate.

The gross disparities in the record suggest someone was completing the logbook very quickly and without caring where U-234 actually was when the entries were being made. In fact, completion of the log seems to have been done with little concern for ensuring the two forms of naviga-tion would validate one another at all! And as the journey progressed, the errors became greater.

Importantly, at the same time, other unexpected but seemingly inten-tional changes occurred.

For example, in the opening hours of 1 May – about the same time Martin Bormann was trying to make it to a rendezvous with a giant U-boat in Hamburg, according to Stalin's report – U-234 broke from its planned journey and turned due east, back toward the North Sea, from a southwesterly course. The logbook records this diversion lasted only one hour. Review of the events leading up to and after this strange change of course, however, may be revealing.

Beginning in the early morning hours of 30 April, about the time Bormann was concluding a series of deceptive dispatches to Doenitz to arrange the final details of his escape from Berlin, a series of changes in U-234's operation were entered in the log. First, and perhaps most tell-ing, the logbook records that Fehler now chose to run submerged in the Atlantic for six days straight, even though the U-boat was now breaking into safer open water; whereas the Bundesarchiv log and Hirschfeld both agree that previous to this time, but after the near miss with the steamer, U-234 had run surfaced almost every night, despite being in the much more dangerous Iceland/Farroe Island narrows.[17]

According to the logbook, U-234 sailed submerged continuously with-out surfacing from the early hours of 30 April until late 5 May – the same crucial time span between Bormann's disappearance from Berlin starting on 30 April, to Doenitz's capitulation on 5 May. If true, running for almost six full days either fully submerged or at snorkel depth was a rare event for any U-boat, much less one running free and undiscovered in the open sea.

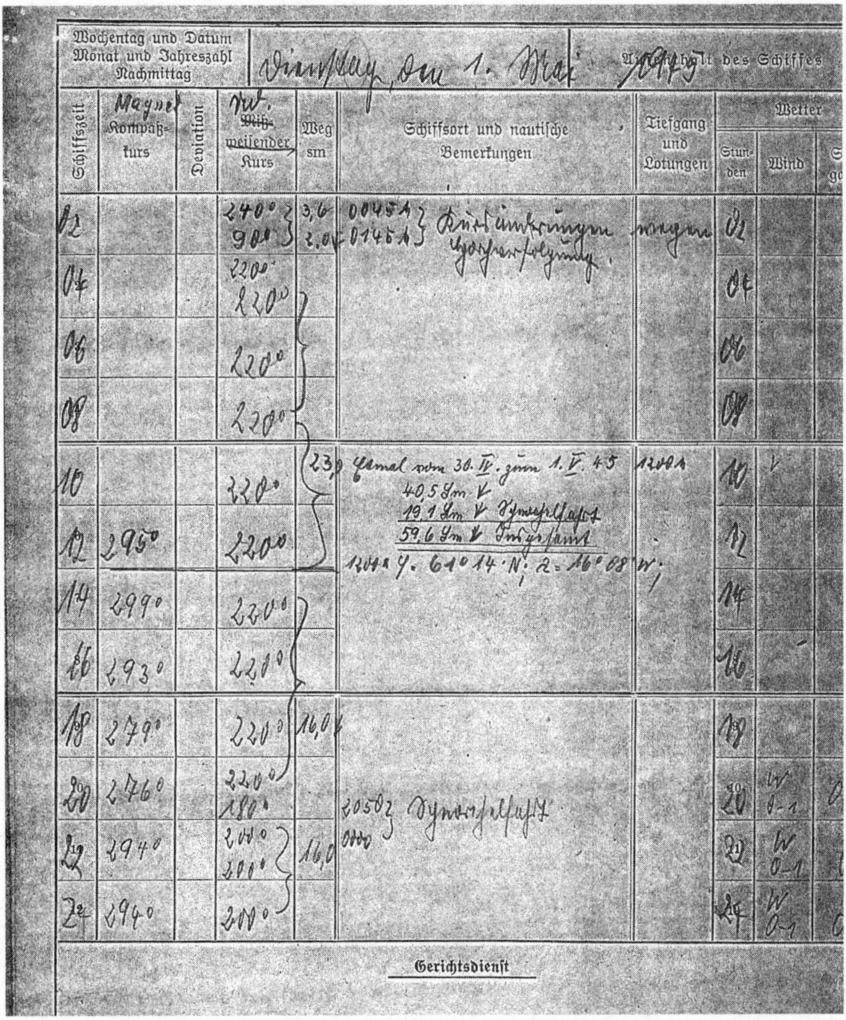

Figure 6: The page for 1 May 1945 from U-234's second logbook. Page shows the first of five full days recording the U-boat ran submerged continuously, either with or without snorkel. The U-boat was alledgedly now in the open sea with little need to run submerged, especially at night. This is the timeframe that Martin Bormann, Hitler's top lieutenant, was picked up in Hamburg by a "large submarine."

Importantly, Hirschfeld's account – proven extremely accurate thus far – conflicts with the logbook, saying U-234 continued to proceed "submerged by day and surfaced at night under the protection of our radar."[18]

Even Fehler identified the logbook is only partially true when he later wrote that on the first two nights after passing through the strait his efforts to surface were thwarted by unidentified aircraft on his radar.[19] He affirms, however, that on the third night the U-boat was able to remain surfaced "for several hours." This is in direct conflict with the logbook. He gives no account of the fourth, fifth and sixth nights.

Hirschfeld's account of these critical first days in the Atlantic, while brief, differs markedly from Fehler's. He states "during the first night we were obliged to dive twice because of aircraft,"[20] the connotation being that during the rest of the first night and on the remaining nights, the boat ran surfaced, not submerged as the log reported. Later analysis of Fehler's account will prove even greater disparity between what was written in the logbook and what appears to have actually occurred.

In trying to decide which record is true, Fehler's, Hirschfeld's or the seemingly faulty logbook, the operational situation of the U-boat must be considered. Fehler was now on the open Atlantic where U-boat interdiction was considerably leaner than on the North Sea and where he had much more room to maneuver, whether surfaced or submerged with thousands of feet of water beneath his keel, and the benefit of the best radar. Additionally, in this part of the Atlantic where it was more difficult to support antisubmarine activity from land bases, U-boat detection by the enemy was usually made only when a U-boat attacked an enemy ship. Therefore a U-boat that did not attack was all but safe from detection. Fehler admitted as much in an undated letter written to Harry Cooper, president of *Sharkhunters*,[21] in which the captain stated he was little concerned about being discovered; he had no intention whatever of attacking anything. In fact, he intended to steer clear of all contact.

These accounts also document the superior protection provided by the special radar with which the U-boat was equipped; which could search the ocean and skies for miles in all directions within a split second without giving away the U-boat's location.[22] The cutting-edge radar system had already saved the U-boat from serious incident once, having early in the voyage detected anti-submarine airplanes, allowing Fehler to evade danger long before the planes could get a fix on the U-boat.[23]

Considering his superior radar and all of these favorable conditions and the greatly improved fuel economy and speed of running surfaced as opposed to snorkeling, Fehler had comparatively good reason to run surfaced, at the very least during the dark of night, as he had done in the much more dangerous North Sea. He even wrote in his letter to Cooper,

"Later on in the open ocean, staying submerged during daytime offeres (sic) a fair chance to pass through undetected," [italics added for emphasis]. He thus inversely infers that he did, indeed, sail surfaced at night during this time, despite the logbook's entries that he was submerged, and again corroborating Hirschfeld's account.[24]

Perplexing and contradictory as it seems, however, the logbook does, apparently deceitfully, record that Fehler had, in essence, "gone to ground" and remained submerged for the entirety of those six days. Considered against the U-boat's operations in the Iceland/Farroe Island Narrows, its superior radar equipment, Fehler's supposed concern for fuel economy, and other conditions as have been outlined, sailing submerged in the open Atlantic for six days straight seems unlikely.

In the context of a second, more intriguing escape scenario, however, as we shall see, such entries in the Bundesarchiv logbook make good sense – as a cover-up for the period of time between 30 April through 5 May, when U-234 detoured on a secret side trip.

A second set of telling data regarding such a detour arises from conflicts within the Bundesarchiv logbook itself, and between the logbook and another record, as well. During the voyage, General Kessler was entering in his diary the U-boat's position coordinates taken at noon each day.[25] His postings match exactly those of the logbook until 30 April – the very day U-234's mysterious movements begin – when Kessler, for the first time during the journey, failed to post coordinates. On that day, the Bundesarchiv logbook showed coordinates at noon of 61° 58' N, 14° 49.5 W, a distance from the previous coordinates that roughly represented the U-boat's average speed thus far. The following day, Kessler posted the exact same coordinates for 1 May that the logbook showed for 30 April – 61° 58' N, 14° 49.5 W.

For 1 May, the logbook in its turn showed 61° 14' N, 16° 08' W, indicating U-234 had traveled the average daily distance again. Up to this point, because the logbook entries remain consistent with daily distances covered, one would assume the logbook is correct and Kessler had just made a mistake. That Kessler's diary exactly matched the coordinates in the logbook until 30 April and then, even when it differed, it showed only an apparent error in transcription of being off by one day, supports this idea. Perhaps Kessler was receiving his data secondhand as provided by the sophisticated radio navigation system U-234 deployed. But Kessler never corrected his 1 May entry, even though it obviously would have been wrong when he updated his diary the following day, on 2 May.

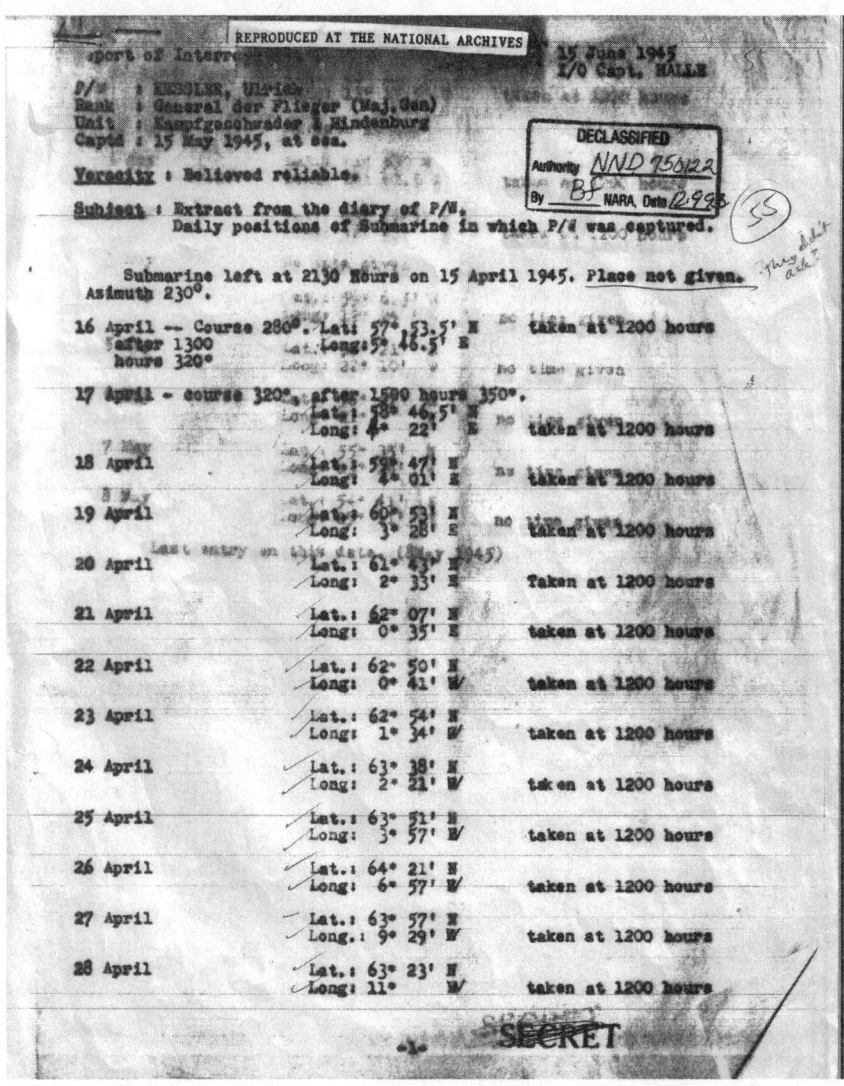

Figure 7: An English-language translation of a transcript of coordinates of U-234 that General Kessler kept in his diary. Two conflicts between General Kessler's entries and U-234's logbook may reveal unknown movements of the U-boat.

Reviewing the coordinates given in the Bundesarchiv log beginning the very next day, however, on noon of 1 May and again on noon of the following day, shows the U-boat traveled barely half of its already slow average daily distance. Such an adjustment considered with Kessler's erratic entries of the days prior may suggest a correction of some sort was made in the log instead. This suggestion becomes more plausible when one considers that the distance traveled for the same 24-hour period, as directly written in the log, is over 60 statute miles – five statute miles above the daily average, not half of it, as the distance between coordinates show for that day. One minor course change recorded during that period would hardly have impacted the overall distance traveled and therefore could not be accountable as a mistake for the discrepancy between the two entries. The record therefore suggests uncertainty about where U-234 was, beginning on 30 April.

The stream of conflicts within the logbook consistently increased from this point forward. A second inaccuracy occurred two days later, between 3 and 4 May, when the posted coordinates recorded the U-boat again moved only a handful of miles, certainly fewer than ten. But the logbook stated 54.7 statute miles had been traveled with no course changes significant enough to account for the difference. Perhaps tellingly, General Kessler's diary on 3 May fails for the second time to record any daily coordinates at all. This herky-jerky motion of the U-boat from day to day as recorded in the logbook, which is out of phase with Kessler's also herky-jerky data, but which Kessler seems to try to account for, seems on the face of it to indicate uncertainty as to where the submarine actually was after 30 April. It is almost as if someone after the fact went back through the log and, ignoring the direct reckoning recorded as direction and speed, entered the series of coordinates they wanted reported regardless of what actually occurred.

A third conflict that continues the fantasy – the largest of the three – occurs in the 24 hours between 5 and 6 May. The coordinates posted show distance traveled of about half of the 55-statute mile daily average, for a total travel distance of about 30 statute miles. But the actual distance reported records a whopping 99 miles – close to twice the average daily distance and three times that calculated by the coordinates reported!

This is the furthest distance in a single day U-234 traveled, by far, recorded to that point in the journal. What compounds this truly significant and very obvious discrepancy is that course bearings given throughout the 24-hour period are almost consistently 220°, a straight line west-south-west. But the end coordinates show U-234's position was about 30 miles *southeast* of its position 24 hours prior. In other words, both the distance *and the direction* traveled are in serious discrepancy within the log *itself!* The distance traveled, as entered directly in the logbook, differs not only by about 60 miles, or three times the distance calculated from the coordinates, but the logbook is off by almost 90° in direction, or one-quarter the arc of the compass, as well.

As noted earlier, marginal differences in a course tracked by coordinates compared against a course tracked by bearings and distance are to be expected. Winds, currents and human error of just fractions of a degree will create variances in position when navigating a submarine. But the size and number of the discrepancies listed above are hardly explainable by anything but the most profound errors, which were very unlikely; suggesting instead that someone recorded a duel record in the same pages.

The variance between the coordinate positions posted on 5 and 6 May and the direction and distance plotted are so extreme, if one does not believe U-234 actually sailed southeast – per the coordinates posted – but rather sailed according to the dead reckoning information, or vice versa – an entire day is lost. This is relevant given General Kessler's trying to account for one. But there is nothing in the document to suggest a correction was made for a lost day. In our scenario of U-234 making a secret side trip, unaccounted for days are central to understanding what the U-boat may actually have been doing during this time. The only other answer for the 'lost day' would be if the U-boat came to a complete standstill for 24 hours, which runs counter to all accounts. But even if it had, why would Fehler have recorded U-234 was traveling in two directions at once?

Endnotes: Chapter Fourteen – Riddles

1   U.S. National Archives II, intercepted radio transmission 18 April, 1945, RG 38 – 370 01/04/07 box 113

2   U.S. National Archives II, intercepted radio transmission 16 April, 1945, RG 38 – 370 01/04/07 box 113

3   U.S. National Archives II, interrogation report of General Ulrich Kessler #1540, p.4 (date unknown), RG 165 – 390 35/10/05 box 495

4   U.S. National Archives II, intercepted radio transmission 12 April and 13 April, 1945, RG 38 – 370 01/04/07 box 113

5   U.S. National Archives II, intercepted radio transmission 18 April, 1945, RG 38 – 370 01/04/07 box 113

6   U.S. National Archives II, intercepted radio transmission 16 April, 1945, RG 38 – 370 01/04/07 box 113

7   Bundesarchiv 24/82 RM 98/676

8   United States Library of Congress, Manuscripts Division, Captured German Documents, Microfilm Roll 18

9   The author personally witnessed a telephone call between Harry Cooper, Sharkhunters president, and Mr. Herbert Werner, on 21 January 1999, in which Mr. Werner confirmed to Mr. Cooper Rosing's presence at Bergen throughout the time span in question.

10   Letter from Sharkhunters president Harry Cooper to the author dated 11 May, 1999

11   Geoffrey Brooks and Wolfgang Hirschfeld, *The Story of a U-boat NCO 1940-1946*, p. 201

12   U.S. National Archives II, *Report On the Interrogation of the Crew On U-234 Which Sur-*

rendered to the USS Sutton on 14 May, 1945, In Position 47°-07'N - 42°-25'W., RG 38 – 370 15/09/01 box 2

13   Geoffrey Brooks and Wolfgang Hirschfeld, *The Story of a U-boat NCO 1940-1946*, pp. 195, 201; also, compare with speeds of U-234 calculated from radio intercepts reported direction finding on May 12, 1945, RG 38 – 370 01/04/07 box 113

14   Undated letter from CaptainLeiutenant Johann Heinrich Fehler to Harry Cooper, president of Sharkhunters International, p. 1

15   U.S. National Archives II, NSA Records, secret German transmissions from Marine Special Forces to Penang, Shonan, Djakarta, Tokyo, 13 February 1945 and March 18 1945, RG 457-190-32-2-7

16   Geoffrey Brooks and Wolfgang Hirschfeld, *The Story of a U-boat NCO 1940-1946*, p. 192; Undated letter from CaptainLeiutenant Johann Heinrich Fehler to Harry Cooper, president of Sharkhunters International, p. 1

17   Geoffrey Brooks and Wolfgang Hirschfeld, *The Story of a U-boat NCO 1940-1946*, pp. 206, 206

18   Geoffrey Brooks and Wolfgang Hirschfeld, *The Story of a U-boat NCO 1940-1946*, p. 207:see also Wolfgang Hirschfeld, *Feindfahrten*, p. 361

19   Undated letter from CaptainLeiutenant Johann Heinrich Fehler to Harry Cooper, president of Sharkhunters International p.2

20   Geoffrey Brooks and Wolfgang Hirschfeld, *The Story of a U-boat NCO 1940-1946*, p. 207; Wolfgang Hirschfeld, *Feindfahrten*, p. 361

21   Undated letter from CaptainLeiutenant Johann Heinrich Fehler to Harry Cooper, president of Sharkhunters International p.1

22   U.S. National Archives II, *Report of Interrogation of U-234 passenger Kay Nieschling, 24 May 1945*, RG 165 – 390; a copy also exists in RG 38 – 370 15-09-04 box 13

23   Geoffrey Brooks and Wolfgang Hirschfeld, *The Story of a U-boat NCO 1940-1946*, p. 207; Wolfgang Hirschfeld, *Feindfahrten*, p. 361

24   Geoffrey Brooks and Wolfgang Hirschfeld, *The Story of a U-boat NCO 1940-1946*, p. 207; Wolfgang Hirschfeld, *Feindfahrten*, p. 361

25   U.S. National Archives II, *Report of Interrogation of General Ulrich Kessler, extract from POW's diary*, RG 165 – 390 35/10/05 box 495

The prize crew of the USS *Sutton* prepares to raise the U.S. flag over U-234, with Lt. Commander Thomas Nazro (second from left), commander of the *Sutton* looking on. Nazro commanded his crew to block Canadian radio messages to U-234 when Canada rightfully ordered the U-boat to surrender to Halifax; then had his crew communicate with U-234 only by Morse code using lights to complete the surrender to the *Sutton* instead.

Chapter Fifteen

Surrender

12 May 1945
From: U234 (Fehler)
To: GZZ 10
Position 50.00 N – 30.00 W. Surfaced, course 260, speed 8.

D/F [Direction finder fix] 51.00 N – 27.00 W 0623Z [6:23 a.m.]
Transmission sent from U-234 at 6:23 a.m., 12 May 1945

12 May 1945
From: U-234 (Fehler)
To: Comsubs Op
Surfaced at 0800B/12/5/45
Position 50.00 N, 30.00 W.
Course 260. Speed 8.

D/F Position 50.00 N. – 34.00 W 2340Z [11:40 p.m.]
A second transmission from U-234 sent over 17 hours later. Despite the reported unchanged position, direction finder fixes showed U-234 was traveling westward at a high rate of speed. In the first transmission, U-234 actually was well east of its reported position; and it actually was well west of its reported position in the second transmission.

O bviously, U-234 could not sail in two directions at once, as indicated in the conflicting dead reckoning entries versus the coordinates posted in the Bundesarchiv logbook. And U-234 did not stand still either – quite the opposite. A last, and also substantial, series of conflicts between the Bundesarchiv logbook and other sources occurs on 11 through 12 May, which included the final 24-hour period recorded in the log – which ended on the 12th, when Fehler first radioed his intent to surrender.

The final coordinates entered in the logbook – 49° 20' N, 31° 51 W – were for noon 12 May. This is perfectly aligned with the south south-westerly course plotted from 7 through 12 May, which adhered to an average bearing of 220°, or south southwest. Surprisingly, however, according to the logbook actual bearings on the 11th and 12th swung widely, from 180°, or straight south – a course reportedly pursued throughout most of 11 May – to a course change to 260°, almost due west, which turn occurred at 2:35 a.m. on the morning of the 12th.

The distance covered during 7 through 10 May, as calculated from the daily coordinates, was about 60 to 70 miles per day, again about the average. But on the 11th, the coordinates showed a doubling of distance to about 120 miles. The total distance sailed entered in the logbook for 12 May is 201 statute miles, which, while a great increase in speed, given the drive south then dogleg west, calculates closely enough to match the 120 miles represented by a straight line from start coordinate to end point.

((PLAIN T[...] OP SECRET ULTRA MR EHB

12 MAY 1945

Ø623/12

FROM: U-234 ((FEHLER))

TO : GZZ 1Ø

POSITION 5Ø.ØØ N - 3Ø.ØØ W. SURFACED, COURSE 26Ø, SPEED 8.

D/F 51.ØØ N - 27.ØØ W (L-2Ø) CASE 12/597Ø/Ø7X 8-25
Ø623Z

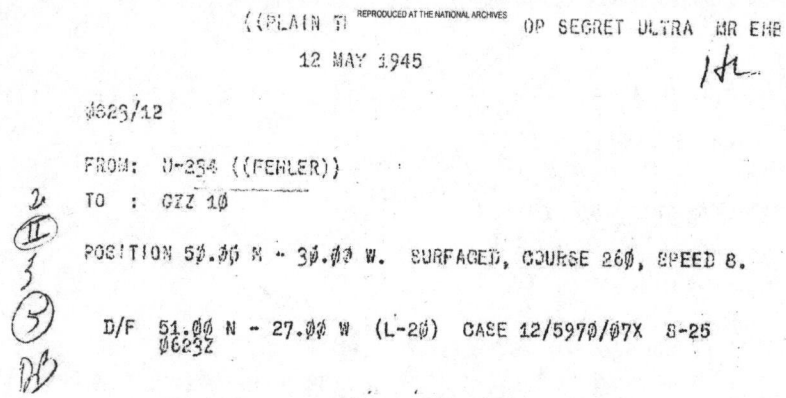

Figure 1: U-234's Captain Fehler falsely signals its location to Halifax. Direction finder coordinates ("D/F" at bottom of transmission) show its actual location.

((PLAIN TEXT MESSAGE)) TOP SECRET ULTRA VG CCN

12 MAY 1945

2356/12/Ø

FROM: U-234 ((FEHLER))

TO : COMSUBS OP

SURFACED AT Ø8ØØB/12/5/45

POSITION 5Ø.ØØ N, 3Ø.ØØ W.

COURSE 26Ø. SPEED 8.

D/F POSITION 5Ø.ØØ N - 34.ØØ W (U1-2Ø) Case 12/77ØØ/2ØX

234ØZ 77ØØ KCS 2145/12

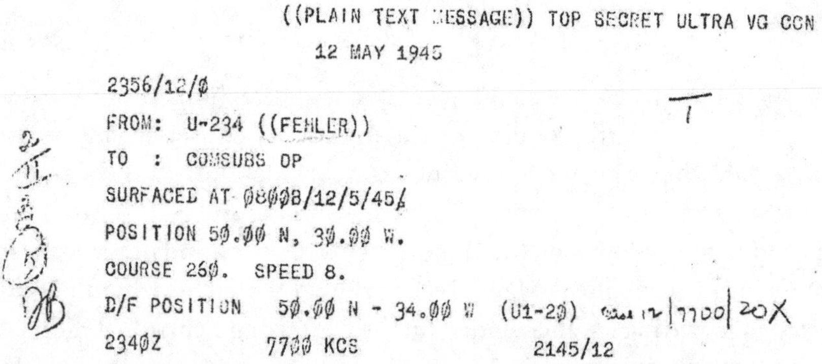

Figure 2: A second transmission almost 18 hours later claims an unchanged location, but direction finder shows the U-boat traveling faster than at any time during the entire journey, and, once again, at different coordinates than reported.

At the outset these entries appear to be accurate, although, as noted, representing a great, and unexplained, increase in U-234's speed as well as an unexplained an illogical course stretching south, followed by a turn west, to bearing 260°. A straight run south southwest would have been consistent with its reported previous path and much more efficient. There is no reason given within the explanations of the traditional history for the doubling of speed, nor for the huge dogleg other than a general desire not to be captured before getting the U-boat positioned for surrender to the United States, despite already transmitting surrender intentions.

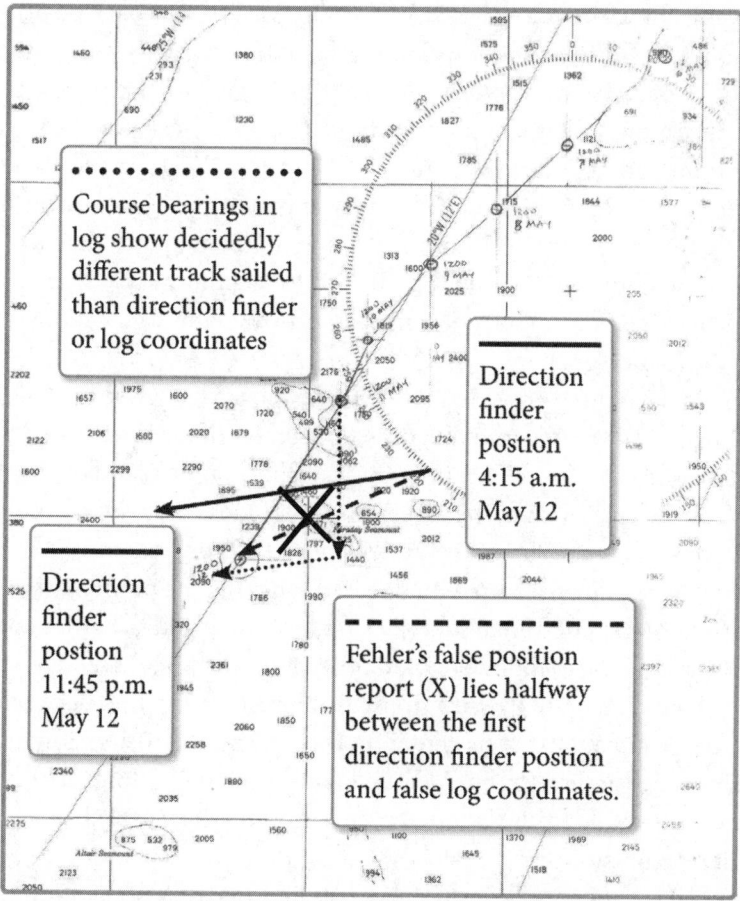

Figure 3: U-234's movements 12 May 1945, the day it signaled surrender intentions. The log coordinates (penciled line with small circles) falsely show it was on a course consistent with its planned route to Japan. Allied direction finders show it was steering a much more westerly course, however, at twice the speed reported. The "X" marks the false position Fehler reported during the 4:15 a.m. transmission. The falsified log and Fehler's admittedly intentional false position report appear intended to hide U-234's actual movements.

Transcripts of radio transmissions from U-234 to Halifax show Canadian direction-finding locations of the submersible that record the actual locations of U-234 at the time it sent the messages. The direction finder in this case revealed locations that corroborated the U-boat was sailing toward 260° on the compass. The direction-finder also showed, however, that on the morning of 12 May U-234 was actually at a position 70 to 80 miles north, and more importantly, 150 miles east of the position calculated from the times and bearings recorded in the logbook. At 4:15 a.m. on 12 May, the intercept's direction-finder coordinates put the U-boat at 51º 00' N, 27º 00' W. The U-boat's position as extrapolated from the logbook's speeds and bearings, however, indicate it should have been at 49° 20' N, 31° 00'W, give or take a few miles. Therefore, as noted, it was trailing about 150 miles east-northeast of the position recorded in the logbook.

Fehler further complicated things – apparently with a plan in mind – by falsely transmitting during that same 4:15 a.m. radio message that U-234's position was 50º 00' N, 30º 00' W, significantly differing both from the direction-finder coordinates and from the calculations made from the logbook data.

So we have three considerably different positions given for U-234 at 4:15 a.m. 12 May: One from the direction-finder of the intercepted transmission, one calculated from the Bundesarchiv logbook, and one that Fehler reported to Halifax. The direction-finder location is by far the most likely to be accurate, since it is the only objective source.

Drawing a line on a map with one end touching the point marking the direction- finder coordinates for 4:15 that morning and the other end the point marking the logbook position posted for noon later that day, shows that Fehler's radioed false position falls almost on the line between the two points and almost at its center. In the 4:15 a.m. transmission he also gave his speed as 8 knots, but he was actually sailing 16 knots according to calculations made from the intercepted transmissions as well as Fehler's own later admission.[1]

From U-234's actual position as revealed by the direction-finder, at the 16-knot speed it was sailing, U-234 would make the position falsely recorded in the logbook just about noon – the time of the daily logbook posting. Doing so would bring it in conformance with the fabricated logbook scenario that suggested U-234, generally, was continuing on its southerly course. Thus Fehler would continue an illusion he had created that U-234 was sailing a conventional course to Japan on the Great Circle.

And U-234's position would be directly in line with the previous six days of posted coordinates, further validating the false entries in the logbook and hiding the fact U-234 was actually racing on a very fast westerly run from roughly the direction of the mouth of the English Channel.

Fehler was caught in the lie, however, when a Canadian plane discovered U-234 out of its reported position.[2] To cover the deception, Fehler then entered the dogleg in the speeds-and-course portion of the logbook, making it look as though both the course steered and the coordinates posted were in accord. Thus the lie was tied up in a neat little bow.

Fehler then does a strange thing. Having carefully set up the deception to the point of entering false data in the logbook, instead of heading straight for those noon coordinates already posted to complete the illusion, Fehler proceeded on a more westerly route.

He also abruptly discontinued his second logbook. There are no more entries in the Bundesarchiv log after noon 12 May.

Why? Possibly because Fehler's deceptive first transmission to Allied forces reported a false position not only calculated to camouflage his movements, but those coordinates would show U-234 proceeding on a course that passed exactly through the intersection of the boundaries of Allied naval control of the Atlantic Ocean. East of this point was the jurisdiction of the British and French, which he was trying to escape but had not, quite yet. West and north of it was the jurisdiction of Canada; and west and south of it was the jurisdiction of the United States. At 4:15 a.m. – probably the soonest Fehler felt with confidence he could reach the desired coordinates by noon – he falsely radioed the information to the Allies, implicitly reporting he was at that moment breaking into the American-controlled sector of the Atlantic. He would be expecting to capitulate to the United States and surrender his boat, passengers and deadly cargo, and then his mission would be over.

Instead of the United States responding to Fehler's surrender message, however, Canada's Halifax station first hailed the U-boat and commanded it to sail for Nova Scotia. Canadian forces seem to have uncovered the fact, either by direction-finder or by air observation – Hirschfeld references that some time during the surrender phase an airplane located and circled the U-boat just prior to Canada signaling to Fehler that he was acting deceitfully[3] – that Fehler was still in their territory. For Fehler, surrender to Canada, apparently, was unacceptable; demonstrated by the fact he began or continued a series of activities designed to avoid Canadian capture, which he later admitted to, and which Hirschfeld chronicled.

From this point on, following only the logbook information, it becomes difficult to know exactly where U-234 was located at any given time. The direction-finder information on the 12th offers the last dependable location. We know the coordinates listed in the logbook during this time were patently false so we can't depend on them. And if one uses only the dead reckoning data in the logbook, the U-boat was already over 150 miles beyond where the direction-finder located it, and, according to Fehler's and Hirschfeld's later accounts, would travel at top speed another 200 to 300 miles in the following two days, placing it several hundred miles from the surrender point on the evening of the 14th, the day it surrendered, unless it traveled a ridiculously convoluted path.

But given Fehler's and Hirschfeld's report of traveling at near top speed throughout the night of the 12th, and possibly the 13th, and knowing the U-boat's 12 May locations as provided by the direction-finders, as well as knowing the direction it was headed and its location at surrender, the only truly plausible answer for U-234's location during this time was that it continued on a dead run westward, not southward as reported.

Realizing general knowledge of his duplicitous maneuvers would reveal a hidden agenda and invalidate the decoy – though disturbingly disparate – logbook he had created, Fehler concluded he should leave no record at all henceforward. Thus Fehler discontinued his semi-fictional second logbook on 12 May. Beyond this point, he probably realized it would be more difficult to create a viable cover story for the convoluted logbook, than to explain his reasons for failing to complete the diary. He could easily say the war was over for his boat, passengers, and crew and so there was no need for further entries.

At 9:45 p.m. 12 May, Fehler again reported U-234's position, 50° 00' N, 30° 00' W, the same position he claimed to hold at 4:15 that morning even though he also reported a speed of eight knots in both dispatches, inferring the U-boat was on the move the entire time. The direction-finding coordinates for this second transmission, however, placed the U-boat at 50° 00' N, 34° 00' W. He had now moved from being well east of his reported position, as indicated in the morning transmission, to well west of that same position, now reported in the late evening. According to calculations from the morning and evening direction-finder coordinates, U-234 had traveled approximately 200 miles in 18 hours – or at least three times the average speed recorded in the logbook. The U-boat was running at over 16 miles-per-hour, as noted earlier, or at over 90 percent of its top surfaced speed.

Captain Fehler admitted to making this mad dash over the Atlantic at "16 or 16½" knots on "the night of 12 or 13 May" in his letter to *Sharkhunters* already mentioned. Hirschfeld confirmed, as well, that Fehler had ordered him to report false speeds and directions to Halifax,[4] which the direction finder data verifies. Thus, once again, the Bundesarchiv logbook and Fehler's uncoded transmissions are proven to be lies. It seems certain, therefore, the dubiously-marked Bundesarchiv logbook now can be accepted virtually as a ruse to cover covert activities.

Fehler had an explanation for these mysterious machinations, though. In his letter to *Sharkhunters*, which is a response to *Sharkhunters* President Harry Cooper's own suspicions about the activities of U-234, the Captain described how he met with his officers and German passengers after hearing about the Reich's capitulation on 8 May to discuss what he should do about surrendering. While many opinions were openly voiced in this meeting, according to radio chief Wolfgang Hirschfeld, Fehler never divulged his own opinions or intentions.[5] General Kessler independently confirmed this in a post-war interrogation.[6] But Fehler later reported in the *Sharkhunters* letter that he and General Kessler had decided prior to leaving Norway that if capitulation was necessary he would not surrender to British forces, allegedly because the food, conditions and treatment of POWs would not be as good as in the United States. As a result, according to the letter, Fehler had determined to get out of the British quadrant of the Atlantic and so was racing toward the setting western sun in the direction of America.

Nothing in any review of General Kessler's statements and interrogations, however, supports this claim. Quite the opposite. Besides never mentioning such a meeting with Fehler, Kessler consistently held to his opinion and his version of events that he felt the U-boat should continue its voyage to Japan or sail to Argentina.

Fehler's explanation of U-234's desperate dash does not hold up under close scrutiny, either. He asserted in his *Sharkhunters* letter that on the night of 12 or 13 May, when Tomonaga and Shoji, the Japanese officers on board, heard the high revolutions of the propeller shafts, that they deduced Fehler had decided to surrender the U-boat.[7] According to Fehler, rather than be captured alive, which would be disgraceful for the two Samurais, the Emperor's officers committed suicide by each taking an overdose of Luminal, a sleeping drug. They were left to this substitute for hara kari because Fehler had confiscated all of the passengers' weapons when they boarded the U-boat in Kiel. Tomonaga's ceremonial samurai sword had, in fact, been turned over to the Captain by Ambassador Oshima at the departure ceremony.[8]

Fehler stated in the letter he sent to Cooper that he planned, had the Japanese not taken their lives, to drop the officers either on the Spanish or Portuguese coast or on the Canary Islands. The discrepancies in this prevarication are obvious and two fold. First, the radio intercepts of 12 May are in plain text, they are not coded messages, indicating Fehler's intent with the messages was to open communications with the Allies – albeit either very cautiously or with a hidden agenda in mind, as indicated by his false coordinates. Regardless of his agenda, with the German capitulation having already occurred, Fehler knew better than to think he could open communications with the Allies and then just sail about the sea wherever he pleased. He knew the Allies would demand a swift capitulation of U-234, which they did.

Thus the evidence demonstrates Fehler was already in the beginning stages of surrendering when he was informed the Japanese had taken the poison, which contradicts his story about planning to take the two officers to safe harbor prior to learning of the suicide.

And second, Fehler freely admitted that at the time the Japanese poisoned themselves, U-234 was headed west at high speed, and direction finder data shows it had been doing so for about 200 miles – supposedly since 2:35 the morning of the 12th according to the logbook, but certainly since 4:15 a.m. that day. He added that his reason for this run was to escape the area of British control. But the three locations Fehler claimed he planned to drop off the Japanese – Spain, Portugal or the Canary Islands – were in very different directions than the one he was racing toward, and they all required crossing vast expanses of Allied jurisdiction. The fact he was racing west at the time the boat's doctor told him the Japanese had taken the poison contradicts his explanation of intending to take them to the Iberian Peninsula or the Canaries, again revealing he had no intention of taking Tomonaga and Shoji to safe haven.

The evidence shows Fehler was bent on surrendering to the United States regardless of the consequences – to Tomonaga and Shoji or to anyone or anything else. In fact, according to his own account, upon hearing the Japanese had taken the poison, Fehler even refused to stop his U-boat to submerge below what he claimed was stormy surface weather long enough for U-234's doctor to recuperate from an alleged case of seasickness so he could treat the poisoned men.

Hirschfeld's account, contrary to the Captain's, while not stating whether Dr. Walter was or was not sick, described Walter's activities in ways that demonstrate he was active and participating in the events underway throughout their entire span, contradicting Fehler's report that

Dr. Walter was ill.[9] And Fehler's assertion the seas were so heavy as to cause a seasoned U-boater like Dr. Walter to become seasick does not jive with the account provided in the USS Sutton's day log that described the weather as clear and that the seas were moderating on that day.[10]

In fact, on the contrary, the doctor was healthy enough according to Hirschfeld, that Fehler ordered him to oversee the Japanese's deaths. "Tonight we must get the Japanese overboard," Fehler explained to Walter. "If the Americans get to them, they'll do everything they can to bring them round. See to it that they die peacefully,"[11] he ordered the doctor.

Granted, Fehler may have been trying to fulfill Tomonaga's and Shoji's last wish to die in peace and with honor. But why would he tell a series of lies in his letter written forty years later, in justification for not reviving them, rather than tell the simple, honorable truth? The evidence suggests that whatever Fehler's hidden purpose, its end would be better served if Tomonaga and Shoji were dead.

Fehler later asserted, again in his letter to *Sharkhunters*, that following his talks with General Kessler before leaving Kiel, and after at least two discussions with his officers and non-Japanese passengers while on the high seas – including Kessler again – that he had decided to surrender to the United States. This he said he did with Kessler's support.

But Hirschfeld wrote that in the surrender discussions at sea, Kessler was in favor of completing the mission to Japan or of heading for Argentina, as were most of the other officers in U-234; a few of whom favored returning to Germany.[12] But Fehler never tipped his hand to reveal what he would do. Kessler unknowingly corroborated Hirschfeld's claim in a post-war interrogation, also stating Fehler never expressed an opinion about where to surrender.[13]

Argentina as a surrender option was a covert ally of Germany's and surrender there allowed the expectation among the passengers and crew of U-234 that they would have a short, uncomplicated stay in South America before a quick return home to family and friends – and rebuilding lives in Germany. A few dissenters, however, preferred to land on some South Sea island paradise instead of ending their journey in Argentina. Hirschfeld reported that only Party Judge Kay Nieschling and the boat's doctor, Dr. Walter, voiced their support for surrendering to the United States. Importantly, as noted, and despite Fehler's later claims, neither Fehler nor Kessler were in that small group.

Thus we have yet another conflict in the record – one of several that crop up between Fehler and Hirschfeld: Was Kessler in favor of surrendering to the United States or not? Again we must try to determine who is telling the truth. Given Fehler's obvious prevarications regarding his intentions toward Tomonaga and Shoji, and his intentional misrepresentations in the Bundesarchiv logbook, as well as the admitted deceptions in his transmissions to Halifax compared against Hirschfeld's consistently provable and accurate accounts, Hirschfeld's version is undoubtedly most correct. Thus Kessler's preference to go to Japan or Argentina is more probable than Fehler's later assertion that Kessler had agreed to surrender to the United States, especially considering Kessler said so under interrogation.

Fehler's uncoded radio transmission of 12 May, intended to open the way to surrender, compared against his immediate refusal to surrender to Canadian or British forces, clearly leaves only the United States as Fehler's intended surrender objective. He later tried to support this objective by cunningly suggesting Kessler agreed to it before leaving Kiel. Hirschfeld's writings and Kessler's statement that Fehler never actually revealed his intentions to any of his passengers or crew about where he would surrender, in view of the fact almost all of the passengers and officers desired a course other than surrender to the United States, further supports the premise of this hidden agenda.[14] The evidence suggests Captain Fehler stubbornly continued to quietly manipulate events until U-234 was "captured" by the *USS Sutton*.

As has been shown, Captain Fehler had demonstrated by both word and action he was bent on surrendering U-234 to the United States. His determination to do so even before the U-boat left Germany – his alleged, though now dubious, discussions with Kessler while still in Kiel to achieve this end, if true – indicates he already may have been laying the groundwork even then. Or perhaps he was simply maintaining the illusion some forty years later when he wrote this account in the *Sharkhunters* letter. At any rate, his mad dash across the Atlantic, carefully manipulated to reach American controlled waters at a critical point in time, drives home his determination to surrender only to the United States. So does his silent decision to land there against the desires of a large majority of his officers and high-ranking passengers. Fehler's intentional deceptions to Halifax combined with his determination to sacrifice his Japanese passengers rather than off-load them in Spain, Portugal or the Canary Islands – or to even make an effort to save their lives at all – all testify of a personal commitment on Fehler's part to surrender only to the United States. This fixation seems far out of keeping with a reasonable assessment of the situation he was in.

Even more shocking – and revealing – is the fact the United States Navy aided and abetted Fehler in his efforts to surrender to it. Hirschfeld recorded that while Fehler was in contact with Halifax, sending deliberately false reports about his position and movements, U-234's radio communications suddenly were jammed by very powerful transmissions.[15] Apparently somebody did not want U-234 in communication with Halifax. Each time Hirschfeld tried to transmit to the Canadian station, regardless of which frequency he used, the jamming would begin anew, which suited Fehler just fine;[16] the overrunning of his radio communications kept the Captain from having to continue his deceptions to Halifax. Soon, the USS *Sutton* could be seen cresting the horizon.

The *Sutton* reached U-234 shortly before dark. Using Morse Code from a lamp, rather than radio signals even though the war had been over for a week, the destroyer ordered U-234 to "head for the Gulf of Maine and to *ignore all further communications from Halifax* [italics the author's].[17] From this Hirschfeld deduced the *Sutton* had done the radio jamming. But few if any United States Navy destroyers had jamming capabilities,[18] so probably the jamming was done from a land-based station that was monitoring events and supporting the *Sutton's* mission. Soon the *Sutton* slipped alongside the U-boat just a few hundred yards to port, but waited until morning to send a boarding party. In the meantime, Hirschfeld witnessed Dr. Heinz Schlicke throw several small tubes of microfilm overboard from the conning tower into the ocean.[19] "There goes the rocket that could fly the Atlantic," remarked Schlicke.

In the morning, a heavily-armed prize crew from the *Sutton* crossed the distance between the two vessels in a small craft and boarded U-234.[20] Nerves were on edge as the outnumbered but well-armed *Sutton* contingent chained the hatch open to ensure Captain Fehler did not try a last-minute dive. Documents were given to Fehler instructing him in the procedures for surrendering his boat and crew; then a skeleton crew of German sailors was left on board to operate the vessel while the remaining passengers and crew were ferried from U-234 to the *Sutton*.

Hirschfeld, one of the few German crewmen left onboard the U-boat, noted U-234 was ordered to make for the Gulf of Maine. Later, this order was changed to direct U-234 to head for the Naval Yard at Portsmouth, New Hampshire.[21] Once again Hirschfeld's near-impeccable account is verified, this time by the *Sutton's* activity report,[22] which recounted how the order was changed for the *Sutton* to escort U-234 to Portsmouth instead of the previous order that it report with the U-boat to Cascoe Bay, Maine.

The *Sutton*, for its part, before locating U-234, had been working along-side two Canadian ships that were also trying to find the U-boat. According to *excerpts* prepared from the *Sutton's* war diary[23] – the diary itself, perhaps significantly, apparently is not available – during the operation the *Sutton* had broken away from the Canadian vessels. The *Sutton* head-ed south on a trajectory that allowed it to intercept the U-boat based on direction finder coordinates the destroyer had received. The *Sutton's* war diary notes that the Canadian ships apparently realized they had "missed their target," but continued to head off in an east-northeasterly direction. Nothing is said in the war diary excerpts of the *Sutton's* having jammed the U-boat's radio transmissions or of ordering U-234 not to respond to Canadian radio communications.

So, having unwound the circular puzzles and unlocked the riddles of U-234, how do we interpret the web of information, disinformation and contradictions surrounding the U-boat's voyage and surrender?

The first step is to determine which evidence is sound and which is not; an objective we have undertaken throughout this and other chapters. Of the five sources of information about U-234's movements – the direc-tion-finding coordinates, the Bundesarchiv logbook, the accounts record-ed in Hirschfeld's two books, Fehler's letters to *Sharkhunters*, and Fehler's position reports to Halifax – the direction finding coordinates are by far the most objective and therefore reliable.

This evidence and a great body of other known facts appear to strongly favor, if not outright prove, large portions of the Bundesarchiv logbook are fabrications. And they appear to be so at least from the first mis-matched coordinates copied by Kessler on 30 April, if not earlier. They extend through the three series of unaligned bearings and speeds versus coordinates recorded from 1 to 6 May, when U-234 allegedly sailed six full days without surfacing, and continue through the outright lies record-ed by Fehler from then through 12 May.

Possibly the logbook was counterfeited after the fact, but it appears to have been "jointly" kept – the fake version and the real version laid down on the same paper. The daily coordinates would have been written to show the fictitious course it was desired the world believe, Fehler as-suming analysts later would choose the easier process of just plotting on a map the coordinates posted to discover U-234's "route." The alternative the future analysts would have to choose from, dead-reckoning, required calculating the combinations of speeds and directions to work out the journey, a much more tedious and time-consuming process. Fehler knew the odds were high investigators would ignore the math and just lay down lines from coordinate point to coordinate point on a map to see where the boat "had been." Especially since they would already have his chart for this cruise, which would also have been falsified, and the coordinates would match what he had entered on the chart.

At any rate, much in the logbook was obviously entered to maintain an illusion for later investigators, which was that U-234 had never varied from its intended mission to Japan until Fehler made the decision to surrender.

Now knowing his consistent deceits, nearly all of Fehler's accounts can be discarded.

On the other hand, as has been highlighted throughout these chapters, Hirschfeld's accounts have all been faithful to what is known from the valid documentation and other sources. With one exception.

The single point of contention between Hirschfeld's account and the theory posited in this text lies in the fact, despite referencing the mysterious message from Hitler's Berlin Bunker, Hirschfeld never mentioned any activity while U-234 was at sea that can be construed as picking up and/or delivering a secret passenger. He detailed activities that occurred on the boat throughout the relevant timeframe, however, and once stated that U-234 "continued to head south at full speed" during a time when, according to the secret detour scenario, it must have been heading west. If this event happened just a day or so earlier, however, it may be a clue as to what the U-boat was really doing. Remember, while Hirschfeld's accounts in all other aspects have been consistently accurate his one weakness has been his recollections of dates.

So in reporting U-234 sailed south at full speed Hirschfeld may suggest a fragment of the premise the U-boat actually was doing something beyond what he was reporting.

It seems implausible Fehler could have secretly picked up a mystery passenger without his chief communications officer knowing about it. Certainly Hirschfeld, as were all others involved in the maiden voyage of U-234 who would have known its secrets, was sworn to silence if it had, indeed, carried an important enigmatic passenger to safety in the final days of the war.

A commitment of silence from those privy to secret information – perhaps including the whole crew – would have been extracted not only from the German government before departing the Reich, but from the

United States government before freeing the prisoners, as well. Perhaps Hirschfeld's writing that the boat was racing south is a small lapse conceding to the weighty burden of carrying such a secret for so long a time. Uncharacteristic as it may seem, his statement that the U-boat was racing southward appears to be one small detail – but a detail that can be correct if the date is wrong – that would stay what is otherwise significant evidence favoring the theory that U-234 made a detour to pick up a mysterious passenger who made his escape on board the U-boat.

While fundamentally reliable, it must be remembered Hirschfeld protected whoever Bubbi was. And although in his memoirs he described Schlicke as an expert on radar, direction finding and high frequency transmissions,[24] he later recounted how Schlicke tossed the microfilm containing plans for the "rocket that could fly the Atlantic" overboard,[25] suggesting Hirschfeld knew Schlicke was more than Hirschfeld originally let on, also. So protecting certain aspects of the secret mission of U-234 was not wholly beyond the capacity of Wolfgang Hirschfeld.

In an effort to resolve these questions and to learn more about the journey of U-234, in late 1998 I sent Mr. Hirschfeld a letter through the *Sharkhunters* organization, requesting an interview. As I had been advised probably would happen, Mr. Hirschfeld chose not to respond to my request.

With all this in mind, combining the information we have learned in this and in previous chapters, what picture can we assemble of U-234's activities and surrender, and Bormann's escape? Is any image becoming clear that would indicate the U-boat's mission?

Taking everything we know about U-234 into account – the messages to it from the Fuehrer Bunker; the wrestling for command of the U-boat; the profoundly slow reported travel speed throughout most of the journey; the mysteriously truncated Library of Congress logbook and secret visit to Bergen; the carelessly doctored Bundesarchiv logbook; the coincident timing as recorded in the Bundesarchiv log of Fehler's alleged but illogical decision to run submerged during the six critical days between 30 April and 6 May compared with the reported escape of Martin Bormann during that same time period; considering the little-known but seemingly reliable series of reports that Bormann escaped in a U-boat; Bormann's connections and control of U-234's cargo, and probably, although covertly, his control of Doenitz himself; as well as U-234's mysterious dash westward apparently from points unknown east of its professed position before surrender; and Fehler's determination at all costs to capitulate to none but the United States – considering all this, it seems probable U-234 was the "large U-boat" reported by Soviet intelligence that had the secret mission of rescuing Martin Bormann from Germany, delivering him safely to Spain, and delivering the cargo to the United States in exchange for Bormann's freedom.

Keeping in mind both the above and the following are conclusions based upon the best evidence as detailed, the most probable scenario that can be reconstructed appears to look something like this: With a struggle over chain-of-command of U-234 raging between Doenitz and Berlin, and having already received communications from Hitler's bunker to stay put, Fehler departed Kristiansand according to Doenitz's order, but at very slow speed in order to remain close at hand when the time came to respond to an expected dispatch to pick up a powerful passenger from Berlin.

Apparently, the chain-of-command issue was still being contested on 18 April, when Fehler secretly altered course to Bergen to check for further communications via BdU North Commander Rosing and the U-boat communication center there. Realizing upon his decision to detour to Bergen that his logbook later would reveal his surreptitious movements and potentially expose his secret mission, he abruptly discontinued keeping this log from the 18th forward. Fehler would later begin a new log designed to camouflage U-234's movements. At Bergen Fehler apparently did not receive the communication from Rosing he anticipated so he continued his slow crawl across the North Sea. He proceeded extremely slowly – no faster than a man walks, just fast enough to maintain steerage of the U-boat – so he would be close at hand when U-234 was needed for the secret pick-up of his mysterious passenger.

In the early morning hours of 30 April, at about the same time Martin Bormann was escaping Berlin by light aircraft, U-234 began a quick six-day cruise back to Germany and out to sea again under cover of a reported six-day submerged voyage in the Atlantic. The falsified "submerged voyage" would in effect make U-234 "disappear" during the deceptive detour. Thus a cover story was provided should she be seen in a location she should not have been, or should another vessel fail to spot her in a location she should have been in.

No record has been found to date of U-234 receiving a message to return to Germany to pick up its passenger, but the author believes such a message was sent and received. The author suggests it was at this point Fehler turned his U-boat east, submerged during the day and surfaced at night, and at high speed headed back into the heavily patrolled North Sea through the strait between Scotland and the Shetland Islands. He then turned south – straight for Hamburg. U-234 made Hamburg in under three days, sailing at top snorkeling speed when submerged (this is where Hirschfeld may have been right about U-234 sailing south at high speed, after all – but in the North Sea, not the Atlantic). She would have sailed with intermittent radar checks from her cutting-edge, single-pulse radar, and probably with covert support and protection from well-placed Western Allied sources – remember the planes that did not attack in the Kattegat.

Quickly picking up Bormann, the large U-boat described by Stalin's intelligence reports then made way, again under surreptitious Western Allied protection, through the English Channel and into the Bay of Biscay, where it rendezvoused with an unknown craft to offload Martin Bormann for his stay in Spain.

Racing west and needing to maintain a cover story that would stand as the official history of the vessel, Fehler realized he was running out of time to surrender following the German capitulation order on 8 May. He needed to be in a credible location along his previously planned journey before surrendering, in order to keep his cover story intact, or else his wayward movements might be revealed. In fact, and more important, he also needed to ensure he was in the American sector of enemy surrender, to guarantee his cargo would be received by the pre-agreed upon country, the United States – and its Manhattan Project. By 12 May, he felt he could report falsely a position in the American Zone that he could reach before it was discovered to be false, and so he duly reported that position by radio.

But calamity nearly ensued when Canada, through Halifax, received U-234's first surrender transmission and ordered Fehler's capitulation before the United States responded. To maintain his cover and avoid surrendering cargo and passengers to an unintended party, Fehler was forced to report inaccurate bearings and speeds – and for a period of time not report at all – until the USS Sutton was able to decoy Canadian ships away and jam U-234's transmissions. The Sutton then located and took possession of the U-boat and her fugitive invaluable cargo and passengers, and escorted her to Portsmouth.

Endnotes: Chapter Fifteen – Surrender

1   Undated letter from CaptainLeiutenant Johann Heinrich Fehler to Harry Cooper, president of Sharkhunters International, p. 3

2   Geoffrey Brooks and Wolfgang Hirschfeld, *The Story of a U-boat NCO 1940-1946*, p. 211

3   Geoffrey Brooks and Wolfgang Hirschfeld, *The Story of a U-boat NCO 1940-1946*, p.

4   Geoffrey Brooks and Wolfgang Hirschfeld, *The Story of a U-boat NCO 1940-1946*, pp. 210, 211

5   Geoffrey Brooks and Wolfgang Hirschfeld, *The Story of a U-boat NCO 1940-1946*, p. 210

6   U.S. National Archives II, General Ulrich Kessler interrogation Report #5899, RG 165 – 390 35/10/05 box 495

7   Undated letter from CaptainLeiutenant Johann Heinrich Fehler to Harry Cooper, president of Sharkhunters International, p. 3

8   Undated letter from CaptainLeiutenant Johann Heinrich Fehler to Harry Cooper, president of Sharkhunters International, p. 3

9   Geoffrey Brooks and Wolfgang Hirschfeld, *The Story of a U-boat NCO 1940-1946*, pp. 210-212

10   U.S. National Archives II, USS Sutton activity report titled Capture of U-234 – Events Leading to, p. 2, 18 May, 1945, RG 38 – 370 15/09/01 box 2

11   Geoffrey Brooks and Wolfgang Hirschfeld, *The Story of a U-boat NCO 1940-1946*, p. 212

12   Geoffrey Brooks and Wolfgang Hirschfeld, *The Story of a U-boat NCO 1940-1946*, p. 210

13   U.S. National Archives II, General Ulrich Kessler interrogation Report #5899, 38 – 370 01/04/07 box 113

14   Geoffrey Brooks and Wolfgang Hirschfeld, *The Story of a U-boat NCO 1940-1946*, p. 210

15   Geoffrey Brooks and Wolfgang Hirschfeld, *The Story of a U-boat NCO 1940-1946*, pp. 211, 212

16   Geoffrey Brooks and Wolfgang Hirschfeld, *The Story of a U-boat NCO 1940-1946*, pp. 211, 212

17   Geoffrey Brooks and Wolfgang Hirschfeld, *The Story of a U-boat NCO 1940-1946*, p. 212

18   Dr. Alan Bath, Rice University Fellow, during personal interview with the author 30 March, 2001

19   Geoffrey Brooks and Wolfgang Hirschfeld, *The Story of a U-boat NCO 1940-1946*, pp. 212, 213

20   Geoffrey Brooks and Wolfgang Hirschfeld, *The Story of a U-boat NCO 1940-1946*, pp. 212, 213; U.S. National Archives II, USS Sutton activity report titled Capture of U-234 – Events Leading to, p. 3, 18 May, 1945, RG 38 – 370 15/09/01 box 2

21   Geoffrey Brooks and Wolfgang Hirschfeld,*The Story of a U-boat NCO 1940-1946*, p. 216

22   U.S. National Archives II, USS Sutton activity report titled Capture of U-234 – Events Leading to, pp. 3, 4 (unnumbered), 18 May, 1945, RG 38 – 370 15/09/01 box 2

23   U.S. National Archives II, USS Sutton activity report titled Capture of U-234 – Events Leading to, pp. 2, 3, 18 May, 1945, RG 38 – 370 15/09/01 box 2

24   Geoffrey Brooks and Wolfgang Hirschfeld, *The Story of a U-boat NCO 1940-1946*, pp. 200, 201

25   Geoffrey Brooks and Wolfgang Hirschfeld, *The Story of a U-boat NCO 1940-1946*, p. 212

Gero von Gaevernitz (left) a German national, was Allen Dulles' (right) principle aide in the Office of Strategic Services intelligence arm Dulles ran throughout Europe, but especially in Nazi Germany. The two also masterminded the "unconditional surrender" of Italy in Operation Sunrise.

General Karl Wolff (right) was the Nazi plenipotentiary in Italy, commanding all occupying forces in the country, who negotiated Operation Sunrise. Wolff, Dulles and Gaevernitz appear to have made a secret deal outside of the "unconditional surrender" agreement. Wolff previously had overseen the construction of a vast processing plant that had every earmark of being a uranium enrichment facility.

Chapter Sixteen

Occam's Razor

"It is axiomatic that you keep your eye on the number two man – the one who does the work"[1]
– Allen Dulles, commenting during the war about Martin Bormann

"Thyssen was [Bormann's] ace in the hole if he ever needed a personal pipeline to Allen W. Dulles."[2]
– Paul Manning, author *Martin Bormann, Nazi in Exile*

There is no hard, documentary, conclusive evidence the secret escape mission of U-234 described in the chapters above rescued Martin Bormann. There is no "smoking gun" or proof beyond reasonable doubt. In fact, it may not have happened. But if the United States was in collusion behind the scenes dealing with the Nazis to obtain the cargo of U-234 in exchange for Bormann's freedom and protection, every effort would have been made to make it look like it did not occur. A revelation of Bormann in American hands and under United States protection would have made an extremely negative impact on American moral authority and its treaty obligations worldwide. Whether even a paper trail of such dealings would have been left at all is highly questionable. The rarefied powers that oversaw such negotiations would surely be careful not to leave telltale signs in this most singular of diplomatic dealings.

On the contrary, they would be certain to leave as few tracks as possible, and to cover or make ambiguous any tracks that may have been left behind. The most proof we can hope for in this vacuum, therefore, is circumstantial evidence – anomalies and telltale signs of some unexplained event. The simplest explanation that includes all of the evidence would be the most likely answer for what occurred. So says Occam's Razor, the scientific principle that defines any reputable theory. It states: *Entities must not be multiplied beyond what is necessary.* In other words, the simplest theory that fits all of the known facts of a problem is the one that should be selected as the most plausible theory. If history is an objective science, Occam's Razor should apply. Surprisingly – or maybe not so surprisingly – it is not often applied in the traditional history at all.

Despite the lack of irrefutable proof regarding his fate, the quantity and quality of circumstantial evidence suggesting Bormann successfully escaped far exceeds that in evidence of his death. This evidence includes Hitler's order that Bormann be flown out of Berlin, which matches Soviet intelligence reports that he was, in fact, flown out within 24 hours of that order. In turn, the report of this escape flight aligns remarkably well with many details of a singularly unique, but true, actual escape flight that was documented separately from the Soviet account and separate of those who revealed Hitler's order for Bormann's escape flight. There is the report the second stage of Bormann's escape was made in a large U-boat, which meshes well with the details of the mammoth U-234 having received radio messages that were interpreted by General Kessler to mean a senior official from Hitler's bunker was on his way to U-234. The convoluted record of the U-boat's travels and the excess of effort expended to hide those actions add veracity to this report. And there is the evidence suggesting Captain Fehler was determined to surrender his important passengers and cargo to the United States – even at the cost of the lives of the Japanese officers onboard – rather than complete his important mission to Japan, surrender elsewhere or return to Germany.

Conversely, the United States appears in advance to have known about and been determined to obtain the U-boat and its cargo. On at least one occasion, and possibly others, Allied warplanes easily could have sunk U-234 but did not, apparently opting just to monitor the U-boat's movements. In the end, the United States jammed U-234's radio transmissions to Halifax, thus ensuring the U-boat would fall into American hands, not Canadian.

Combined, the primarily objective, disparate facts recounted in this volume, and all the detailed evidence supporting them, create a scenario for Bormann's escape far more likely to have occurred than the traditional history recounting his death. There is no solid evidence for Bormann's death, and what evidence there is is composed almost entirely of the suspect, often irrational, eyewitness accounts of Nazi sympathizers and Hitler henchmen. The witnesses for the traditional history all potentially had reasons for ensuring Bormann was presumed dead, as do many others who would like the world to come to the same conclusion.

The traditional history leaves many crucial events unexplained, while the theory advanced within these pages resolves almost all – and certainly all of the critical – previously ignored anomalies and mysteries surrounding the events. Even the only supposedly "hard evidence" for Bormann's death, the DNA-tested skull, is fraught with inconsistencies ranging from whether the body was actually buried where the one tested was dug up, to the provenance of the skull itself – a provenance that surely could not pass chain-of-evidence muster in a legitimate court of law.

By applying Occam's Razor to the evidence, far more of the anomalies and mysteries are explained with the new scenario than by the old, disjointed account. So one must suggest it is time the traditional history give way to the new, more congruent one; or that at least a serious, deeper study be completed that includes careful review of all of the newly revealed information.

To believe a great portion of the actions outlined in this book actually occurred, one must believe the United States government, in some form and at some high level, was in league with Martin Bormann and those involved in his escape. These government entities would probably have assisted in the escape by ensuring safe passage for the U-boat by "pulling strings" where necessary, as demonstrated in the "non-attack" of the warplanes overflying U-234 in the Kattegat, and possibly by allowing U-234 to sail unimpeded to Hamburg and back to the Atlantic through the English Channel.

Certainly, jamming U-234's radio transmissions to break contact between Halifax and the U-boat – and then signaling Fehler by lamp not to respond to Halifax's transmissions – appears to be direct intervention on behalf of the United States government to exclude its ally, Canada, from receiving the U-boat's surrender under its rightful jurisdiction.

And to believe the United States took part in such events is to admit it also maintained a clandestine relationship of some nature with Martin Bormann after the war, protecting him from a distance. Such an affiliation with the second-ranking kingpin of the Nazi Empire would be anathema to the American people and also to the majority of Europeans who suffered under his Nazi Party regime. Most especially, the Russians would be enraged. If a connection between the United States and Martin Bormann became known, Joseph Stalin immediately would have suspected treachery by his ally the United States – which, in fact, he did. In September 1945, Stalin broadcasted the assertion that Bormann was in Allied hands.[3] If his accusations were genuine, Stalin would have wondered what Bormann had given to receive such rarefied assistance, rather than be incarcerated and tried at Nuremberg with the rest of his cronies.

Whatever the ransom, to get the treatment from the United States Bormann wanted it must have been of utmost importance to the geo-political equation, and would have been in direct violation of the Allies' unconditional surrender treaty requirement. The participating American leaders knew this, therefore evidence suggesting a relationship between Bormann and the United States would need to be carefully avoided, if possible, or destroyed or buried deep, if not.

And so proof of an arrangement between Martin Bormann and the United States, if there was one, does not appear to exist. What is apparent, however, is that the United States went to considerable trouble to *ensure* evidence of such a relationship does not exist!

During my research in the National Archives, I tried to locate all of the documentation about Bormann that I could find within State Department and, specifically, Office of Strategic Services files. I located several second-party reports notifying these agencies of sightings and meetings with Bormann, suggesting the survival and whereabouts of a very alive Martin Bormann following the war.

I also located a key report that identified all of the top Nazi fugitives still unaccounted for immediately after the war, in Record Group 457 file 190-37-11-1 box 192. The report does not list Martin Bormann or Heinrich Mueller as still missing, even though at that point in time, according to the traditional history, their whereabouts were unknown.[4] Apparently the OSS knew where they were, while everyone else involved was searching high and low for them.

Many of these reports are substantive[5] and were provided by sources the agencies labeled as reliable, such as a State Department report I found that indicated Bormann was living in Spain with a certain Leon DeGrelle, and was running a Nazi escape operation from there.[6] This documentation is impressive. In addition, many other authors, including Manning and Farago, have revealed compelling governmental documentary evidence of Bormann's survival.

But while researching the evidence, the same index that led me to these documents also contained cards referencing mysterious other files about Bormann within Record Groups 226 and 190.[7] Instead of being reports about sightings, the index descriptions seemed to suggest the documents were agency records regarding personal information about Martin Bormann. These included details about his apartment in Munich, found in Record Group 226 file number 122640. His headquarters in Pullach is referenced in Record Group 226 file number 123900; and, most stunning, in

Record Group190-3-32-3 box 1022, resides an apparent OSS evaluation stating that Martin Bormann was "the most powerful man in Germany."

When I searched for these records the index referenced, however, they were not in their files. There were no placeholder cards substituted for the missing documents telling researchers the records were checked out to someone else. There were no slip-sheets indicating the files were still classified and therefore not available. There was nothing: Just missing numbers in the sequence of the files.

As mentioned in a previous chapter, I have spent many hours researching in the National Archives I and II, the Library of Congress and the Southeast Regional Archives in Atlanta, Georgia. During these research sessions I have reviewed thousands – probably tens of thousands – of documents that at the time of their origins were highly classified. These included presidential records, extensive Manhattan Project Records, captured German records, war crimes trials records and the records of U-234 and her captured passengers and crewmembers, as well as records from other U-boats and the State Department and OSS. Only twice in ten years of research did I come across documents missing from their files with absolutely no explanation. Such an omission is almost unheard of in the well-protected archives, which has a stringent procedure for the handling of documents to ensure they are not lost or damaged. In every other instance I encountered, when a document was not in the file as it should have been, a card was left in its place explaining the document was at that time checked out to an archivist. Or a sheet of paper was in the file stating the document was still classified due to its importance to national security and therefore was not available for review.

The only exceptions I have found are these three missing documents about Martin Bormann and the absent "Reactor Shipments" documents mentioned in Chapter Seven – all potentially crucial documents that may substantially rebut the traditional history. Certainly these documents are not required for national security seventy years after the events. Even if they were, there should have been an information card signifying this distinction.

When I described the missing Bormann files situation to an archivist, I was at first greeted with mild disbelief. When he had looked through the file boxes and not found the files, however, he shook his head and exclaimed someone had either refiled them incorrectly or the State Department had removed them. He offered no further explanation. Thinking they may have been incorrectly filed, I carefully searched every folder in each of the deficient boxes, but could not find the missing documents.

One of the hard and fast rules in the archives is that a researcher may have only one box on a research table at a time, and all other boxes must remain closed and on the cart provided for the transport of the document boxes. The box on the table is the only one allowed open at any time and only one document is allowed out of the box at a time. All documents must be returned to that box and the box returned to its cart before another box may be removed from the cart to the table and opened. Boxes are not allowed opened at all while on the cart.

This system is designed to ensure documents are not incorrectly filed or lost. That three files from various boxes, and even from different record groups, all concerning the same subject – Martin Bormann – were accidentally misplaced, while the records of virtually every other subject I queried within the archives seem to be immaculately kept, therefore, seems highly improbable. The more likely event is the State Department or OSS – or its successor the CIA – which, like all contributing agencies maintain control of their documents while in the archives, intentionally removed the missing files about Martin Bormann.

Why would the State Department or OSS/CIA remove the files without explanation? The reason seems obvious: there was information in the files the agency did not want revealed; quite possibly information proving Bormann was alive and the OSS, CIA or State Department had helped with his escape and freedom. Any conceivable information about Bormann different than this should not require unexplained removal from the files at this time in history. If the documents were sensitive to national security, certainly those who removed the files would have used the national security dispensation to cover the otherwise unexplained missing documents, rather than allow them to be conspicuous by their absence. The documents' unexplained disappearance certainly seems to indicate somebody is stonewalling.

Despite the traditional history, the overwhelming preponderance of particulars appears to demonstrate Bormann survived, seemingly with American collusion. Besides the evidence provided in these chapters, this evidence is supported by a plethora of reliable reports of Bormann's being alive and well following the war, advanced by a broad variety of observers, many of whom had nothing to gain from such revelations. I have personally reviewed many such reports – possibly as many as fifty. While some reports are fraudulent or specious at best, many others, when carefully scrutinized, continue to withstand the tests of time and concerted efforts to debunk them. Some are so sound in their details and the integrity of their sources as to seem unimpeachable – although many people have tried to prove them wrong – such as the extensive account given by Dr. Otto Biss, who provided medical services to Bormann in 1959.[8]

The only substantive evidence Bormann did not survive Berlin is the reported positive DNA identification of the remains unearthed at the Lehrter Fairgrounds Station. As has been noted previously, these findings must be viewed with skepticism since, according CIA investigating agent James McGovern, the body supposedly tested and positively identified was no longer buried at the location where the remains were disinterred. And by May 1998, when the testing was done, Martin Bormann almost certainly had finally died and the remains tested may, in fact, have been his – proving nothing except they may have been substituted for those of the person exhumed at Lehrter Station.

The last possibility is the remains tested may have been a relative of Bormann's who was recruited as his double, in the same fashion Hitler's double was a distant cousin of the Fuehrer's. Depending upon the level of testing, DNA tests would possibly be a close enough match to mistakenly identify Bormann instead of his missing relative. Ultimately, the provenance of the remains compared against these possibilities makes the DNA tests far from conclusive.

If Martin Bormann escaped onboard U-234, one piece of information regarding his escape – one very important piece – remains unexplained. How could Bormann, at the seat of the Nazi Party and the Third Reich, Hitler's top lieutenant and a mortal enemy of the United States, have negotiated secretly with the top leadership of American intelligence, politics and the military to arrange the surrender of U-234 and its potent cargo? Through what conduit could he have made a secret peace overture, a proposal that would not jeopardize him personally but would be taken seriously by the United States? The answer is not conclusive, as is little about Bormann's fate. Considering Occam's Razor and the facts outlined above, it may not be possible nor is it critical to this study to prove with certainty Bormann found a pipeline to, and negotiated with, the United States. The requirement, for now, in order to pique continued research, is to show only that such capabilities were available and that such events are the most plausible explanation of the evidence.

Looking at the possibilities concerning these negotiations and history as it unfolded, it is not surprising another suspicious string of events and personalities with Bormann's stamp on them seem to be "coincidentally" connected. If certain events are, indeed, linked, as they appear to be, they solve more historical anomalies that previously have been dismissed or ignored by the traditional history.

Allen Dulles, President Roosevelt's personal envoy[9] in continental Europe and leader of the OSS on the continent,[10] was operating an intelligence apparatus from Berne, Switzerland in February 1945 when an emissary of SS General Karl Wolff secretly approached him.[11] General Wolff was Wehrmacht Plenipotentiary for Italy, which meant he was responsible for all German occupation troops not fighting on the Italian Front, and he was head of the Security Police and Secret Police in Italy.[12] Interestingly, prior to this assignment he had been Himmler's personal chief of staff, SS adjutant to Hitler and liaison between the SS and I.G. Farben[13] – especially for the buna plant at Auschwitz during its construction.

Through these offices and responsibilities Wolff was fully privy to the mysterious workings of the I.G. Farben plant at Auschwitz and its apparently enriched uranium product; and he was well connected with Martin Bormann and his inner circle of bureaucrats and industrialists. For many years, Wolff also held the purse strings to Himmler's personal funds, most of which were garnered from Himmler's "Circle of Friends," a small but powerful cartel of business magnates that included I.G. Farben industrialists Buetefisch and Duerfeld.[14] Both men were central figures in the I.G. Farben plant at Auschwitz, and both were connected to Bormann through Farben's chairman, their boss Hermann Schmitz. Sixty percent of the funds Wolff managed for Himmler's personal interests and projects was provided by the Circle of Friends, while the other forty percent was provided to Wolff by Bormann,[15] either directly from party coffers or through the Party's Adolf Hitler Fund, which Bormann also controlled.

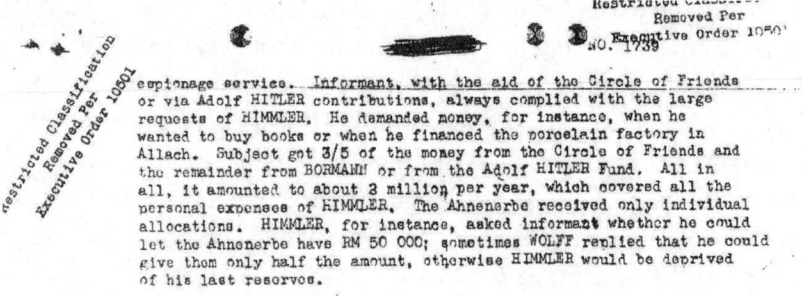

Figure 1: Extract of General Karl Wolff interrogation in which he describes his passing of funding from Bormann to Himmler. Martin Bormann controlled all such funding.

The traditional history of the surrender of the German troops in Italy holds that Wolff suggested to Dulles through a secret emissary that they open negotiations for a separate capitulation of the German armies in Italy. Dulles listened to the envoy with interest and on 8 and 9 March[16] met with General Wolff in person at Dulles' apartment in Zurich. According to Winston Churchill,[17] and supported by the official reports of the negotiations,[18] Dulles told Wolff that the only acceptable capitulation was full and unconditional surrender. The American, British and Soviet Governments were then notified of Wolff's query, according to Churchill.

The traditional history asserts Wolff then agreed to "pave the way" for the unconditional surrender of Germany's southern army, which he appears to have done, although possibly not unconditionally. In the process of developing what was to be called Operation Sunrise, several more meetings were held between Dulles or his envoys and Wolff or his envoys over the span of the next two months.

According to this traditional historical account, on the surface all seems well and good; but it contains incongruities. First, according to Churchill's statement – although Dulles' official report and the files of Operation Sunrise, which are vague on the subject, do not necessarily support it – Stalin had been informed of the initial talks, and efforts were made to get the Soviets involved.[19] But they never participated in the Swiss discussions.[20] The reason given was the difficulty on the Western Allies' behalf of smuggling a Soviet representative into neutral Switzerland, with which the Soviet Union had no diplomatic ties.[21] More difficult challenges, however, did not keep the operation from smuggling an Allied radio operator straight into Wolff's chief of staff headquarters in German-occupied Milan while the war was still raging, to provide communications to complete the surrender details.[22] Nor did it keep them on multiple occasions from smuggling general staff-level English and American intelligence and military officers across several borders in and out of Switzerland, to manage the planned surrender.[23]

More importantly, the surrender of Italy was very much in both Russian and Swiss interests. It seems unlikely the two countries could not work out a covert agreement if their sole and mutual objective was to conclude the Italian surrender. Given such considerations, the excuse for excluding the Soviets appears hollow.

Soon the perpetually paranoid Stalin, stirred up by Nazi innuendo[24] – the Germans were playing for both a separate peace and an Allied break of ranks, whichever they could achieve[25] – was angrily accusing the Western Allies of secretly negotiating with the Germans. Stalin pestered the Anglo-Americans until the West eventually decided to end the contact with Wolff rather than find a solution that allowed the Soviets to participate.[26] Of course, by this time the talks had gone on for two months. At the very last minute the program was saved, but still Russian observers were not allowed to be present until the very final details of the surrender document were being completed.[27]

Such kibitzing indicates perhaps there was more happening surreptitiously than Churchill and Dulles admitted. The United States consistently denied Stalin's accusations, and the official record of the operation appears to support this stance; Dulles and his envoys and Allied leaders clearly stated in their communications the importance of not giving impressions that could be construed as negotiating. All talks referenced in written or signal communications were characterized as discussions opened for the purpose of arranging full and unconditional surrender. In a cable sent on 5 April,[28] Roosevelt denied to Stalin that agreements had been reached or that negotiations were even ongoing. He wrote:

> I have complete confidence in General Eisenhower, and know that he certainly would inform me before entering into any agreements with the Germans. He is instructed to demand, and will demand, unconditional surrender of enemy troops that may be defeated on his front.... I am certain that there were no negotiations in Berne at any time, and I feel that your information to that effect must have come from German sources, which have made persistent efforts to come between us...

> Finally, I would say this: it would be one of the great tragedies of history if at the very moment of the victory now within our grasp such distrust, such lack of faith, should prejudice the entire undertaking after the colossal losses of life, material, and treasure involved.

> Frankly, I cannot avoid a feeling of bitter resentment toward your informers, whoever they are, for such vile misrepresentation of my actions or those of my trusted subordinates.

While it is true the Germans were trying to play the Allies against each other, Roosevelt could not know when he blasted Stalin's "informers" that the chief man he was denigrating was none other than Kim Philby, the Soviet master spy.[29] Philby would later defect to the Soviet Union and a communist hero's welcome following three decades of service as a Russian spy who intrigued throughout the top echelons of British intelligence.

Philby had been the source of Stalin's information in an incident that reportedly occurred several months earlier, when Dulles secretly met with another shady emissary suing for peace for Germany – a Herr Langbehn.[30] Himmler, notably Wolff's boss at the time, ostensibly had sent Langbehn, but to Dulles Langbehn described himself as connected to the German Foreign Ministry. To show Dulles he was acting in good faith, Langbehn presented certain Foreign Ministry documents that were compelling to Dulles in their value and in proving Langbehn's bona fides, and that his negotiation query was in earnest. Dulles later described in enthusiastic tones the impact and value of the goods "in all their pristine freshness."[31]

Dulles had the papers copied and sent to OSS headquarters in Washington and London. In London, Kim Philby received the papers and promptly forwarded them to Stalin. Moles at OSS headquarters in Washington confirmed to Stalin Philby's findings.

According to the traditional history, the Langbehn "peace initiative" set in motion by Himmler, purportedly with Hitler's blessing, was actually planned as a form of political sabotage – part of the process of breaking up the Allies. The intent was to weaken the Allies' East/West alliance with artificial documents that would put the Soviets at odds with the United States and Britain. Counter to Himmler's plan, however, the documents Langbehn presented to Dulles were very real, not the specially forged papers Himmler thought were being offered. As noted, they were very compelling to Dulles. And Hitler was furious at the "mistake."

What, or who, caused the important switch of the documents from fake to real papers may prove interesting when considered against ensuing developments. The information within the documents, the actual timing of the meeting and its results, I have been unable to ascertain other than that it was initiated in the summer and fall of 1943. In fact, certain information around this negotiation appears to actually have been connected with Operation Sunrise, too, or perhaps the two are one, with the timing confused. I have been unable to untangle the two using the information I have discovered.

The timing, government services involved and personalities participating in this and the Operation Sunrise affair are all aligned, however, to suggest a possible connection between the Langbehn and Wolff negotiations. In fact, Langbehn's name was mentioned by Wolff when he was interrogated as a witness for the Nuremberg trials,[32] inferring he knew the man and worked with him as one of Himmler's industrialist contacts, mentioning specifically Langbehn's connections with Swedish Banker Raoul Wallenberg.

At any rate, Langbehn had approached Dulles on behalf of Himmler with very real and compelling "Foreign Ministry" documents, which one must assume were important papers relating to Germany's relationship with at least one other country, or more. The papers would either have been military or commercial intelligence in nature, or a combination of these; and would have been important enough to get Dulles' rapt attention and a quick dispatch up the chain of command. They could have been any documents that fit this bill, but it is reasonable to assume the documents dealt with the recent agreement for technology exchange between the Third Reich and Japan. This agreement certainly fits the criteria of all the requirements above and would have been an eye-popping revelation to Dulles. A portion of this material would become the cargo of U-234, including the enriched uranium from the I.G. Farben plant at Auschwitz.

Himmler thought the documents being compromised were the faked papers. But for Langbehn, or anyone else for that matter, to have made a simple mistake of accidentally exchanging intentionally fraudulent documents created only for this political sabotage, in place of real, very important, Foreign Ministry documents that one must believe were well guarded, seems highly improbable. More likely, someone behind the scenes got the real documents into Langbehn's hands and was playing Himmler for the fool, apparently in a very real, but guarded, communication to the West through him. In this scenario, Himmler served as an unwitting front man and buffer, thus saving the unidentified arbiter from exposing himself.

From the outset, the ploy looks like a classic Bormann intrigue. By the spring of 1943, with Stalingrad fallen, Bormann had concluded the war was all but lost and he had already begun his secret campaign to export as much of Germany's economy as possible outside of the Third Reich. To ensure he would be around after the war to control that fortune, he needed to guarantee his post-war freedom and protection with those who would then be in control. Naturally, he would have begun looking for a conduit to the West, and through his broad range of dealings with Himmler possibly found Himmler's ruse and then co-opted it; using Wolff to send the technology exchange papers to Switzerland through Langbehn in place of the fraudulent documents.

Unfortunately, agents in Switzerland reported back to Hitler that real documents had been leaked and Hitler, livid, held Himmler to account. Himmler was only able to save himself by arresting his emissary to Berne – who was presumably Langbehn.

Bormann now would have needed to find another pipeline to the West. Enter General Wolff. Or, as noted, possibly Wolff already had served as the contact that got Bormann's technology exchange papers into Langbehn's hands in the first place. As has been stated already, Wolff and Lang-behn shared a working relationship through Himmler. And as also noted previously, Wolff had connections with Bormann as well. Wolff had been Himmler's personal chief of staff, Himmler's SS adjutant to Hitler, and SS liaison to the I.G. Farben plant, all of which required significant interfacing with Martin Bormann. In addition, Wolff was now the master of all of occupied Italy.

These positions and the experience gained from them would have made Wolff perfect for Bormann's negotiation needs. As Italian plenipotentiary Wolff had a degree of autonomy and physical distance from Berlin and close proximity to Switzerland that allowed him to relatively easily contact, and even meet with, emissaries from the West. He also commanded the occupying troops in Italy and maintained good relations with the commanders of the fighting troops there. Thus he had the capacity to bring a surrender to fruition – or the surrender could at least play the role of cover story, possibly for the real negotiation at hand, that of exchanging the enriched uranium and other cargo of U-234 for Bormann's, and by extension now, Wolff's, freedom.

As an officer in Hitler's court, Wolff had learned the tricky political landscape and how to engage in sophisticated high-level negotiations while watching his back, which Bormann would be well placed to protect anyway. As Himmler's personal chief of staff, Wolff had been responsible for collecting and distributing Bormann's multi-million reichsmarks-per-year contributions to Himmler's personal accounts. This made Wolff a tool of Bormann as well, and exposed him to a healthy appreciation for Bormann's power and modus operandi. And as a key player in Germany's enriched uranium production project, Wolff was singularly knowledgeable about its secret purpose and value, and therefore its use as a bargaining chip with the United States. For Bormann, Wolff was perfect for handling the delicate matters of the secret negotiations and to address the questions and details the Americans surely would have regarding the ransom being offered.

In turn, Wolff could gain much from this symbiotic relationship. With Bormann in Berlin to watch his back – and possibly even by then to have convinced Hitler secret negotiations with the West might be prudent – Wolff could win his own freedom along with Bormann's by practicing his discrete diplomacy with a fair level of safety – as it appears he did. In fact, according to Wolff's post-war interrogations, as early as 6 February 1945 Wolff had discussed with Hitler that if the "secret weapon" was not completed in time, whether or not he should approach the West with surrender options.[33] He indicated in the interrogation Hitler not only did not forbid him from pursuing contact with the West, but that he, Wolff, interpreted this to be Hitler's unspoken approval of such a program, which Wolff then followed.

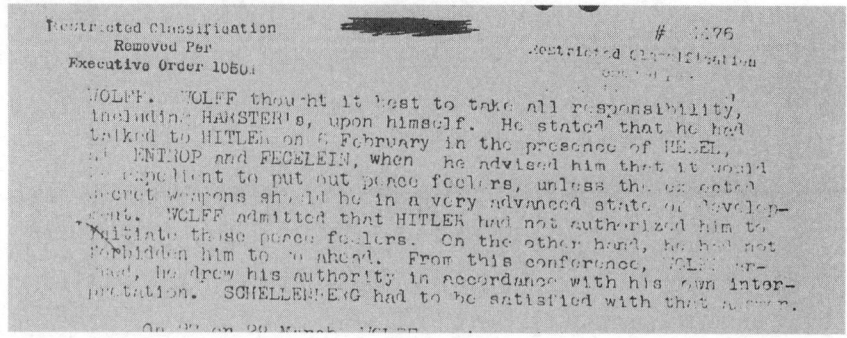

Figure 2: An extract of a Wolff interrogation references discussions Wolff had with Adolf Hitler regarding the secret weapon and its importance to successful peace negotiations.

In a full report on Operation Sunrise that Allen Dulles and his chief assistant Gero van Gaevernitz prepared at the end of the war, according to Wolff, Hitler had even issued a "secret order to seek any possible contact with the Allies."[34]

It seems doubtful given Hitler's penchant for getting even with traitors, that Wolff would have gone forward on the basis of Hitler's no-comment alone, without Bormann, or someone, assuring him of Hitler's approval.

Later, when Wolff's actual surrender efforts were revealed to Hitler, the Fuehrer complimented Wolff on following this course and on its apparent success, and thanked him for his initiative.[35] Hitler's approval came despite the fact he had ordered there would be no unconditional surrender, which would indicate something was being innocuously offered in exchange for Germany's capitulation. Hitler's approval also came despite the fact Wolff had been threatened by his detractors – including Himmler by this time – who were going to reveal his surrender activities to the Fuehrer, and who assured Wolff the Fuehrer would take drastic measures against him.[36] Himmler, trying to save his own post-war hide, did not want Wolff's negotiations conflicting with secret talks he was conducting with the Allies through the Red Cross, nor did he want word of his negotiations getting back to Hitler and having Hitler squelch all such talks.

Again, Bormann's influence appears to have been present in these events, for who else had the weight with Hitler to garner his support for Wolff to pursue peace negotiations with the West, while the opportunity was denied to Himmler? In fact, Himmler's arrest was ordered by Hitler when the Fuehrer learned Himmler was parlaying with the West through the Red Cross. Why would Hitler have applauded Wolff and denounced Himmler for pursuing surrender, unless a different, secret desired result was expected of Wolff's efforts, which Himmler's negotiations might ruin? The openly known fact it was Bormann who stirred Hitler to order Himmler's arrest, probably in part to stop Himmler's negotiations from interfering with Wolff's, suggests the master of the Wolff plan was Bormann, himself.

Add to this the fact Hitler, despite forsaking his own survival, had ordered that Bormann be rescued from Berlin in order to preserve the political paperwork testifying of the Fuehrer's consent to preserve the Nazi cause after his death, and it seems the secret weapons discussion between Hitler and Wolff may have held greater importance than at first review.

General Kesselring had complained to Wolff, "Our situation is desperate, nobody dares tell the truth to the Fuehrer, who is surrounded by a small group of advisors, who still believe in a last specific secret weapon which they call the 'Verzeiflunga' weapon."[37] Interestingly, the report goes on to explain Kesselring did not doubt the existence or viability of the secret weapon – in fact, he believed the weapon would "prolong the war [but] could not decide it." But the General stated he would refuse to order its use, fearing the bloodbath it would cause. Kesselring's belittling accusation of the weapon being controlled by a small group of advisors rings of Speer's charge of Bormann's "Sunday supplement" reporting of the atomic bomb project. And, as has been shown in a previous chapter, whenever the secret weapon was mentioned, particularly during the last days of the war, it was always tied to Bormann, who appeared to be its overseer.

In reality, Hitler probably was so exhausted and dazed by his imminent downfall that Bormann probably had to do little more than make the suggestion for this scenario and Hitler, weary and desperate for a chance at some sort of victory, would have accepted it.

One must ask, given the secret weapon was not used by the Germans, what happened to it? Since Wolff's discussion with Hitler started with the disposition of the weapon, it is logical to assume and reasonable to believe the discussion continued along the lines of what to do with this very important asset.

If Hitler was willing to entertain and even encouraged Wolff to pursue an agreement with the West on his behalf, certainly the objective of such negotiations, and whatever currency was available with which to negotiate, was also discussed. Hitler made clear during his last interview with Wolff that, while he approved of the dialogue with the West, unconditional surrender was out of the question. But according to the traditional history this is exactly what Wolff was negotiating. On the other hand, Hitler, Bormann and Wolff had to know they were not going to get something for nothing. They almost certainly would not have left the outcome of negotiations open, but they would have recognized that whatever was agreed to could not appear to fly in the face of the Allies' very public commitment to unconditional surrender.

Given the outcome of events, as described throughout this book, Hitler's purpose for the negotiations, it seems, was to get Bormann to freedom and ongoing protection, carrying Hitler's final orders and his last will and political testament, in order to provide a breath of hope that some form of Nazi control would survive. Bormann apparently had convinced Hitler that his plan to export and rebuild Germany's economy after the war, and thus ultimately win the conflict for Germany by economic means, still had potential. Indeed, as noted in a previous chapter, Hitler appears to have supported the plan from its inception. Likewise, it is reasonable to believe Bormann convinced Hitler such an outcome would post-humously justify the Fuehrer's life's work and eventually honor his legacy.

The exchange currency for facilitating this agreement with the West would be the secret weapon. With that exchange came the added benefit of giving the United States, which Hitler much preferred over the Soviet Union and communism, a tool to keep communism in line or even eliminate it – atomic bombs.

If the Wolff/Dulles negotiations went further than a simple unconditional surrender – as Stalin's insistence and other indications suggest – and the secret mediation originated as an overture from Bormann to Dulles, upon hearing Bormann's name Dulles most likely would have been "all ears." For Dulles had identified Bormann years earlier as the Hitler minion most worth watching.

"It is axiomatic that you keep your eye on the number two man – the one who does the work," Dulles once said of Bormann, whom he had met at a pre-war reception.[38] Dulles' older brother, John Foster Dulles, who would soon be Eisenhower's Secretary of State, also had connections to Martin Bormann – through Bormann's old consort, I.G. Farben chairman Hermann Schmitz[39] – whom he had met during the Versaille Treaty negotiations.

Bormann, in his turn, recognized the value of Allen Dulles as a conduit to Roosevelt and had already gone to great lengths to create a pipeline to Dulles if he ever needed one. Industrialist Fritz Thyssen and Allen Dulles had met and hit it off following World War One, when the pair represented their respective countries in the industrial reparations negotiations following that war. Thyssen became an ardent supporter of Hitler in the early years of the Nazi Party, but later withdrew his support and openly criticized Hitler in a public letter in protest of The Fuehrer's human rights violations. Hitler, enraged, threw Thyssen into a concentration camp. Bormann, however, "felt Thyssen was his ace in the hole if he ever needed a personal pipeline to Allen W. Dulles," wrote Paul Manning.[40] And so Bormann ensured Thyssen and his wife were kept in a private home outside the main camp. Although it is unknown whether Bormann ever used Thyssen to contact Dulles, his foresight and actions in case the need ever arose speaks volumes regarding his understanding that Dulles would be the right person to contact when the critical moment arrived.

Thus a remarkable concentration of connections to Bormann was centered within Operation Sunrise – the Allied code name for the Wolff/ Dulles talks to surrender occupied Italy. There is no evidence within the documentation of the secret talks that proves Bormann was involved. Nor is their evidence, on the other hand, that precludes his having used the meetings as an opportunity to negotiate his freedom in exchange for the enriched uranium and other components on board U-234.

Proving Bormann actually took part in or influenced Sunrise may be impossible. In light of Bormann's apparent connections to U-234 and the U-boat's activities, including Fehler's determination to surrender its important cargo and its high-profile passengers, even at the expense of Tomonaga's and Shoji's lives, it seems probable, nonetheless, secret agreements were being pursued. If, indeed, this was the case, the most logical place for these agreements to have been completed was during the talks of Operation Sunrise.

Three additional points are worth considering in support of the above scenario. First, although the unconditional surrender agreement was written out in detail, during postwar discussions and interrogations Wolff often referred to the "oral agreement" he had made with Allen Dulles. Why would he specify verbal agreements rather than the surrender in whole unless he was trying to infer a separate importance to his discussions from the actual surrender itself, and thus that some of the agreements he and Dulles had concluded were not part and parcel of the instrument of surrender? These private oral agreements inferred, alone, would have violated the unconditional surrender command.

Having reviewed much of the Operation Sunrise files in the National Archives, I was unable to find any notes actually taken during the meetings between Dulles and Wolff or their parties. All available documentation concerning these discussions are either reports summarizing the conclusions of the talks or indices of wireless transmissions that record the working out of logistics and reporting in broad terms on their progress. The lack of actual minutes or personal notes recording the proceedings may indicate a sanitizing of the record to eliminate proof of actual agreements made.

Supporting this suggestion further is the fact immediately following the Italian capitulation, Wolff spent three days while still secured in his own headquarters in Fasano, Italy sequestered with Allen Dulles' right-hand man, Gero von Gaevernitz. Only Gaevernitz and Wolff, ostensibly to help Wolff compose his memorandum of events, attended these meetings. Not even Gaevernitz's OSS companion, a man who was sent specifically to monitor the surrender process, was allowed to participate.[41]

The unidentified OSS agent – thought to be Donald Jones, Dulles' man in Lugano[42] – recorded, however, Wolff had twice requested the meeting immediately following the "unconditional" surrender "to discuss the settlement of certain urgent matters."[43] What *conditions* were left to settle in private following the *unconditional* surrender? What could not be discussed in the presence of Gaevernitz's colleague, the surrender monitor, much less be included in a report about the supposedly above-board 'unconditional surrender?' There seems to have been no basis for such secret conferences if the unconditional surrender was actually implemented per the traditional history.

In addition, certain passages about the negotiations alluded to in the secret Operation Sunrise report, filed by Allen Dulles and Gero von Gaevernitz after the close of the European war, have been censored.[44] Again, this was done despite the fact the talks were supposedly based solely on unconditional surrender, which would seem not to require such mystery. The section introducing Wolff's report within that same document also admits that "one or two items" of Wolff's report had been eliminated because they were "not pertinent to this phase of our story."

A second possible proof-point suggesting Wolff provided secret concessions as part of the surrender includes the fact General Wolff was not tried at Nuremberg immediately after the war with the other key defendants, despite his clear complicity in crimes against humanity. Although he was Himmler's direct intermediary with I.G. Farben at Auschwitz, and he was SS leader and secret police chief in Italy, Wolff seems to have been immune from war crimes prosecutions at the Nuremberg Trials. Of course he denied his complicity, as did virtually all others involved in such activities. But he admitted the original idea of using forced labor for SS profit was his brainchild,[45] and as commander in charge of providing forced laborers for Auschwitz, he was personally and directly responsible for 25,000 deaths.[46]

In addition, in Italy his troops massacred hundreds of Italian partisan prisoners on at least three occasions.[47] Certainly he was as guilty as Farben's Krauch, Ambros and Buetefisch, and Auschwitz's Commandant Hoess, or Bormann himself, and less guilty than Grand Admiral Doenitz, all of whom were tried and convicted immediately after the war. Doenitz was convicted solely on the basis he did not countermand a direct order from Hitler turning captured crewmembers of an Allied torpedo boat over to the SS, who executed them.[48] Certainly this set a precedent under which Wolff should have been held responsible for the slaughter of hundreds of helpless partisan prisoners under his command, not to mention the tens of thousands who died at Auschwitz.

Wolff was not only ignored as a possible defendant at the initial Nuremberg Trials but he was released in August 1949 following his usefulness as a witness to the crimes of others. He was later sentenced to four years by a denazification court, but was released after only one week! Twenty years after the war, Wolff drew attention to himself by granting an interview during the Adolf Eichmann trial. His comments in the interview aroused public anger to the point he could no longer be protected, and his past finally caught up to him. He was tried and sentenced to 10 years for providing Jews to the death camps.[49]

Despite the eventual conviction, such protection of an obvious war criminal suggests collusion on behalf of the United States in shielding Wolff from going to trial – and in not serving his time, in the case of the denazification conviction. Long experience with the likes of Klaus Barbi and Reinhard Gehlen has shown the United States has only provided such protection when it had received powerful intelligence, political or military assets from the men it was protecting. It seems logical involvement in the U-234 surrender negotiations contributed to Wolff's protection, as well.

Some will argue Wolff secretly was granted immunity in exchange for initiating the Italian surrender discussions. These assertions may, in fact, be true, but doing so also would have violated the terms of the unconditional surrender. The war crimes case of Admiral Doenitz demonstrates how diligent the Allies were in pursuing suspected war criminals. Doenitz, who was responsible for ending the European war by surrendering all of Germany within one week of Hitler's death, was not only tried but convicted on charges far less serious than those of which Wolff was admittedly guilty. The basis for Doenitz's trial was so slight even American and British military commanders were appalled that Doenitz was tried, much less convicted.[50]

In comparison, Wolff's admitted complicity as the originator of the forced labor idea for I.G. Farben, compared to the specious charges against Doenitz, should have been a gauge ensuring Wolff would be tried with the others. And in comparison to Doenitz's surrendering all of Germany and thus ending the war within five days of taking charge, the Italian surrender – despite Wolff's documented efforts to expedite and facilitate it – took over two months to complete and did not actually occur until the same time the Reich's core armies in Berlin were capitulating and Doenitz himself was preparing to surrender. Therefore, in its final context, Wolff's surrender of the Italian front was meaningless. Why should the Allies have given any special treatment to Wolff on that account?

Finally, the shadowy Herr Langbehn, who first revealed those enigmatic, extraordinary documents that were so compelling to Allen Dulles, bears a name of striking likeness to a Captain Lieutenant Langbein of the German Navy's foreign bureau, the Marine Sonderdienst Ausland Commission. Bormann biographer William Stevenson wrote that Martin Bormann had overall responsibility for the cargo of U-234,[51] which seems to be corroborated by General Wolff's comments during interrogations as a witness for the Nuremberg Trials, in which he stated Bormann and Walter Schellenberg, one of Himmler's toadies, were responsible for the Ausland, or foreign, commissions.[52]

Interrogations of U-234's prisoners and captured German records indicate Langbein, under command of officer K.K. Becker[53] of the Marine Sonderdienst Ausland Commission, actually facilitated the collection and loading of the secret documents and materiél[54] onto U-234 before its departure from Kiel.[55] In fact, Langbein is the name signed at the end of the freight manifest. It might be a long shot, but could Langbehn and Langbein have been the same man? Could the "attorney" who ostensibly was responsible for the Foreign Ministry documents shown to Dulles, which may have included records pertaining to U-234's cargo, also served as an officer of the navy's foreign bureau who oversaw the documents and cargo loaded on to U-234?

The connection seems too compelling to ignore. Might Dulles have unwittingly misspelled the name upon hearing it? What is known is far from conclusive; and a positive answer may be too much to expect, but the possibility certainly should be explored further. The author encourages others who are interested in the answer to pursue it.

Endnotes: Chapter Sixteen – Occam's Razor

1   Paul Manning, *Martin Bormann: Nazi In Exile*, p. 253

2   Paul Manning, *Martin Bormann: Nazi In Exile*, p. 254

3   William Stevenson, *The Bormann Brotherhood*, p. 293

4   U.S. National Archives II, *Intelligence Summary Report* – Red, RG 457-190-37-11-1-Box 192

5   U.S. National Archives II, letter titled *Officer in charge of the American Mission*, RG 59 862.20200/9 – 2247; letter from Uruguay Ambassador Ellis O. Briggs to Secretary of State titled "Further Drew Pearson Article on Uruguay," 20 January, 1948, RG 59 833.021/1 – 2048; letter from J. Edgar Hoover to Assistant Chief of Staff, G2, War Department, 24 April 1946, RG 59 740.00116EW/4-2446

6   U.S. National Archives II, *State Department Telegram from Madrid to Secretary of State*, 29 August 1947, RG 59 862.20252/8 - 2947

7   U.S. National Archives II, *Index for the Office of Strategic Services*, RG 226-190-3-32-3 box 1022; RG 226 #123900; RG 226 #122640; RG190-3-32-3 box 1022

8   Ladislas Farago, *Aftermath*, pp. 243, 244

9   Paul Manning, *Martin Bormann: Nazi In Exile*, second photo section caption

10   Louis L. Snyder, *Encyclopedia of the Third Reich*, p. 75; Paul Manning, *Martin Bormann: Nazi In Exile*, second photo section caption

11   U.S. National Archives II, various files of Operation Sunrise, RG 238, M1019 Roll 80; RG 238 M1270 Roll 22; RG 226, Entry 110 Boxes 1 and 2; Dollinger, *The Rise and Fall of Nazi Germany and Imperial Japan*, p. 188

12   U.S. National Archives II, War Crimes Records, *Interrogation Summary #1795, of General Karl Wolff, Nuremberg, 12 April, 1947*, RG 238 – M1019 Roll 80; OSS cable from Berne #538 AFHQ for G-2 From (General) Airey, date thought to be 21 March 1945 but unsure, RG 226 Entry 110 Box 2; Report on the Sunrise – Crossword Operation Feb. 25 – May 2, 1945, p. 8, by Allen W. Dulles and Gero von Gaevernitz, RG226 Row 9/24/4 Entry 190 Box 24

13   Joseph Borkin, *The Crime and Punishment of I.G. Farben*, p. 117

14   U.S. National Archives II, War Crimes Records, *Interrogation Summary #1739, of General Karl Wolff, Nuremberg, 8 April, 1947*, pp. 1, 2

15   U.S. National Archives II, War Crimes Records, *Interrogation Summary #1739, of General Karl Wolff, Nuremberg, 8 April 1947*, p. 2; and *Interrogation Summary #2797, of General Karl Wolff, Nuremberg, 25 June, 1947*, RG 238 – M1019, Roll 80

16   U.S. National Archives II, untitled 76-page report about *Operation Sunrise*, pp. 10-15, RG 226 Entry 110 Box 1;also various other documents in same file; Hans Dollinger, *The Decline and Fall of Nazi Germany and Imperial Japan*, p. 188

17   Hans Dollinger, *The Decline and Fall of Nazi Germany and Imperial Japan*, p. 188

18   U.S. National Archives II, untitled 76-page report about *Operation Sunrise*, pp. 16-32, RG

226 Entry 110 Box 1

19   U.S. National Archives II, untitled 76-page report about *Operation Sunrise*, pp. 16-32, RG 226 Entry 110 Box 1; untitled 76-page report about *Operation Sunrise*, p. 20, RG 226 Entry 110 Box ; Hans Dollinger, *The Decline and Fall of Nazi Germany and Imperial Japan*, p. 188

20   U.S. National Archives II, untitled 76-page report about *Operation Sunrise*, p. 20, RG 226 Entry 110 Box 1; cable from Berne, 20 March, 1945, p. 2, RG 226 Entry 110 Box 2; untitled 76-page report about *Operation Sunrise*, p. 23, RG 226 Entry 110 Box 1

21   U.S. National Archives II, *Report on Operation Sunrise – Crossword, Feb 25 – May 2, 1945* by Allen W. Dulles and Gero von Gaevernitz, pp. 15, 16, RG 226 Row 9/24/4 Entry 190 Box 24

22   U.S. National Archives II, report titled *Sunrise Radio Operator*, RG226 Entry 110 Box 1; *Crossword, Feb 25 – May 2, 1945* by Allen W. Dulles and Gero von Gaevernitz, p. 21, RG 226 Entry 110 Box 2; OSS cable from Berne #540(9), 20 March 1945, RG226 Entry 110 Box 2; *Report on the Sunrise – Crossword Operation Feb. 25 – May 2, 1945*, p. 8, RG226 Row 9/24/4 Entry 190 Box 24

23   U.S. National Archives II, *Report on Operation Sunrise – Crossword, Feb 25 – May 2, 1945* by Allen W. Dulles and Gero von Gaevernitz, pp. 12-16, RG 226 Row 9/24/4 Entry 190 Box 24; OSS cable from Berne #540(9), 20 March 1945, RG226 Entry 110 Box 2

24   U.S. National Archives II, OSS cable from Berne #626, 11 April, 1945, RG 226 Entry 110, Box 2

25   U.S. National Archives II, untitled 76-page report about *Operation Sunrise*, pp. 33, 35, 47, RG 226 Entry 110 Box 1

26   U.S. National Archives II, untitled 76-page report about *Operation Sunrise*, p. 42, RG 226 Entry 110 Box 1; OSS cable from Berne #540(9), 20 March, 1945, RG 226 Entry 110, Box 2

27   U.S. National Archives II, *Report on Operation Sunrise – Crossword, Feb 25 – May 2, 1945* by Allen W. Dulles and Gero von Gaevernitz, p. 46, RG 226 Row 9/24/4 Entry 190 Box 24; untitled 76-page report about *Operation Sunrise*, p. 42, RG 226 Entry 110 Box 1

28   Hans Dollinger, *The Decline and Fall of Nazi Germany and Imperial Japan*, p. 188

29   Paul Manning, *Martin Bormann: Nazi In Exile*, pp. 91, 92

30   Paul Manning, *Martin Bormann: Nazi In Exile*, p. 91

31   Paul Manning, *Martin Bormann: Nazi In Exile*, p. 91

32   U.S. National Archives II, War Crimes Records, *Interrogation Summary #3722, of General Karl Wolff, Nuremberg, 1 October, 1947*, p. 1, RG238 – M1019, Roll 80

33   U.S. National Archives II, War Crimes Records, *Interrogation Summary #4476, 1 December 1947*, RG 238 M1019, Roll 80; also *Interrogation Summary #4453, 16 December, 1947*, RG 238 M1019, Roll 80

34   U.S. National Archives II, *Report on Operation Sunrise – Crossword, Feb 25 – May 2, 1945* by Allen W. Dulles and Gero von Gaevernitz, p. 25, RG 226 Row 9/24/4 Entry 190 Box 24

35   U.S. National Archives II, War Crimes Records, *Interrogation Summary #4476, 1 December 1947*, RG 238 M1019, Roll 80; also *Interrogation Summary #4453, 16 December, 1947*, p. 2, RG 238 M1019, Roll 80

36   U.S. National Archives II, War Crimes Records, *Interrogation Summary #4476, 1 December 1947*, RG 238 M1019, Roll 80; also *Interrogation Summary #4453, 16 December, 1947*, p. 2, RG 238 M1019, Roll 80

37   U.S. National Archives II, *Report on Operation Sunrise – Crossword, Feb 25 – May 2, 1945* by Allen W. Dulles and Gero von Gaevernitz, p. 25, RG 226 Row 9/24/4 Entry 190 Box 24

38   Paul Manning, *Martin Bormann: Nazi In Exile*, p. 253

39   Joseph Borkin, *The Crime and Punishment of I.G. Farben*, pp. 168, 169

40   Paul Manning, *Martin Bormann: Nazi In Exile*, p. 254

41   U.S. National Archives II, *Operation Sunrise* official reports, unidentified report, pp. 10, 12, 15, 22, RG 226, Entry 11, Box 1

42   U.S. National Archives II, *Operation Sunrise* official reports, *List of persons...of the Sunrise Operation*, pp. 2, 3, RG 226, Entry 110, Box 1; also *Report on Sunrise-Crossword*, p. 42 and *Report on events from 27 April to 2 May* (apparently prepared by General Karl Wolff), p.1, RG 226, Entry 110, box 1

43   U.S. National Archives II, multiple *Operation Sunrise* official reports, unidentified report, p. 1, RG 226, Entry 11, Box 1

44   U.S. National Archives II, *Report on Operation Sunrise – Crossword, Feb 25 – May 2, 1945* by Allen W. Dulles and Gero von Gaevernitz, p. 32, RG 226 Row 9/24/4 Entry 190 Box 24

45   U.S. National Archives II, War Crimes Records, *Interrogation Summary #769, 16 December 1946*, p. 3, RG 238, M1019, Roll 80

46   Peter Hayes, *The European Strategies of I.G. Farben, 1925 – 1945*, p. 63; Joseph Borkin, *The Crime and Punishment of I.G. Farben*, p. 3; Paul Manning, *Martin Bormann: Nazi In Exile*, p. 153

47   U.S. National Archives II, War Crimes Records, *Interrogation Division Summary of General Karl Wolff interrogation, Nuremberg, 26 October 1945*, pp. 1-3, RG238 – 1270, Roll 22

48   Louis Snyder, *Encyclopedia of the Third Reich*, p. 72

49   Christopher Ailsby, *SS: Roll of Infamy*, p. 183

50   Louis Snyder, *Encyclopedia of the Third Reich*, p. 72

51   William Stevenson, *The Bormann Brotherhood*, p. 64

52   U.S. National Archives II, War Crimes Records, *Interrogation Division Summary of General Karl Wolff interrogation, Nuremberg, 31 August 1945*, p. 3, RG238 – 1270, Roll 22

53   U.S. National Archives II, memorandum from H.T. Gherardi titled *Interrogation of Niesch-ling, 27 July 1945*, RG 165 – 390 35/11/05 box 540

54   U.S. National Archives II, *Report on the Interrogation of the Crew of U-234 Which Surrendered to the USS Sutton on 14 May, 1945, In Position 47°-07' N - 42°-25' W*, 27 June, 1945, RG 38 – 370 15/09/01 box 2; also memorandum titled *In regard to: Freight and Supplies of U-234*, 18 March, 1945, signed: Langbein, Korvettencapitan, RG 457 – 190-32-2-7,

55   U.S. National Archives II, *intelligence report of interrogation of Managing Director Saudel Aircraft Works of Kahla, Germany, 8 May 1945*, RG 38 – 370 01/04/07 box 113

Hiroshima destroyed. As part of a technology exchange agreement between Germany and Japan, U-234's original destination was to bring its load of technology, including the bomb-making materials, to the Island nation for use against the United States.

Epilogue

There was much written of a postwar "foreign trade offensive" and of a "European Economic Community" in which Germany would act merely as the 'flag bearer' and predominate by 'elastic political methods... not with brutal force.[1]*

             – Peter Hayes, *Industry and Ideology: I.G. Farben in the Nazi Era*

For a secret concern, the ramifications of the surrender of U-234 had far-reaching effects. Shortly before U-234 landed at Portsmouth Naval Yard, a leading Japanese scientist reported to the Japanese House of Peers he was about to introduce a weapon "so powerful that it would require very little potential energy to destroy an enemy fleet within a few moments."[2] According to Robert Wilcox, author of *Japan's Secret War*, "the reference was clearly to an atomic bomb." By extension, the reference actually appears to have been toward the cargo on board U-234, and possibly from other U-boats, as well. The evidence is strong the Japanese program had neither the technical capacity nor the needed uranium stocks to make such a bomb on its own. On the other hand, information exists that suggests at least one U-boat carrying nuclear components besides U-234,[3] and possibly more,[4] left German soil destined for Japan. It is unlikely, however, these vessels carried all of the workings necessary to make a bomb; and it is especially unlikely they carried enriched uranium.

With the surrender to the United States of U-234 and the nuclear components that were no longer going to Japan, Japanese possession of an atomic bomb to use against its enemies was unlikely. And yet immediately after the attacks on Hiroshima and Nagasaki, a Japanese broadcast claimed they had "similar weapons and will retaliate."[5] Perhaps this was a bluff, or perhaps it was true, but more likely, they had not yet realized the weapon they were prematurely claiming to possess had already been turned over to their enemies. While there are reports the Japanese tested an atomic bomb,[6] certainly they never used one in battle.

The leaders of Japan were not the only ones left wondering what had happened to their bomb. A few months after the United States dropped the bombs on Japan, leaders in certain Latin American countries began complaining the bombs had been stolen from Germany.[7] How these leaders may have discovered that fact is fascinating, considering Martin Bormann probably continued his escape from U-234 through Spain and on to Latin America, with those leaders' collusion. He probably told them a politically sanitized version of what had happened. The Latin American leaders' revelation is therefore not only another piece of evidence suggesting Bormann's escape, but indirectly, possibly of the uranium being enriched, as well.

Whatever the case, the surrender of U-234 certainly caused a commotion. According to former naval intelligence officer Bruce Scott Old, General Groves "almost had apoplexy when the Germans launched a submarine called U-235."[8] Old asserted Groves thought the U-235 designation referred in some way to a cargo of enriched uranium the U-boat was carrying. Then Old explained he thought Groves had confused U-235 with U-234.

Intelligence Officer Old must have been right, there must have been some confusion, because it is highly improbable Groves' excitement was caused solely by the surrender of a U-boat designated U-235, for two reasons. First, because U-235 was not surrendered. The U-boat had been a training boat throughout most of the war and then was sunk in the Baltic Sea on its maiden combat mission.[9]

And second, because it was widely known, even by the American public, that U-boats were designated with consecutive 'U' numbers. Certainly there was a U-235, and there was absolutely no reason to believe its 'U' designation was in any way related to any cargo it may have carried or mission it was intended to perform. Therefore, there is no reason whatever to believe a U-boat designated U-235 would cause any anxiety in General Groves; he would have expected it and thought nothing of it. Old went on to say Groves was concerned the report on the mysterious U-boat indicated it had been heading for Argentina.

Certainly despite the confusion, the details of this story match exceedingly well with those of U-234. If Groves was aware of a U-boat carrying U^{235}, it would almost certainly have been U-234. And given the German transmission intercepts decoded using the captured Enigma, and presumably other intelligence, as well, Groves probably knew the entire story behind U-234.

Apparently, the story Old recited was a skewed account of the surrender of U-234. The story, importantly, also adds support to the argument the Manhattan Project knew early on about the surrender of the U-boat and the uranium on board was, indeed, uranium enriched in U^{235}.

All of these rumblings pale in importance, however, compared to the larger picture of the impact upon our world of U-234 and its strange cargo. Looking back comfortably from the vantage point of nearly three quarters of a century since the end of World War Two, it is easy to presume that throughout the last half of the war its outcome and the race for the atomic bomb would reach a predetermined conclusion.

The evidence now available about U-234's cargo and passengers paints a frighteningly different picture, however.

The evidence, taken in whole, shows the United States was not necessarily leading in the race for the atomic bomb, as has been claimed. The evidence shows Germany was very near having all of the components for a bomb; and that the Nazis were dealing their bomb to the Japanese to use in the Pacific. The evidence, in fact, shows atomic bombs may have been ready for use by both sides at a frighteningly close point in time. The consequences could have been abysmal.

A key question is, if the German program had the components for a bomb, why did it not use one? The answer is simple: by the time enough enriched uranium was available to complete a weapon, the Germans had lost control of the skies over Europe. Since the Luftwaffe had lost control of the skies, there was little that could be done to transport the bomb to a strategic target. Any German bomber approaching Allied territory would be attacked mercilessly, and therefore had little chance of reaching a viable target objective.

Other transport systems were impossible as delivery options, or highly problematic at best. Trains traveling in and out of the Reich were carefully searched – when they were allowed to cross the frontier at all – as were all other forms of ground transportation. How would a weapon the size of a cannon be hidden? Surface ships, likewise, were tightly controlled. A submarine delivery was possible, but was very problematic and too risky. To deploy the bomb by U-boat meant the vessel would have to sneak undetected into the harbor of an enemy major city or military installation and either sacrifice itself and crew or leave the bomb in the harbor with a mechanism to detonate it hours after the U-boat had departed.

Detection of the U-boat approaching or trying to enter the harbor – a high probability after the half-way point of the war – meant failure and possible loss of the weapon, a risk too high to accept given the great expense and potential of the bomb. In addition to the great risk involved in a non-air delivery, up to

75 percent or more of the destructive capacity of the weapon would have been lost in a surface or sub-sea explosion. The ultimate in damage efficiency for the bomb was detonation about 1,500 feet directly above the center of its target. Without the capacity to deliver the bomb to a target of commensurate value, preferably by air, use of the weapon would have been a waste.

But on board U-234 were not only enriched uranium, but plans, parts and personnel to build V-4 rockets, Messerschmidt 262 jets and even the Henschell 130 stratosphere plane.[10] Although the ME262 was designed as a jet fighter, Hitler had ordered that it be redesigned and deployed as a small bomber.[11] That idea was taken one step further when a plan was hatched by General Kreipe to have a small bomber, armed with an atomic bomb, piggybacked across the Atlantic to New York.[12] At the distance limit of the mother plane the small craft would be launched in-flight to finish the bomb run. Once the payload had been dropped, the pilot would ditch the jet, parachute into the ocean, and then be retrieved by a U-boat.

The plans and components for a first high-altitude cockpit were also on board U-234.[13] The cockpit may have been part of the Henschell 130 stratosphere plane or the ME262. As bizarre as it may seem, the cockpit may, in fact, even have been designed for the V-4 rocket.

Interrogations of some of the prisoners of U-234 may shed interesting light on what possibly was planned for these V-4 rocket, high-altitude cockpit and atomic bomb components. Both General Kessler and Party Judge Nieschling, who were passengers on board U-234, answered questions during their interrogations about cockpits that had been installed in V-1 flying bombs and Japanese rocket planes.[14] Indeed, Hanna Reitsch, the brave German aviatrix already mentioned within these pages for flying Bormann out of Berlin, was awarded the Iron Cross by Hitler himself for test flying the V-1 bomb, which had been modified for a pilot.

Nieschling indicated that in the hands of the Japanese, the intent of such a weapon was to have it piloted by kamikazes.[15] The Japanese were already using kamikazes to pilot their small, wooden, one-man, rocket-propelled bomb-planes the Americans disparagingly called Baka bombs.[16] Baka means 'foolish' in Japanese. The very short-range Baka bomb was piggybacked to its destination by a four-engine plane, and carried a charge of high explosive. The Baka bomb was relatively ineffective, however, compared to its cost to produce, to deliver to a target, and especially in its steep cost of human kamikaze lives.

The specially designed V-4 rocket U-234 was carrying, on the other hand, was a powerful weapon that could carry a substantial payload across several thousand miles, if a comment Colonel Schlicke made to radioman Hirschfeld regarding it being the rocket that could cross the Atlantic is true.[17] Armed with an atomic warhead, which the Germans were already working on,[18] it would become the ultimate weapon of war, and in fact became the model for today's intercontinental ballistic missiles. The V-4 also had the advantage of traveling at great speeds – no airplane could catch it to shoot it down.

17-Minute Oversea Rocket Plane Among Germany's War Secrets

Special to THE NEW YORK TIMES.

WASHINGTON, Aug. 26—American and British technicians, closely and quickly following the Allies' military advances across France and Germany, have taken possession of a wealth of information about German "secret weapons" on which the enemy counted so much but that he did not have time enough to develop.

Besides an atomic bomb, on which, as has been made known, the Germans had made considerable progress, German scientists and engineers had developed a defense against radar and experimented on piloted rocket missiles that it was thought would be capable of crossing the Atlantic in seventeen minutes. These and many other German war secrets were disclosed today by the Office of War Information in reporting on the operations of a combined American and British intelligence organization that made daring forays on targets containing vital war information. The work of this group, called the Combined Intelligence Objectives Subcommittee, was itself, as disclosed today, a secret weapon of our own.

The CIOS, according to the OWI, has been operating inside Germany for many months, tracking down Germany's inner war secrets, and all the information uncovered has been channeled to both London and Washington, from where it was directed first to the war with Japan and now toward our post-war technical and scientific planning. In the United States the work is performed by the Technical Industrial Intelligence Committee, which functions under the Chiefs of Staff.

Instructions Were Specific

The CIOS' teams moved into Germany with plans and instructions as specific as those carried by a Flying Fortress crew or a party of Commandos. They concentrated on the "targets" believed to be richest in vital information on weapons, oil production, raw materials, synthetics, new engineering and chemical processes, inventions, patents and machinations in finance, economics and politics.

More than 2,000 missions to such

"targets" have already been made, and the information obtained was estimated by the receiving authorities as being worth "millions of dollars" in research and scientific development. The findings indicated, the OWI reported, that "German invention was far ahead of her capacity to translate theory into industry.

"The rapid advances of the Allied armies prevented her from putting into practice many of the technological advances evolved in the laboratories of her scientists," the OWI said. It added that some of the unlocked secrets might soon make some American technical processes "obsolete and outmoded."

Some Secrets Unrevealed

Not all the secrets have been disclosed, but the most startling ones were said to pertain to the development of the atomic bomb and the production of "heavy water," used in one method of making the bomb. The defense against radar was a system of radar camouflage consisting of anti-radar coverings and coatings. It would be employed, presumably, on submarines and other weapons.

The Germans contemplated a piloted missile with a possible range of 3,000 miles. The designer envisioned for it a commercial application for flying passengers across the Atlantic in a little more than a quarter hour.

Other Finds Listed

Other discoveries were:

The Germans were working on a formula for new war gases that they hoped, would prove more deadly than any chemical agent yet developed.

They had specifications and construction details for naval vessels of advanced design, including submarines with high underwater speeds and apparatus for sustained underwater operations. They had highly advanced jet engines, rocket-assisted take-off and aerodynamics designs.

They had found new uses for many staples. From coal the Germans were making a synthetic butter as well as alcohol of both beverage and industrial types, aviation lubricants, soap and gasoline. *New York Times 8/27/45*

PARLEY TO STRESS CONTROLLING REICH

Eisenhower Calls Military-Government Conference— Rehabilitation Minor

By DREW MIDDLETON
By Wireless to THE NEW YORK TIMES.

FRANKFORT ON THE MAIN, Aug. 26—The control of Germany and not her rehabilitation will be emphasized at a military-government conference that Gen. Dwight D. Eisenhower will open here tomorrow.

Conflicts within the military government, some of the weaknesses in the application of the American policy for the occupation zone and the necessity for greater liaison between headquarters here and field units will be discussed in the three-day conference. This is the first meeting of its kind and it is all the more important in view of the difficult and almost critical phase in which the military government now finds itself. New problems apparently unforseen six weeks ago require immediate attention, among them the changing attitude of the American occupation and military-government forces toward the Germans.

Moreover, the character of the command is changing and thus necessitates a shift in methods and allegiances. Lieut. Gen. Lucius D. Clay has supplanted Lieut. Gen. Walter Bedell Smith as General Eisenhower's principal lieutenant now that military government and not operations is paramount, and this necessitates a readjustment all around.

The biggest question—control vs. rehabilitation—is inherent in the apparently contradictory terms of the four-power occupation of Germany, which aims at political decentralization while maintaining Germany as an economic unit. During the past three months, the military government, in pursuit of the latter aim, has emphasized the rehabilitation of Germany to a considerable extent. According to an unimpeachable source the Americans' rôle as a controlling power rather than as an active participant in rebuilding Germany in excess of attaining the economic necessities, as laid down in the Big

the electoral registers would be inevitable.

Figure 1: A report in the *New York Times* documents a German rocket plane that could cross the Atlantic Ocean, presumably with an atomic warhead, if available.

The rocket's only shortcoming was lack of a guidance system. The kamikaze could solve that, too. All the rocket needed was a cockpit that would allow the pilot to survive and control the weapon in the rarefied atmosphere of near-space on its way to its target. Was this the purpose of the high-altitude cockpit? Were there plans to adapt the kamikaze strategy of the V-1 and Japanese Baka rocket-plane to the exponentially faster, more powerful, greater-distance capabilities of the V-4, or even the Henschell stratosphere jet?

The German/Japanese strategy might have looked something like this: Upon Germany supplying V-4 or stratosphere jet components, technology and expertise to Japan, the Japanese would build the aircraft equipped with controls to be operated by a kamikaze pilot placed in the specially-equipped high-altitude cockpit. The aircraft would be armed with a uranium warhead that would be detonated at the appropriate time by the ill-fated pilot, saving the program the considerable additional expense and development time of designing a guidance system that would lead the rocket over thousands of miles to within a few hundred feet of its target.

The speed and high-altitude characteristics of the V-4 and, presumably, the stratosphere jet, were indefensible by the Allies. And the long range of the rocket – which would allow the pilot to fly the weapon from the Japanese mainland to the closest Allied-controlled islands – had the double benefit of providing the element of surprise to the attack (the operational distance of the stratosphere plane is unknown). Once over the target island – perhaps the first would be the Allied-held land closest to the Japanese homeland, Okinawa or Iwo Jima – the kamikaze pilot would detonate the bomb, completely eliminating the Allies' strategic outpost and huge numbers of the enemy and his war-making materiéls.

With this sacrifice the kamikaze would achieve the highest possible honor among his people, and, should the war be won by his bravery, he was sure to be a national hero – posthumously of course. With this island base of operations won back, the following suicide rocket would be launched from that location to the next strategically held enemy island, and so on back across the Pacific, roughly in reverse order of how the Allies had won the islands from the Japanese.

Presumably as many as 8 to 10 bombs would be required before the United States, Britain and Russia – the Soviets would be in the Pacific War by then, and would have been bombed by similar attacks from China and Manchuria – would surrender. Japan would win the war and Nazi Germany, as Japan's ally, though once defeated, would rise like a phoenix from the battlefield ashes to control Europe, while Japan lorded over the Eastern Hemisphere.

It is easy to imagine the consequences such an outcome would have meant to the United States and the rest of the Americas. Certainly Japan and Germany could not allow American sovereignty to continue unchecked in the Western Hemisphere. The United States had the economy and resources to support a significant military defensive from its shores, or a substantial guerrilla resistance force. The Japanese and Germans would have had a difficult challenge controlling the vast enemy territories they already held by virtue of the V-4 offensive on their own continents, much less maintaining over-stretched command and communications and supply chains across the Atlantic and Pacific.

Probably a stalemate would have resulted between the Japanese and German juggernaut versus the United States, constructed of dubious treaties enmeshed in ultimatums – a Cold War with an enemy other than the Soviets and with an entirely different complexion.

Or perhaps during the months the Japanese and their imported German technicians were completing their bomb program, the Manhattan Project, between November and the end of 1945, would have solved its challenges triggering the plutonium bomb. The Japanese, had U-234 not surrendered to the United States, easily could have received the German goods from U-234 as early as July. With that cargo they may have concluded their atomic bomb and V-4 rocket preparations by November – roughly the same timeframe the Manhattan Project's bombs earliest could have been ready. Who would have used the first atomic weapons? And what would the response have been?

Perhaps as early as late 1945 or early 1946, nuclear war would have seared our collective experience as a family of beings mutually inhabiting this planet. What would be the outcome? What would each of our lives be like? On equal atomic terms, would the mission of one group of nations to assure self determination to all countries, confronted by the perceived requirement of other nations to sustain their own people by forcibly annexing the land and resources of other sovereignties, have dictated an unimaginable ending to the conflict?

Or would the leaders of two social systems so diametrically opposed to one another, for the sake and at the cost of the marginalized existence of many billions of people, have overlooked each other's immoralities to find life, of its own virtue, a more justifiable objective. Could the two sides agree to disagree, treating the subjugation of millions or billions of people as inconsequential compared to the alternative?

Fortunately, these questions never were answered because U-234 did not achieve its Japanese objective but was surrendered to the United States; which then used the resulting weapons on their intended original destination: which spawned a whole new battery of questions.

Should the bombs have been used at all as unbelievably destructive as they were? Was there any excuse for destroying hundreds of thousands of innocent lives along with the few guilty ones, in the single blink of an eye, and condemning many of the survivors to a slow death to radiation sickness and various cancers for hundreds of thousands more?

Or was using the bombs justifiable to save what some say certainly would have been a million or more American and Japanese lives that would have been expended in the final battle to seize and occupy Japan. Others argue Japanese surrender was imminent. Or that surrender would have been forthcoming with just a show of the weapons' power and therefore, they argue, use of the weapon was actually undertaken to reset the geo-political game board in America's favor.

Certainly American war planners felt compelled to consider the ramifications if Stalin completed plans begun when he declared war on Japan 8 August, 1945. Having promised his allies he would indeed declare war on Japan in this timeframe, he had already amassed a one-million-man army on the Manchurian border, with which he attacked on the following day and had conquered Manchuria, Inner Mongolia and northern Korea within two weeks.

With the remaining plan to move a vast portion of his army from defeated Europe and turn them east to join in the attack on Japan, the United States planners had to consider what would be the outcome; especially because Russia and Japan had been warring almost continually over this part of the world since the turn of the century. Here was an opportunity for the Soviets to decisively win the final victory.

With the Battle of Manchuria holdings to protect his right flank and consolidate his position on his east coast, and another 3 million men shortly available to attack from there to Japan's main island of Honshu, Stalin had the very real potential of ending the war quickly if he chose to do so. Such a quick resolution would be very much in his favor.

If he could take Honshu before the United States could from its much less potent position spread across a smattering of islands over a thousand miles away, and in a timeframe the Americans could not hope to emulate, the whole Asia Pacific Rim would be his. American negotiating power then would be almost nothing if Stalin chose to ignore previous agreements. And in the position of strength he would then be in – a much greater position than the Soviet Union had ever been in, or would ever be in again – it certainly seems plausible he would have done so.

The result would be a world with only one Super Power, and that would have been the Soviet Union under Stalin. The United States would not have had the opportunity to demonstrate it had the game-changing bomb, and would have been relegated once again to the background. The growth and strength of democracy would have been very much weakened, and communism would have become the reigning world political paradigm.

The world, in so many, often unfathomable, ways, would have been a markedly different place were it not for the historic outcome of the mission of U-234. Beyond altering what our world would look and feel like had U-234 delivered its cargo and passengers to Japan, the surrender of U-234 also has had a weighty and long-lasting direct influence on the lives we each lead today. The surrender of U-234 has helped define our present-day world. The quick and deep revival of the West German economy appears to be the fruits of Martin Bormann's Flight Capital Program – made possible by Bormann's apparent escape and post-war freedom – and guaranteed by the United States in return for U-234's surrender.

The Flight Capital Program that fueled the swift post-war resurrection of the West German economy – probably with the covert support of the United States – therefore, appears to have had a profound impact on the European and world economies in their turn. Bormann's plan for continued German dominance in Europe after the war apparently was so well structured, so deeply entrenched in the fabric of the many companies, cartels, industries and national economies co-opted, and so rich in those assets, that its permutations easily can be seen up to today. The plan especially can be seen in the European Economic Community that was confederated around the Euro, with Germany at the heart of the initiative. According to author Peter Hayes' book *Industry and Technology: I.G. Farben in the Nazi Era*, that confederation was planned for by Bormann, this author believes, in 1943.[19] Hayes wrote:

> There was much written of a postwar "foreign trade offensive" and of a 'European Economic Community' in which Germany would act merely as the "flag bearer" and predominate by "elastic political methods ... not with brutal force."[20]

The survival of and economic power generated by such multi-billion-dollar titans as Bayer, Hoescht, Volkswagen, AGFA-ANSCO and a long list of others, can all be traced to Bormann's Flight Capital Program. And their cumulative influence can be felt throughout the world economy, affecting each of us intimately, though unnoticeably, as we live our lives day to day.

The world, of course, continues to turn in the present as it has in the past. Approaching three-quarters of a century after the last global conflict ended, the echoes of its orators and ordnance are reverberating in ever-softening tones as we dash away toward new destinies – which too often are being defined by ever more meddlesome technologies and increasingly intractable amorality. At times it behooves us to stop a moment and look back. To try to wave clear the obscuring smoke of the past and discern through that awful mist, what caused the pall; so that new methods may be found to resolve the critical questions upon which our mutual peace and security lie. As we look back, we should not be shocked to find that "great doors sometimes swing on small hinges." That an eclectic handful of men and women – some of them great, but as often people of middling mien – stand at the center of enormous events and knowingly or unknowingly pull the levers and turn the knobs that define our world.

So it was with U-234.

THE END

Endnotes: Epilogue

1   Peter Hayes, *Industry and Ideology: I.G. Farben in the Nazi Era*, p. 368

2   Robert Wilcox, *Japan's Secret War*, p. 170

3   *Sharkhunters*, KTB 104, p.4

4   *Sharkhunters*, KTB 103, p.7; Geoffrey Brooks and Wolfgang Hirschfeld, *Hirschfeld: The Story of a U-boat NCO 1940-1946*, pp.189, 241

5   Glenn T Seaborg, *The Plutonium Story: The Journals of Professor Glenn T. Seaborg*, p.745

6   Robert Wilcox, *Japan's Secret War*, pp. 15, 16

7   David Irving, *The German Atomic Bomb*, p.294

8   Robert Wilcox, *Japan's Secret War*, p. 160

9   Telephone interview with *Sharkhunters* member Michael Koss, researcher and author of unpublished paper about the history of U-235; Geoffrey Brooks and Wolfgang Hirschfeld, *Hirschfeld: The Story of a U-boat NCO 1940-1946*, p. 199

10   New York Times, 27 August 1945;London Times, 27 August 1945

11   U.S. National Archives II, *Interrogation Report of Luftwaffe General Ulrich Kessler #5399, 25 June 1945*, RG165 – 390 35/10/05 box 495

12   David Irving, *The German Atomic Bomb*, p. 236

13   U.S. National Archives II, a green, hardback ledger book with the title *"CASH"* printed on the front lists a multitude of drawings and parts carried onboard U-234, and is marked as such inside, including "pressurized cabin parts," various pages, RG 38/370-15-05-07 box 3; Robert Wilcox, *Japan's Secret War*, p. 141

14   U.S. National Archives II, *Report of Interrogation of Kay Nieschling, 24 May 1945; Report of Interrogation of Luftwaffe General Ulrich Kessler #5399, 25 June 1945*, RG165 – 390 35/10/05 box 495

15   U.S. National Archives II, *Report of Interrogation of Kay Nieschling, 24 May 1945*, RG 165 – 390

16   Hans Dollinger, The Decline and Fall of Nazi Germany and Imperial Japan, p. 335; U.S. National Archives II, *Report of Interrogation of Kay Nieschling, 24 May 1945*, RG 165 – 390, also RG 38 – 370 15-09-01 box 2

17   Geoffrey Brooks and Wolfgang Hirschfeld, *Hirschfeld: The Story of a U-boat NCO 1940-1946*, pp. 212, 213

18   David Irving, *The German Atomic Bomb*, p. 185

19   Peter Hayes, *Industry and Technology: I.G. Farben in the Nazi Era*. P. 368

20   Peter Hayes, *Industry and Ideology: I.G. Farben in the Nazi Era*, p. 368

Sources

Aftermath, Ladislas Farago

Alvarez, Luis Alvarez

An American Genius, Herbert Childs

Anatomy of the Auschwitz Death Camp, Yisrael Gutman and Michael Berenbaum

Atomic Energy Desk Book, The, John F. Hogerton

Atomic Energy For Military Purposes, H.D. Smyth

Bormann Brotherhood, The, William Stevenson

Boston Globe, The, May 19 and 20, 1945

Brighter Than A Thousand Suns, Robert Jungk

Bunker, The, James P. O'Donnell

Complete Encyclopedia of World Aircraft, The, David Donald

Concise Encyclopedia of Nuclear Energy, Interscience Publishers

Crime and Punishment of I.G. Farben, The, Joseph Borkin

Daily Mail Online, 13 July, 2011, Nazi nuclear waste from Hitler's secret A-bomb programme found in mine, Allan Hall

Danger and Survival: Choices About The Bomb In The First Fifty Years, McGeorge Bundy

Decline and Fall of Nazi Germany and Imperial Japan, The, Hans Dollinger

Der Spiegel, 14 February 1964

Deutsche U-boote 1906-1966

Divide and Prosper, Raymond G. Stokes

Dopplegangers, Hugh Thomas

Encyclopedia of the Third Reich, Dr. Louis L. Snyder

Enola Gay, Max Morgan Witts and Gordon Thomas

European Strategies of I.G. Farben, 1925 – 1945, The, Peter Hayes

Explaining Hitler, Ron Rosenbaum

Feindfahrten, Wolfgang Hirschfeld

Fliegen, Mein Leben, Hanna Reich

Forbes, For Your Information, 25 November 1991

German Atomic Bomb, The, David Irving

Germany's Secret Weapons, Brian Ford

Gestapo Chief: The 1948 Interrogation of Heinrich Mueller, Gregory Douglas

Graphite Reactor-A Historical Landmark at the Oak Ridge National Laboratory, The, Oak Ridge National Laboratory brochure

Griffin, The, Arnold Kramisch

Himmler, Peter Padfield

Hirschfeld: The Story of A U-boat NCO 1940-1946, Geoffrey Brooks and Wolfgang Hirschfeld

Hitler's Tabletalk, Dr. David Picker

Industry and Ideology, Peter Hayes

Inside The Third Reich, Albert Speer

Into The Atomic Age, Chapman Pincher

Japan's Secret War, Robert Wilcox

"Last Dive," *The, National Geographic Magazine,* October 1999, Priit Vesilind

Letters of Martin Bormann, The, Trevor Roper-Smith

London Times, 27 August 1945

Los Alamos Primer, The, Robert Serber

Making of the Atomic Bomb, The, Richard Rhodes

Management of the Hanford Engineer Works In World War II, Harry Thayer

Manhattan Project, Stephen Groueff

Martin Bormann: 100,000 Marks Reward, James McGovern

Martin Bormann: Nazi In Exile, Paul Manning

Murderers Among Us, The, Simon Wiesenthal

New York Times, 27 August 1945

New York Times, March 3, 1973, Paul Manning

Now It Can Be Told, Leslie Groves

PhD, B Reactor Museum Association-History of 100-B/C Reactor Operations, Hanford Site, Michele S. Gerber

Plutonium Story: The Journals of Professor Glenn T. Seaborg, The, Glenn Seaborg

Project Alberta, Harlow Russ

Quest For Economic Empire, Peter Hayes

Secret History of the Atomic Bomb, Anthony Cave Brown and Charles B. MacDonald

Secretary, The, Jochen von Lang

Sharkhunters KTBs

SS: Roll of Infamy, Christopher Ailsby

Survival In Auschwitz: The Nazi Assault On Humanity, Primo Levi

Uranium People, The, Leona Libby

Wall Street Journal, Greg Steinmetz

Wiesenthal Files, The, Alan Levy

Index